The Life, Death, and Afterlife of the Record Store

The Life, Death, and Afterlife of the Record Store

A Global History

Edited by
Gina Arnold, John Dougan, Christine Feldman-Barrett,
and Matthew Worley

BLOOMSBURY ACADEMIC
NEW YORK • LONDON • OXFORD • NEW DELHI • SYDNEY

BLOOMSBURY ACADEMIC
Bloomsbury Publishing Inc
1385 Broadway, New York, NY 10018, USA
50 Bedford Square, London, WC1B 3DP, UK
29 Earlsfort Terrace, Dublin 2, Ireland

BLOOMSBURY, BLOOMSBURY ACADEMIC and the Diana logo are trademarks of
Bloomsbury Publishing Plc

First published in the United States of America 2023

Copyright © Gina Arnold, John Dougan, Christine Feldman-Barrett,
and Matthew Worley, 2023

Each chapter copyright © by the contributor, 2023

Cover design by Louise Dugdale
Cover image © lechatnoir/Getty Images

All rights reserved. No part of this publication may be reproduced or transmitted
in any form or by any means, electronic or mechanical, including photocopying,
recording, or any information storage or retrieval system, without prior
permission in writing from the publishers.

Bloomsbury Publishing Inc does not have any control over, or responsibility for, any
third-party websites referred to or in this book. All internet addresses given in this
book were correct at the time of going to press. The author and publisher regret any
inconvenience caused if addresses have changed or sites have ceased to exist,
but can accept no responsibility for any such changes.

Whilst every effort has been made to locate copyright holders the publishers would be
grateful to hear from any person(s) not here acknowledged.

A catalog record for this book is available from the Library of Congress.

ISBN: HB: 978-1-5013-8450-9
PB: 978-1-5013-8451-6
ePDF: 978-1-5013-8453-0
eBook: 978-1-5013-8452-3

Typeset by Deanta Global Publishing Services, Chennai, India

To find out more about our authors and books visit www.bloomsbury.com and
sign up for our newsletters.

Contents

List of Figures   vii

Introduction   *Gina Arnold, John Dougan, Christine Feldman-Barrett, and Matthew Worley*   1

Prologue: The Record Store That Saved My Life   *Mark Trehus*   7

Part I   Record Stores as Community

1. "We 'Bout It 'Bout It": The Independent Record Store in Post-Katrina New Orleans   *Jay Jolles*   17
2. Firecorner: The Importance of the Reggae Record Shops in Black London and the Cultural Confluence of West Indian Music   *Kenny Monrose*   26
3. Journey of a Girl in a Plaid Skirt and Knee Socks   *Holly Gleason*   40
4. The Cult of the Record Bar   *Stephen Shearon*   50
5. Magic in Here: Brisbane's Alternative Record Stores from the 1970s to the Digital Age   *Ben Green*   60
6. *High Fidelity* across Twenty-Five Years: Record Shops, Taste, and Streaming   *Jon Stratton*   72
7. Reflections from the Girls behind the Counter   *Lee Ann Fullington*   85

Part II   Cultural Geography of Record Stores

8. "Ways of Living": Touristification, Gentrification, and Curatorship in Spanish and Portuguese Record Stores   *Fernán del Val*   97
9. Living Popular Music in "High Fidelity": Portugal's Independent Record Stores, 1998–2020   *Paula Guerra*   109
10. Music on the Turntables When the Tables Are Turning: A History of Record Stores in Romania from Late Socialism to the Present   *Claudiu Oancea*   120
11. Jazzhole: How a Record Store Became the Lone Priest of Nigerian Oldies' Pop Culture   *Eromo Egbejule*   134
12. The Influence of Imported Records and Their Stores on the History of Popular Music in Japan   *Ken Kato*   141

13 Recording the Irish Experience: The Record Shop and Fair as
    Archive *Paul Tarpey*   153
14 The Revolution Will Not Be Televised, It Will Be Taped: Western Music
    Acquisition in Pre- and Post-Revolution Iran *Lily Moayeri*   164

Part III Sites for Fandom and Performance of Subcultural Capital

15 Making Indie Noises in the Corporate Outlet: How Hanging Around
    and Working in Small Record Shops in Aotearoa New Zealand Changed
    My Life *Roy Montgomery*   175
16 Rip Off Records (Hamburg) and the Microhistory of Capitalism
    Karl Siebengartner   186
17 Soul Bowl: Rare Soul Uncovered *Christopher Spinks*   197
18 Lucky Records: Music Makes the People Come Together *Mariana Lins*   209
19 Rough Trade Paris, 1992–9: The History of a Scene *Jean Foubert*   219
20 Musicians in the Record Store: Celebrity Encounters through Amoeba
    Music's *What's In My Bag? Christine Feldman-Barrett*   230
21 "Contents Expected to Speak for Themselves": A Preliminary
    Understanding of North American Self-Service Record Retail
    Tim J. Anderson   241
22 Lost in the Booth: British Record Store Listening Booths as Atmospheric
    Sites of Intimacy *Peter Hughes Jachimiak*   252

List of Editors and Contributors   263
Index   269

Figures

10.1 Advertising pamphlet for the Muzica Store in Bucharest (early 1970s) (author's personal collection) — 123
10.2 Front cover of 1975 Electrecord questionnaire regarding the soon-to-be-introduced format of cassette tapes (author's personal collection) — 125
10.3 Muzica Store in 1979 (Revista Muzica, April 1979) — 126
10.4 Page of "Cooperativa Radio TV" catalog, from the early 1980s (author's personal collection) — 128

Introduction

Gina Arnold, John Dougan, Christine Feldman-Barrett, and Matthew Worley

I have watched independent record stores evaporate all over America and Europe. That's why I go into as many as I can and buy records whenever possible. If we lose the independent record store, we lose big. Every time you buy your records at one of these places, it's a blow to the empire.

—Henry Rollins

Record stores keep the human social contact alive, it brings people together. Without the independent record stores, the community breaks down with everyone sitting in front of their computers.

—Ziggy Marley

The digital economy of the early twenty-first century has proven disastrous for many long-standing bricks-and-mortar businesses and perhaps the most beleaguered of these retail casualties is the record store. Doubly impacted by the technological changes to music dissemination and shifting market habits incurred by internet commerce, the record store is increasingly an obsolete construct. Among the first to shutter were the large chains (e.g., Sam Goody's, Tower, Virgin, HMV, among others), megastores ill-equipped to respond to changing consumer interest in a "clutter-free" future wherein the ownership of physical media would be supplanted by the streaming economy's impermanent access to digital files. Now, vanishing with increasing regularity are independent record stores. Once gathering places for like-minded consumers and nascent band members that provided a forum for innovation and creativity and a fulcrum for social change have been transformed, with some exceptions, into reliquaries where obsessives and eccentrics collect expensive, limited-edition vinyl artifacts. The demise of the local record store is a harbinger of the kinds of problems inherent in such a seismic cultural shift.

This book explores, from a variety of perspectives and methodologies, how record stores became such important locales. As an agora, a community center, and a busy critical forum for taste, culture, and politics, the record store prefigured social media. Once conduits to new music, frequently bypassing the corporate music industry in ways now done more easily via the internet, many independent record stores, in direct opposition to rock radio programmed by corporate interests, championed the most local of economic enterprises, allowing social mobility to well up from them in unexpected ways. In this regard, record stores speak volumes about our relationship to shopping, capitalism, and art. The editors of this anthology believe that record stores are spaces rife for examination because their cultural history is in some ways the story of the best side of capitalism seen in microcosm. To that end, three analytical motifs are utilized: cultural history, urban geography, and auto-ethnography to find out what individual record stores meant to individual people, but also what they meant to communities, to musical genres, and to society in general. What was their role in shaping social practices, aesthetic tastes, and even, loosely put, ideologies? This collection of stories and memories and facts about a variety of local stores not only recenters the record store as a marketplace of ideas but also explores and celebrates a neglected personal history of many lives.

The focus of this volume is on the culture of record stores from roughly the mid-twentieth century through the early 2000s, stores that, in some instances, were essential to the development of local music scenes. Related to this is the record store as subcultural space, how these clubhouses for music fanatics were, at times, genre-specific sanctuaries for "outsider communities" such as punk, metal, soul and R&B, and hip-hop. Independent record stores have often served as a public sphere for such "outcasts," providing a space for them to gather where, to paraphrase Jürgen Habermas, opinions take shape and are circulated, and decisions are made without violence and, from the mid-to-late twentieth century onward, private places where misfits could meet up with other misfits. These accounts fall under the heading of auto-ethnographies in which participants reflect upon the centrality of a specific record store and its impact on a city's cultural vitality. As important cultural spaces, record stores have been portrayed in the fictional films *Empire Records* (1995), *High Fidelity* (2000), *Hearts Beat Loud* (2018), *Mixtape* (2021), and documentaries *I Need That Record* (2008), *Sound It Out* (2011), *Last Shop Standing* (2012), and the Tower Records story *All Things Must Pass* (2015), as well as in books like Graham Jones's *Last Shop Standing* (2015), Eilon Paz's *Dust & Grooves: Adventures in Record Collecting* (2015), Marcus Barnes's *Around the World in 80 Record Stores* (2018), and Garth Cartwright's *Going for a Song: A Chronicle of the UK Record Shop* (2018).

More recently, this idea was made surprisingly literal in the 2022 animated movie *Minions: The Rise of Gru*. In the film, which is set in the early 1970s, a group of supervillains called the Vicious Six have their lair in the basement of a downtown record store called Criminal Records, and it's difficult to imagine a setting better designed to tap into baby boomer memories. Like real record stores of that era, the aisles are lined with racks of LPs, with prominently displayed covers of popular FM rock radio staples like *Frampton Comes Alive*; the whole place is presided over by a snotty clerk called Nefario, who only allows favored customers into the inner circle of villainy if he utters a secret password and plays a Linda Ronstadt 45 on the stereo . . . backwards.[1]

Minions: The Rise of Gru is aimed at the under-ten set, but there is no getting away from the sensation that its true audience is not their parents but their grandparents. Moreover, implicit in the setting is everything this book posits. Record stores in the 1970s *were* lairs of a sort, not for supervillains but for society's gentler outcasts, which is to say anyone who needed to escape from the tight bonds of mainstream culture. The life story of Mark Trehus, for example, epitomizes the phenomenon: herein he relates his postwar upbringing with distant, alcoholic parents, a descent into drugs in the 1970s, then a rebirth through the medium of rock music, vinyl LPs, and the purely American joys of owning a small business. Mark's personal "Criminal Records" was Oar Folkjokeopus in Minneapolis, and it doesn't really deviate that far from the one depicted in *Minions: The Rise of Gru*. The same goes for record stores located in the heart of Berlin and Brisbane and London and Tehran, as well as all the other record stores in all the other cities explored herein. These spaces served as sleeper cells of a sort, where rebellious youth, sickened by the mainstream, could gather with the like-minded. In many of the locations outside of America, record stores served an even more explicitly political purpose, whether as an escape from fascism, communism, racism, or religious fanaticism, but all of them served a similar purpose, allowing their devotees to hide in plain sight.

* * *

The history of the record store as cultural nexus ostensibly begins with the opening of Spillers Record Shop in Cardiff, Wales, in 1894. Officially recognized by the Guinness Book of World Records as the world's oldest, Spillers sold cylinder recordings and phonographs a mere seventeen years after Thomas Edison introduced the world to the technology that enabled sound to be recorded and played back. Edison's invention was altered and improved a decade later by Emile Berliner who patented the gramophone, a device that would become the first commercially successful machine to play another Berliner invention, the flat disc, items Spillers soon added to its inventory. Today, the store is thriving with patrons of all ages forming queues in the early morning for the limited-edition vinyl releases on Record Store Day. "Over-the-counter music sales have consistently been holding their own since I took the business over from my dad in 2010," notes owner Ashli Todd. "But I have absolutely no desire to expand or to do anything massively lucrative, other than sustain the viability of Spillers for more years to come. I just want to run a lovely, sustainable record shop in Cardiff for the people who value that."[2]

In the United States, the emergence of record store culture dates to 1932 and the opening of George's Song Shop in Johnstown, Pennsylvania. Founded by brothers Eugene and Bernie George, it remains the nation's oldest record store that, remarkably, survived during one of the worst periods of the Great Depression, one that saw significant declines in industrial production and gross domestic product, as well as a skyrocketing unemployment rate that, by 1933, had reached 20 percent. After Eugene died from a stroke in 1962, his nineteen-year-old son John took over ownership and runs the shop to this day. The five-story location stocks over one million records and compact discs and its signage proudly proclaims, "If we don't have it, nobody does." While the city of Johnstown has seen a precipitous decline in population since the 1960s (in 2017 it was the third fastest shrinking city in the United States) George's Song Shop, now over ninety years old, remains, thanks in part to the resurgence of vinyl sales. "It started with the young people," George says. "Anybody that walks in here under 35 years old, they're looking for records. If they're older than 35, they're usually looking for CDs."³ There is, however, no talk of retirement. "[This store] has been in my blood all my life," he notes, "I could retire today and not have to worry about anything, but I don't want to, I just love coming to work. I don't want to give it up even though I'm getting up there in age, so to speak. I feel pretty young, and I'll go as long as I can."⁴

The long history of places such as Spillers and George's Song Shop, while noteworthy, is, increasingly, the exception rather than the rule. The recent closure of Boston's Skippy White's Records (2019) and the possible closure of Nashville's Ernest Tubb's Record Shop (2022)—venerable stores combining for 135 years of business—represent not just the loss of bricks-and-mortar retailers but, more significantly, the loss of shared cultural spaces. Fred LeBlanc (Skippy White was his radio DJ handle) opened his shop in 1961 selling primarily soul, funk, gospel, and R&B records. A French-Canadian from Waltham, Massachusetts, whose last remaining store was in the heart of the city's Roxbury neighborhood, White's "decades of devotion to the music and musicians he loves has earned him almost unrivaled respect and love from Boston's black community."⁵ And while the business had been eroding for quite a while, octogenarian White admitted that even the public's rekindled love of vinyl arrived too late after bricks-and-mortar retailers were decimated by digital sales.

> It used to be that if you wanted to find out what was happening, the latest in what was going on, you [went] to the record store to find out. That's what was happening, all those many years . . . a place where those who love the music can pour over the details, the stories, where these conversations can lead to new discoveries.⁶

Despite being known as "Music City USA" Nashville had a dearth of record stores when Ernest Tubb opened his in 1947. As was the case with Skippy White, Ernest Tubb's specialized in selling only two kinds of music: country and western. The iconic shop, which relocated to downtown Nashville's Lower Broadway in 1951, predated the existence of the city's famed Music Row and many of the celebrity-themed honky tonks which now line both sides of the street. Enhancing the shop's reputation as one of the country's most important specialty record stores was its hosting the Midnite Jamboree, the informal, intimate post Grand Ole Opry show that, for over seventy years, featured legendary performances by hundreds of country stars including Hank Williams, Sr., Patsy Cline, and Loretta Lynn. According to music historian Bill DeMain, "Along with [revered honky tonk] Tootsie's Orchid Lounge, Tubb's is one of the last remaining connections that Nashville's Lower Broadway has to old country music."⁷ The store's precarious future is partly the result of it "aging out" its current location. Tourists crowding the nearby honky tonks are disinclined to shop for records, an economic reality exacerbated by Nashville's relentless gentrification of the past twenty years making the building and the lot on which it stands extremely valuable real estate—just not for a record store. Recently, however, the store and lot on which it stands was sold to a small group of local investors that included Tubb's grandson Ernest Dale Tubb III for the astronomical price of $18.3 million ($2,000 per square foot, more than triple what the property sold for in 2020) forestalling its imminent demise. While, as of this writing, there are no definitive plans on how the site will change and function (it cannot be demolished or the exterior changed significantly as it's a protected historical building), current speculation is it will be more than simply a record store—it might become a museum/venue that, however, continues to sell two kinds of records: country and western.[8]

Skippy White's and Ernest Tubb's and several other stores, once significant cultural landmarks of their respective communities for more than forty years, are now places that are gone. The buildings now repurposed or razed by the wrecking ball, their histories slowly slipping below the horizon of recognition. The latter happened in March 2020 to middle Tennessee's most important record store of the mid-twentieth century, Randy's Record Shop in Gallatin, Tennessee, a store that Greg Reish, director of the Center for Popular Music at Middle Tennessee State University notes, "[was immeasurable in] helping to establish middle Tennessee as an epicenter of the American popular music industry in the postwar years and launching the careers of numerous country, bluegrass, rock 'n' roll, gospel, blues, ragtime and pop artists."[9]

"Nostalgia," writes essayist Joe Bonomo, "means a desire to return home, and the special, irreplaceable pleasures there, even if that home is defined by its absence."[10] For generations born in the mid-twentieth century or later, records and the music on them have been the prism through which individual identities and communities were created and expressed. For all the stories of record shops closing, there are still steadfast independent entrepreneurs opening small shops sustained by the recent dramatic increase in vinyl sales and the coveted, exclusive limited-edition releases featured on Record Store Day, an event aiding in the transformation of the twenty-first-century record store into a postmodern haven for vinyl connoisseurs both young and old. From coast to coast, in US cities large and small, record store culture is being maintained and, in some instances, flourishing as retailers maximize their e-commerce presence on social media giving their local stores a national profile. This includes "how-to" videos for those interested in opening a store with little or no retail experience.[11] This is also true globally, with important record stores functioning as de facto community centers supporting local and regional music scenes in places such as Lagos, Nigeria; Dubai, United Arab Emirates; Kingston, Jamaica; Beirut, Lebanon; Havana, Cuba; and Reykjavik, Iceland. And while a sense of loss, itself the residue of collateral damage rendered by the inevitability of change, is the leitmotif of this anthology, the hope is that, in sum, this will be not necessarily an encomium to a bygone era but a nuanced understanding of what has been lost and what significance there is in what remains.

Notes

1 To promote *Minions: The Rise of Gru* in London, a pop-up record store called Despicable Discs opened in Foubert's Place which claimed to "[bring] a '70s explosion back to Carnaby [Street] . . . the one-stop vinyl swap shop will see visitors exchange their own groovy records from the 1950s, '60s, and 70s with all records

donated to Oxfam." https://www.shaftesbury.co.uk/en/media/press-releases/2022/shaftesbury-announces-the-opening-of-minions--the-rise-of-gru-re html (accessed July 30, 2022).

2. A. Todd, "Remixing the Record Store," wales.com/en-us/economy/investment/remixing-record-store (accessed June 30, 2022).

3. L. Harris, "America's Oldest Record Store Is Just as Hip Today as When It Opened in 1932," parade.com/661040/lharris-2/Americas-oldest-record-store-is-just-as-hip-today-as-when-it-opened-in-1932, April 21, 2018 (accessed June 15, 2022).

4. Ibid.

5. B. Coleman, "Boston's Skippy White: A Vinyl Life," medium.com/@briancoleman/bostons-skippy-white-a-vinyl-life-3cd10422dbe April 16, 2018 (accessed June 23, 2022).

6. Christopher Gavin, "Skippy White Has Been Selling Records in Boston for Nearly Six Decades. But It's Time to Close up Shop," boston.com/news/local-news/2019/12/19/skippy-whites-records-closing, December 19, 2019 (accessed July 3, 2022).

7. "Ernest Tubb's Record Shop Is Closing," YouTube.com/watch?v=uvWL-gzRKIM (accessed July 11, 2022).

8. Ernest Tubb Record Shop Property Sold, Hope for Preservation Renewed, SavingCountryMusic.com July 29, 2022. https://www.savingcountrymusic.com/ernest-tubb-record-shop-property-sold-hope-for-preservation-renewed/?fbclid=IwAR34dtX14eF9PEq4C7yZ1kr-4VMbV2zxxmf29XOrhKmrT8hDuUEzlGL47Hw#l6793al3k4hjewws6t8 (accessed July 30, 2022).

9. Randy's Record Shop Demolished, theportlandsun.com April 25, 2020. https://www.theportlandsun.com/news/randy s-record-shop-demolished/article_30e280e6-84fb-11ea-87ee-4be4f7e1de27.html (accessed July 30, 2022).

10. J. Bonomo, "Muscle Memory," nosuchthingaswas.com/2018/02, February 23, 2018 (accessed July 5, 2022).

11. A YouTube search of "record store videos" will lead the curious down a very deep rabbit hole of information from the video series Let's Go to the Record Store, HiFi America Record Store Tours, and the UK's Behind the Counter, as well as videos from individual stores such as Amoeba Records (three California locations), The "In" Groove (Phoenix, AZ), Noble Records (Charlotte, NC), NTX Vinyl (Dallas/Fort Worth, TX), Too Many Records (Portland, OR), Spin Me Round (Easton, PA), among many others.

Prologue

The Record Store That Saved My Life

Mark Trehus

The event jumpstarting this collection was the closing of Minneapolis' Treehouse Records (formerly Oar Folkjokeopus, more commonly known as Oar Folk) in 2017. I'd posted a link to Facebook about the store's final days which started a conversation with Gina Arnold and Christine Feldman-Barrett, the three of us, in one manner or another, having had a connection to the store. It was Christine who suggested we put together an edited collection on "record store culture" past and present. Gina and I agreed, and we thought, after we'd recruited fellow coeditor Matthew Worley, it would be a fun and important project.

Early into the call for chapter submissions, Mark Trehus contacted me about Oar Folk's/Treehouse's inclusion in the anthology. It made perfect sense insofar as his story was unique; he'd been a customer, employee, and, eventually, owner of the location that had occupied the corner of 26th and Lyndale Avenue in South Minneapolis for nearly fifty years. The store's distinctive name was the result of previous owner Vern Sanden who decided to create a quasi-portmanteau of two of his favorite albums: Skip Spence's Oar and Roy Harper's Folkjokeopus. When Mark assumed ownership in 2001, he renamed it Treehouse (earlier having a record label of the same name), but the store never lost the spirit, dynamism, and cultural impact of the original Oar Folk, a record store that, from its inception in 1973, "was a key portal to new music for adventurous fans and musicians in the Twin Cities."[1]

Mark recalls Oar Folk as a "postgraduate rock and roll maniacs' school. The tuition was paid for with my weekly paychecks and the receipts were evidence of edgier record purchases." From the moment he bought the first Modern Lovers album there in the mid-1970s, he was linked to the store, the long-term relationship (and subsequent employment and ownership) cemented after years of addiction ended with sobriety in 1985. It's a story best told by Mark, one that unfolded via email and phone conversations between us in 2021–2, with some material excerpted from Cyn Collins's book Complicated Fun: The Birth of Minneapolis Punk and Indie Rock, 1974-1984. "I was on that corner for 32 years," he notes pensively, "that's a long time."

—John Dougan

I was born in Minneapolis in 1955 the eldest of three boys. At the age of four, I discovered my mother's modest collection of 45s and was drawn to the rock and roll sounds of Elvis, Bill Haley and the Comets, and the Big Bopper. I was fascinated with the process of stacking those seven-inch platters, pulling the arm over the top securing them on the spindle, flipping the switch, watching as one record dropped from the stack onto the turntable, the tonearm swinging over and briefly hovering above the entrance groove, dropping the needle onto the disc—and presto! Big Bopper's "The Purple People Eater Meets the Witch Doctor" came out of those tinny-sounding, cloth-covered speakers. The die was cast.

My father was a high-functioning alcoholic, never missing a day of work as a purchasing agent for a computer company, but he usually checked out after dinner with home projects and other distractions like magazines devoted to his passion for flying and his Colt 45 malt liquor. My mother was a clinically depressed stay-at-home Mom struggling with bipolar disorder, who was often drunk when I came home from school. My parents' favorability barometer during my upbringing was measured by my academic achievement. Staying out of their way and not engaging with them was how I learned to survive. I felt alone and defective.

I was a gifted student, but as time went on, I rebelled, and my priorities changed. The Beatles appearance on *The Ed Sullivan Show* in 1964 altered the way I looked at the world. Soon I graduated from throwing snowballs at cars and stealing cartons of cigarettes from the supermarket to shoplifting records from large department stores. Bullied by older kids in the neighborhood I was desperate for acceptance. Weekend drinking and sleepovers at the homes of my friends turned into daily pot and frequent LSD use by the time I reached high school. I wanted to be removed from the plasticity and shallowness of my empty, dull, suburban existence. I wanted to live in a different reality without the pain of a dysfunctional life at home. Drugs and music were my escape, and I was addicted to both.

By ninth grade I began to frequent rock concerts all the while accumulating an increasingly large collection of LPs. Discovering new music became the principal focus of my existence. My musical palette expanded as I soaked up every live show I could attend—from B.B. King to the Bonzo Dog Band. Tony Glover's midnight radio show on KDWB in Minneapolis and Beaker Street with Clyde Clifford on KAAY in Little Rock, Arkansas; distant sounds transmitted in the dead of night provided an ongoing education in the musical counterculture. I felt different from those who seemed to have prescribed destinies and for the next decade I remained directionless. I tried college, but it didn't stick. After a harrowing brush with the law, I needed to find a more legitimate way to pay the bills besides selling drugs. I couldn't conceive of a life that didn't embrace the counterculture. Rock and roll had always been a refuge, but I now looked to it for salvation.

* * *

Oar Folkjokeopus was more than just a record store. It was a clubhouse for music fanatics. It was a place where people came to listen to music and talk about their

favorite stuff. The exchange of information that happens at record store was essential to building the great community we had here.[2]
—Peter Jesperson, former manager of Oar Folk and the Replacements

Shopping at Oar Folk could be intimidating. My future roommate, Hüsker Dü drummer Grant Hart, referred to it as "a place to be condescended to." My first interaction with the store was to phone it in search of a particular Patti Smith 45. Andy Schwartz, who would go on to edit the influential music magazine the *New York Rocker*, answered. I asked sheepishly, "Do you have the Patti Smith 45 with 'My Generation' on the B-side?" "Of course, we do!" he sneered, slamming the phone into its cradle, hanging up, confirming the brash attitude of the hipper-than-thou record clerk for which Oar Folk was notorious. Despite his brusque tone (or maybe because of it), I immediately headed down to this singularly different record shop on the south side of town and bought that record.

In 1977, Oar Folk was as much a clubhouse as it was a record store. New and used records made their way into the hands of eager listeners via the store's influential tastemakers and gatekeepers like Terry Katzman and store manager Peter Jesperson. Promotional LPs regularly came in, serving the frugal budgets of the shop's steady customers. I went to the store almost daily and plopped on the floor against the front counter to feverishly comb through the newest secondhand arrivals. Nearly everything in those days was uniformly priced: a near mint record might garner the seller as much as $1.75, which in turn would be priced at around $3.50. Older, used records were sold for as much as current promotional arrivals; rarely was there any price difference for these future blue-chip rarities. Records that Peter felt strongly about were bought in large quantities because he knew, as an influencer, he would sell them. When the Sex Pistols' "God Save the Queen" came out, there was a line down the street of people waiting to buy it. Oar Folk was where you mingled with like-minded music fans and developed long-standing friendships. It was a clique of musicians, writers, artists, scenesters, and record buyers where everything revolved around what was new and cool or older and essential. There were no bongs or "head shop" gear sold here. Oar Folk was all about rock and roll.

Soon, Oar Folk and the Longhorn Bar (aka Jay's Longhorn, Minneapolis' version of New York's famed CBGB) were where I spent a considerable amount of my free time. My best friend Thor Lindsay (who would soon move to Portland, Oregon, and eventually start T/K Records with trust fund heir Tim Kerr) and I broke from the blue-collar, booze, and pot-fueled rowdy crew we knew from high school and started hanging out with the artists, musicians, and scene-makers comprising the small group of square pegs at the Longhorn. Here was a place where those who self-identified as misfits connected. And Oar Folk was the mecca to which the scene's musicians and record buyers were drawn; more importantly, these places were ours. Thor and I enthusiastically separated ourselves from the mainstream, trading in our bell-bottoms for ripped jeans, maligned and disparaged by the unenlightened philistines from the Northeast suburbs as a couple of punk rock losers. We wouldn't have wanted it any other way.

Despite the rough shape I was in, by the early 1980s I'd managed to get jobs at other area record stores the Wax Museum and Harpo's. Peter had given me a job DJing between live sets at the Longhorn, but my dream was to work at Oar Folk. However, I was still too messed up, a fact that didn't exactly sway Peter (and, later, Jim Peterson) into hiring me. Hell, I wouldn't have hired me either.

In August 1985 I went into treatment. On October 7, 1985, a fire gutted Oar Folk. Vern Sanden had bought the store in 1973 from Wayne Klayman, who'd opened it in 1970 as North Country Music, its name taken from the Bob Dylan song "Girl from the North Country." Vern was a regular customer, older than Klayman, with previous work experience as an air traffic controller, not exactly the resume you'd expect for a prospective record store owner, but he loved music, especially 1950s rockabilly.[3] While it was Peter's tastes that shaped Oar Folk more than any other single factor, Vern was often behind the scenes, buying cool rockabilly, blues, and soul reissues for the shop. It was his vision that allowed Peter an outlet for his passion and knowledge, which in turn played an inestimable role in fueling the scene the store fostered.[4]

Not long after the fire, Terry Katzman and Jim Peterson left to open their own store, Garage D'or, with financial assistance from recording engineer Paul Stark, one of the cofounders of the indie record label Twin/Tone. It was located six blocks from Oar Folk, across the street from Twin/Tone's offices. Vern took a chance on me and my sobriety. I was hired along with former Uneeda Records owner Bill Melton to comanage the rebuilding store.

A rivalry developed between Oar Folk and Garage D'or. Other shops like Northern Lights and Let It Be had opened and record-buying options were multiplying as the scene splintered and spread. But nothing would fully replace the spirit and relevance of the pre-fire Oar Folk. The fire symbolized the end of an era, and the early 1980s glory of the Twin Cities music scene was starting to fade, with several stores competing for a share of the table scraps.

When Bill decided to leave the shop, I became sole manager. As the so-called grunge movement emerged in the late 1980s, Oar Folk was the first to pick up on bands like Mudhoney and Nirvana. I had also become immersed in running my own record label with my partner Johnny Dromette (né Thompson) who'd recently moved to Minneapolis from Cleveland. We released records from a band formerly "managed" by John (the Pagans), and Minneapolis upstarts Cows and Babes in Toyland. Because I also loved 1960s garage rock the racks were dotted with reissues and new releases from local and regional neo-garage bands as well as those from Australia and Sweden. We still championed old favorites like the Velvet Underground, the Stooges, and Pere Ubu, but we also carried a great deal of blues, R&B, soul, jazz, and country records. Although the pre-fire Oar Folk was an impossible act to follow, I did everything a selfishly obsessive music lover could do to maintain the store's relevancy. It wasn't always easy. Managing the store, the label, and my sobriety was a lot to take on.

The technological reality we faced during this time was the transition from vinyl to compact discs. While other record stores and aging head shops (e.g., local veteran indie stalwart the Electric Fetus) were abandoning vinyl to stock CDs, Vern and I stubbornly held out for as long as we could. We finally broke down and went semi-digital, it was necessary for the store's survival. Plus, at that time, there was a lot of music that wasn't available on vinyl. But Vern and I loved vinyl, that's where our hearts were, and we stuck to our guns. Vern always had my back in that regard and I can't thank him enough.

Vern had been talking for some time about getting out of the record business and asked me if I was interested in taking over the store. The building's landlord, however, wouldn't offer me a long-term lease; he was preparing to sell the property, which included a commercial space next door and a duplex behind the record shop. I was faced with unemployment and an uncertain future. I'd arrived at a career crossroads. I was forty-five years old and had no marketable skills.

I decided to take the plunge and took out a second mortgage on my house, which covered about half of the purchase price. After being turned down by five banks, a local community bank finally gave me an additional loan. Cobbling together additional funds from family members, I'd managed to raise enough money to buy the properties. I was scared and nearly broke. What the hell was I doing? I didn't know the first thing about property management but had just purchased three run-down buildings in order to save my job. There wasn't enough money left for me to buy Oar Folk's inventory or the business per se and after some uncomfortable negotiations, I agreed to help facilitate Vern's going-out-of-business sale. On March 31, 2001, he left with his remaining stock and store name; on April 1, I opened Treehouse Records. I put my life savings of $15,000 into new inventory. With a drained bank account, a mountain of debt, and a handful of used records from my collection, I became a record store owner.

Within a year, Treehouse looked a lot like the old Oar Folk and the immediate community support was uplifting. I hired the latter-day Oar Folk assistant manager to help run the store, with part-time help filling in evenings and weekends. This went on for a couple of years, until I found out I had been carrying the hepatitis C virus for nearly two decades, a consequence of my prior drug use. The disease had progressed, and I had cirrhosis of the liver. After forty-eight weeks of debilitating twice-weekly Interferon self-injections and daily Ribavirin capsules (I had every side effect imaginable), I was unsteady on my feet, largely zapped of energy, and my long-term romantic relationship ended. During this health crisis, much of the daily work had fallen on my employees, and I took a far less active physical role in the store's operation.

Records started disappearing as supervision got lax. I didn't handle it well and things got messy. As the store owner, I'd instructed my employees to give me first crack at the used record spoils. Unfortunately, I'd underestimated the level of entitlement of some of my younger employees, who resented my addiction to the juicer fruits of the business. My longtime assistant manager, whom I relied on to handle the store's day-to-day operation, left to manage a sports collectables shop in a suburban mall. In the meantime, I'd lost touch with the mechanics of running the business. I'd also discovered that he was a "player's coach" who was more concerned with employee popularity than he was living up to the trust I'd placed in him. Still, replacing him and training someone new at this most specialized of record stores was a daunting proposition.

The decision to shutter the doors in 2017 was semi-spur-of-the-moment, even though I'd been thinking about it for a while. It had been a great ride. I had had the greatest job in the world, along with amassing an insane record collection. I hosted in-store performances by everyone from Dan Penn and Clarence "Gatemouth" Brown to the Clean and Rodriguez. Oar Folk/Treehouse had been acknowledged by major media outlets as one of (if not the best) record store(s) in the Twin Cities, even earning a blurb in a Rock and Roll Hall of Fame program. Between my recovery groups and my companions immersed in rock and roll, I had the best friends in the world. Adding to this was my engagement to a woman I met at the New Orleans Jazz & Heritage Festival, who loved me enough to trade in New Orleans for the frozen north to be with me.

But not everything was idyllic. I realized and accepted that I was a terrible manager/owner with poor supervisory skills. I had a great record collection, but life felt empty. I was struggling with depression, complex post-traumatic stress disorder (CPTSD), a lack of equanimity, and too often lashed out at employees or on social media platforms, garnering many younger detractors. I no longer trusted my employees, as several had committed grand larceny. One came to work just to steal a large portion of my personal collection—worth thousands of dollars—that I'd temporarily stored in the shop's basement while moving in with my fiancée. I naïvely assumed that employees still abided by an unwritten code that rendered the thought of ripping off a mom-and-pop store unthinkable—especially one with the status and reputation of Oar Folk/Treehouse. I was wrong and was emotionally devastated by the betrayal.

Compounding this was the conflict I experienced as the liberal, pro-small business, iconoclastic, idealistic champion of a record store with history and integrity at odds with the positive impact gentrification was having on my economic bottom line. I had a real estate lawyer friend tell me not to fight the changes to the neighborhood because it was increasing my property value. But the part of me who wanted to maintain the integrity of the neighborhood and the store was far more important than what was going to pad my wallet when I sold the store and retired. The six-story, chain store-anchored apartment building blocking the sun from my south-facing windows was a constant reminder that I was losing that battle.

I'd arranged an in-store performance in May by Minneapolis punk rock legends the Suicide Commandos to celebrate the release of *Time Bomb*, their first album of new material in thirty-nine years. It was important to me that they do the last in-store appearance. The Commandos were the band that introduced me to a new era of rock 'n' roll and, subsequently, a new way of life. I don't remember exactly what I said as I was introducing them, but I blurted out that I'd be closing the store at the end of the year. There was a deep, collective gasp. No one was prepared for it, including me, but in an instant, I'd made it official. I was simply exhausted, burned out, and knew it was time to move on. Even though there had been some lean years, the store was operating in the black. I arranged a small New Year's Eve retirement party with live music, including a performance by my friend and musical idol Spider John Koerner. When it was over, I locked the doors of Treehouse for the last time.

After the store closed, I went into a funk. I thought, this is what I do, this is who I am. If the store ceases to be relevant, does that mean I'm irrelevant too? So, I let the space sit with my old inventory for well over a year until my wife and my therapists helped me out of my rut and see that there was indeed life after retirement. I did some remodeling to upgrade the space and make it rentable. The last thing I expected was that someone would open another record store. I'd concluded that my business model was no longer relevant or sustainable in my rapidly gentrifying neighborhood. I thought it might become a coffee shop or an upscale deli. I did meet with a few people interested in renting the space, but their business plans weren't particularly compelling. Eventually, I gave in to a persistent woman who reimagined the site as a boutique incorporating the history of Oar Folk/Treehouse emphasizing its importance in local music history. The lease was signed just before the Covid pandemic hit. So far, the place remains shuttered. The build-out continues.

Retirement has been a revelation as my perspective on life has changed dramatically. However, I still have a deep, abiding love of music. My records are important artifacts, like paintings and precious antiques. I still have plenty of music in my personal collection: 50,000 or so albums and singles in my home, with several thousand more spread over five storage units. I am fighting a never-ending battle to consolidate the records, music books, CDs, and assorted ephemera and disperse the rest of it. I still frequent my neighborhood record shop. Music was, is, and always will be my life's passion. The candy store may be closed, but I still have a sweet tooth.

Notes

1 C. Collins, *Complicated Fun: The Birth of Minneapolis Punk and Indie Rock 1974-1984*, 67 (St. Paul, MN: Minnesota Historical Society Press, 2017).
2 Ibid.
3 North Country Music, https://twincitiesmusichighlights.net/north-country-music/ (accessed July 30, 2022).
4 Collins, *Complicated Fun*, 85.

Part I

Record Stores as Community

1

"We 'Bout It 'Bout It"

The Independent Record Store in Post-Katrina New Orleans

Jay Jolles

This past summer, when Hurricane Ida made landfall in New Orleans, barely eclipsed in intensity and damage by Hurricane Katrina, it claimed an integral part of New Orleans music history in its wake. The Karnofsky Tailor Shop, known more famously as the place that Louis Armstrong got his start—and his first cornet—was located on Rampart Street, a home to other historic jazz landmarks central to New Orleans.[1] The main commercial corridor in New Orleans, South Rampart Street is nestled deep in the swampy back end of the city in an area densely populated by Black New Orleanians as a result of the city's post-slavery racial order. In the 1920s, Morris Karnofsky opened Morris Music a few blocks away. It became the first place in New Orleans to sell jazz records.[2] While the Karnofsky building wasn't in great condition before Ida, it had been added to a list of properties and historical landmarks marked to be restored.[3] However, when it collapsed in late August,[4] the destruction of the Karnofsky Shop threw into sharp relief how the impacts of natural disasters are not just limited to infrastructure, demonstrating their potential to erode the cultural fabric of a city.

New Orleans is often hailed as the birthplace of jazz, cementing the genre as inextricable from the founding of the city. Yet, music more broadly is a fundamental aspect of the culture of New Orleans, with a litany of genres, such as sissy rap, bounce, and zydeco, finding storied local repute. While there exists a considerable body of scholarship regarding the relationship of the city to music, and vice versa, there has been little critical examination of the social spaces both produced by and enmeshed in the music. In the context of life in New Orleans following Hurricane Katrina, the genres considered endemic to the city worked to connote a sense of home subject to a particular diaspora in which the vernacular traditions integral to the New Orleans music scene were fundamentally altered. So, while scholars have long written about sound and place in conjunction, particularly when it comes to music, the recovery effort following Katrina complicated that calculus.

In this chapter, following Ana María Ochoa Gautier's notion of the "aural public sphere,"[5] in which music, media, and discourse circulate, I argue that in New Orleans, the independent record store is not merely a locus but rather an engine of culture that both establishes new and reifies old existing social and cultural ties in the wake of Hurricane Katrina. Such a function not only works to elaborate upon the ways local record stores have long championed local interests but also demonstrates how the social spaces produced by the record store in cities or other locales which have experienced sustained large-scale collective trauma have the potential to serve as sites of and for memory.

Bounce: The Sound of Music Post-Katrina

The musical response—both locally and globally—to Katrina was swift and robust. Ranging from the charity songs produced by artists like Green Day & U2[6] to Lil Wayne's incisive critiques of the Bush administration reflected in "Georgia . . . Bush" and "Tie My Hands," the event and its aftermath worked to usher in a new era of music from and within the city. While the city is perhaps most widely and popularly known for its jazz legacy, there is an irony inherent in the fact that it is deceptively difficult to find modern jazz in present-day New Orleans. Most highly successful New Orleans jazz musicians, or at least the ones that one might associate with the city, built their careers elsewhere.[7] One of the city's signature genres, however, is a particular type of rap music known as bounce. A grassroots genre with an "instantly recognizable beat based on two drum machine rhythms—the "Triggerman" and the "Brown" beat—sampled and resampled for use in virtually every bounce song," has a long history.[8]

Bounce emerged in relation to "local themes, communal affirmation, and dance."[9] Sonically, bounce encapsulates the synergy of many elements of not only New Orleans music but also the music of the southern more broadly. Relying upon heavy brass instrumentation, "due, in part, to a sustained dialogue with brass band music," bounce is replete with call-and-response vocals.[10] Aside from its signature rhythm, perhaps the thing that makes a track's sound most legibly bounce is its emphasis on the local. According to Holly Hobbs, founder of the NOLA Hip-Hop and Bounce Archive, the genre is really "the music of . . . the working class and underclass of New Orleans."[11] Given the well-documented assessment(s) that the city's working and underclasses were the most disproportionately affected by Hurricane Katrina, it is perhaps unsurprising that bounce artists took it upon themselves to chronicle this especially painful chapter of New Orleans history. Artists like Fifth Ward Webbie and Mia X in particular lamented the Bush administration and FEMA reconstruction efforts in their respective songs, "Fuck Katrina" and "My FEMA People."

These tunes and many like them demonstrate the importance of hearing a local story authored in a local voice. In bounce songs, which is often referred to as "project music,"[12] there is often an emphasis placed on repeated chanting phrases. In particular, artists will usually shout out or "roll call" specific housing projects, New Orleans neighborhoods, or wards. While intra and interward violence in New Orleans has long been a source of deep tragedy and heightened tensions, "the pan-New Orleans unity expressed in bounce has strong historical roots. This unity expressed in bounce stands in direct contradiction to the tensions and violence that exist in the city."[13] Bounce thus represents a space in which the sonic clashing of these wards, made manifest by the distinct signatures each ward puts on their particular style of bounce, catalyzes a space of creative fusion instead of violence. Primarily recognized as club music, bounce is just one of many genres produced in and associated with New Orleans that serves a functional purpose. Much like the music of the second-line tradition and its corollary the jazz funeral, the main purpose of bounce is to get listeners moving. Bounce thus "has the ability to bind normally antagonistic groups and mentalities through dance, an ability made possible by the sole intent . . . to make people feel good."[14]

Following Katrina, the return to the city was slow and hard-fought. But a year after the storm, popular bounce artist Big Freedia returned to New Orleans, working almost single-handedly to revive the club bounce scene by hosting weekly "FEMA Fridays" at Caesar's in the West Bank. According to Freedia, "It was the only club open in New Orleans at the time."[15] As a result, the centrality of bounce to New Orleans's post-Katrina rebuilding cannot be overstated. After the storm, New Orleanians were forcibly relocated across the country,[16] resulting in the nation's largest internal diaspora since the civil war.[17] Evacuees from the storm thus took the genre with them to other cities. Big Freedia was one such artist, first arriving at an army base in Arkansas before ultimately ending up in Houston.[18] In Texas, Big Freedia and other local artists found new venues and, by extension, audiences for their music. As a result, the popularity of the genre grew, and Big Freedia began featuring on tracks with artists such as Beyoncé, Drake, Kesha, and Icona Pop. While the mainstreaming of the genre in the last decade has meant commercial success for Big Freedia and artists like her, many of the neighborhoods from where bounce primarily emerged are still struggling to rebuild.

Even today, population numbers in New Orleans hover a bit below 80 percent of its pre-Katrina health.[19] Moreover, the reinvention of the city, fueled by disaster capitalism, predatory lending, and a hardscrabble tourist market, has meant rising rent and property costs, displacing longtime citizens and replacing them with younger white gentrifiers. As such, there has been rightful concern about appropriation of the bounce genre and the ways in which it has undoubtedly been complicated by the changing demographics of the city. Unfortunately, this is a fairly common phenomenon, especially when it comes to genres that are bound up in or inextricable from the struggle of Black artists in particular. As Hobbs notes, "Blues artists didn't get any money from blues, or very few of them did. It wasn't until later that people figured out, oh, this music is important and these artists should be preserved."[20] Despite this, or perhaps in spite of it, bounce remains a patently New Orleans genre, continuing to evolve within and driven by the work of independent record stores in the city.

New Orleans Record Stores and the Aural Public Sphere

In her work on aural modernity in Latin America, Ana María Ochoa Gautier writes that an aural region is "constituted by the mediations and (dis)junctures between different practices enacted by sonic transculturations."[21] In much the same way, New Orleans' aural contour is made up of an amorphous collection of sounds and styles, bound together by the relationship produced among race, place, and functionality.[22] And while Ochoa Gautier argues that in Latin America, this sonic constitution ultimately plays a negative role in developing what has been a profoundly unequal modernity, in New Orleans it instead works to invert some of the problematic paradigms that can plague life in the city. As such, I take up Ochoa Gautier's concept of the "aural public sphere" to demonstrate how the independent record store in New Orleans works as an indispensable engine of culture that seeks to fortify cultural ties through the power of rap and bounce specifically.

In particular, I explore two independent record stores integral to the musical fabric of New Orleans: Nuthin But Fire Records alongside Peaches Records and Tapes. These two stores, of the many scattered across the city,[23] are the only ones that sell primarily—almost exclusively—rap and bounce records. In the long aftermath of Hurricane Katrina, these two record stores exemplify the ways in which record stores have the potential to function as what sociologist Ray Oldenburg terms "third spaces," locales that are "informal and often gathering places where patrons could shop, interact, and take part in a communal process."[24] Not only do these record stores serve an economic function in a city still reeling from financial distress, they have also become markers of the city's march toward normalcy, an ever-present reminder of the adaptable and industrious nature of the city's music industry.

Concurrent with the rising influence of new media vis-á-vis the rise of streaming and internet-based music dissemination and promotion, Hurricane Katrina drew members of New Orleans' hip-hop scene to engage in entrepreneurial activities—side hustles, if you will—in order to supplement income. Performers often invested in small-scale operations, such as pursuing ownership of a local record store as one of many options. In his work on the significance of Black-owned record stores in Durham, North Carolina, Joshua Clark Davis writes, "Black merchandisers envisioned the record trade as an arena in which African Americans could pursue a broader strategy of economic self-sufficiency and sustaining black public life."[25] The industrious nature of these entrepreneurs demonstrates, among other things, how the tentacular reach of these stores in other different but similarly oriented industries benefits from and contributes to the aural public sphere. Much like the independent record store owners in Durham, "Many record retailers," particularly in the rap and bounce markets, "were inveterate entrepreneurs and ran a range of businesses, including nightclubs, recording labels, and entertainment management and promotion companies."[26] While Clark Davis's work is chiefly preoccupied with the significance of record stores in postwar North Carolina, such insights are relevant to this project in that they shed light on a similar set of circumstances shared by record stores and their owners in post-Katrina New Orleans.

Nuthin But Fire Records is "a hub of the local scene," owned by Sess 4-5, a rapper well known for putting on some of the biggest rap music events in the city.[27] As a result, Nuthin But Fire has become central to the enduring legacy of rap music in New Orleans, particularly bounce.[28] Opened in 2006, right after Katrina, Sess accounts for his success through a market heavily dedicated to bounce. Nuthin But Fire creates a sort of triangulation among brass band music, bounce, and the clubs, drawing attention to the particular desires for a particular kind of sound. Located on North Claiborne Avenue, Nuthin But Fire is nestled deep in the predominantly Black South 7th Ward neighborhood of New Orleans, a part of town scarcely visited by tourists. As such, Nuthin But Fire represents, like Clark Davis notes of the Black-owned Snoopy's Records in Durham, "a crucial nexus where African American enterprise, consumer culture, community, and of course, music" all meet.[29]

Much like Nuthin But Fire, Peaches Records and Tapes is "independent, NOLA owned & [we] 'bout it 'bout it."[30] These two record stores are devoted to driving support for local rap and DJs, with Peaches being the only independent record store in New Orleans specializing in the distribution of rap records. Opened for the first time in 1975, Peaches was long housed in the old Tower Records building in the French Quarter. That is, until Hurricane Katrina. After the storm, due to rising rent costs and the continuing downward trend of the physical music market, the future of Peaches became a lot more precarious. As Joshua Clark Davis notes, this phenomenon is unfortunately quite common, resulting in the disproportionate shuttering of predominantly Black-owned independent record retailers: "Generally speaking, the last decade has been unkind to music retailers of all kinds, but especially the independents, which include the vast majority of black-owned stores."[31] In lieu of closing completely, or filing for bankruptcy like the other seven original Peaches franchises, the New Orleans location moved to Uptown. While the relocation to its current address on Magazine Street prevented the total loss of an integral part of the New Orleans record store circuit, it fundamentally altered Peaches' role in the landscape.

Eddie Gaspard, longtime Peaches manager, said of the shop's forced move out of the French Quarter, "This was the meet spot. It was the hanging spot. Now it's gone."[32] Gaspard's commentary on the fate of the original Peaches location is indicative of its centrality not only to the music scene but also as a social space. By identifying Peaches as "the hanging spot," Gaspard draws attention to the shop as a site for community. Indeed, these record stores do not merely serve as distributors of records. They are also where many young rap artists get their start, and they therefore introduce many local rappers to the broader pastiche of the hip-hop music scene in New Orleans. "Local rap," according to LeMenestrel and Henry, has "been thriving since Katrina,"[33] which is marketed not only by a robust underground scene but through the local record stores as well. Though Clark Davis notes that independent record stores, particularly ones that are Black-owned, struggle to offer a robust "selection of competitively priced music," such a circumstance actually works in favor of Nuthin But Fire. While the store does provide a limited selection—they only carry local artist titles, particularly in bounce—they are one of, if not the only record store in the area to do so. And as such, their limited selection of titles is actually an asset, especially as the popularity of bounce as a genre continues to surge. This strategy squares with Clark Davis's assessment regarding the means necessary for sustaining the health of an independent record store—they remain in business either by "serving as boutiques for the most devoted of music fans or by selling considerable stock online, and often independent stores do both."[34]

Music, and its ability to not only circulate but be circulated, is central to New Orleans and its culture. In his book *The Power of Black Music*, Samuel Floyd writes of the song of jazz clarinetist Sidney Bichet's grandfather Omar, who died as an enslaved young person. Floyd writes that the song serves as an example of cultural memory, "a repository of meanings that comprise the subjective knowledge of a people,"[35] thus locating within Omar the driving force behind Black music in America. In much the same way, in the face of flood, digitization, and myriad other unfavorable circumstances, bounce music has become a driving force in New Orleans, particularly after the storm. Indeed, we might think of Nuthin But Fire Records and independent stores like them as what are often referred to as *lieux de mémoire*, a phrase made popular by French historian Pierre Nora who argues that these "memory places" are in some ways defined by their complexity. He writes, "at once natural and artificial, simple and ambiguous, concrete and abstract, they are *lieux*—places, sites, causes—in three senses—material, symbolic and functional."[36] While Nora's concept is most commonly used to refer to key historical sites, contested places bearing some degree of commemorative significance, recent trends in memory studies scholarship have moved toward examining *lieux* that exist beyond the confines of heritage places. In many ways, these record stores are "touchstones for the community to connect with the past through a connection with place"[37] in the same way that a battlefield or monument might be.

Indeed, "part of the appeal of sites of memory is that they often take on a symbolic life of their own."[38] As a result, Nuthin But Fire and Peaches thus become examples of what Veit Erlmann calls a "practiced place,"[39] a location that emerges through a set of relationships in space and time. This is significant in that the idea of New Orleans as home to these genres and the musicians who produce and contribute to them is thus anchored not only by musical practices but by networks of interactions. These work to shape not only the meaning of music post-Katrina but the lives of the artists who make it. As such, it becomes clear that some things not only can but will withstand the force of a hurricane.

Conclusion

New Orleans and the particular profound histories it contains are no less susceptible to erasure today than they were sixteen years ago. This is a fact no more crudely exemplified than by Hurricane Ida's landfall on the precise date of the sixteenth anniversary of Katrina. As climate change and rising waters continue to subject coastal communities to unrelenting precarity, it remains unseen how the various communities embedded within the cultural fabric of New Orleans will contend with a most uncertain future. It stands to reason, however, that music will play a large part in it. As Rich Paul Cooper writes of bounce musicians in New Orleans, "Their music is a voice resisting the erosion of their lives, their homes and their histories."[40] And I must admit that I don't find Cooper's assessment hyperbolic in the slightest. As much as New Orleans is recognized for its musical culture, robust social calendar, and tourist hotspots, it is "more than a stage where things happen."[41] Amid the changing tides of the market when it comes to the most popular forms through which to consume music, this chapter has made clear the ways in which two particular independent record stores in New Orleans have worked to serve as loci of culture and engines of future success for local musicians.

Notes

1. John Hasse, a curator of American Music at the Smithsonian, acknowledges that "There is probably no other block in America with buildings bearing so much significance to the history of our country's great art form; jazz."
2. WWOZ, "Morris Music," *A Closer Walk*, July 27, 2017. https://acloserwalknola.com/places/morris-music/ (accessed January 1, 2022).
3. WWL Staff, "New Orleans' Historic Karnofsky Shop Collapses during Hurricane Ida," *WWLTV*, August 30. https://www.wwltv.com/article/news/local/orleans/new-orleans-historic-karnofsky-shop-collapses-during-hurricane-ida/289-f0a6d93a-da3b-4446-9bab-557a1037ed60 (accessed January 1, 2022).
4. N. Clark, "Ida Leveled the Karnofsky Shop, Louis Armstrong's Second Home," *NPR*, August 31, 2021. https://www.npr.org/2021/08/31/1032868655/ida-leveled-the-karnofsky-shop-louis-armstrongs-second-home (accessed January 1, 2022).
5. A. Gautier, "Sonic Transculturation, Epistemologies of Purification and the Aural Public Sphere in Latin America," *Social Identities: Journal for the Study of Race, Nation and Culture* 12, no. 6 (2006): 803–25.
6. The song, entitled "The Saints Are Coming," was released in 2006 on the album *U218 Singles*.
7. Some of these artists include Grammy Award-winning trumpet player Nicholas Payton and pianist-composer Harry Connick Jr.
8. M. Miller, *Bounce: Rap Music and Local Identity in New Orleans* (Amherst: University of Massachusetts Press, 2012).
9. C. Cooper, "Bouncin' Straight Out the Dirty Dirty: Community and Dance in New Orleans," in *Hip Hop in America: A Regional Guide*, ed. M. Hess (Santa Barbara: ABC-CLIO, 2010), 523–48.
10. Miller, *Bounce*.
11. H. Hobbs, "About," *NOLA Hiphop and Bounce Archive*, 2012. https://digitallibrary.tulane.edu/islandora/object/tulane%3Ap16313coll68?sort=mods_originInfo_dateCreated_dt%20desc&islandora_solr_search_navigation=0 (accessed January 1, 2022).
12. H. Hobbs, "Bounce," Music Rising at Tulane: The Musical Cultures of the Gulf South, 2012. https://musicrising.tulane.edu/discover/themes/bounce/ (accessed January 1, 2022).
13. Cooper, "Bouncin' Straight Out the Dirty Dirty."
14. Ibid., 531.
15. C. George, "The Real and Raw History of New Orleans Bounce, as Told by Big Freedia," *INDIE Magazine*, June 1, 2018. https://indie-mag.com/2018/06/bounce-music-big-freedia/ (accessed January 1, 2022).
16. D. Noor, "Bounce Music Helped Put New Orleans Back Together After Hurricane Katrina," *Gizmodo*, August 28, 2020. https://gizmodo.com/bounce-music-helped-put-new-orleans-back-together-after-1844860567.

17 R. Cornwell, "Hurricane Katrina: The Storm That Shamed America," *The Independent*, August 20, 2020. https://www.independent.co.uk/news/world/americas/hurricane-katrina-the-storm-that-shamed-america-2057164.html.
18 D. Noor, "Bounce Music Helped Put New Orleans Back Together after Hurricane Katrina," *Gizmodo*, August 28, 2020. https://gizmodo.com/bounce-music-helped-put-new-orleans-back-together-after-1844860567.
19 A. Plyer, "Changing New Orleans Neighborhoods," *The Data Center*, September 21, 2021. https://www.datacenterresearch.org/reports_analysis/changing-new-orleans-neighborhoods/.
20 Noor, "Bounce Music Helped Put New Orleans Back Together after Hurricane Katrina."
21 Gautier, "Sonic Transculturation, Epistemologies of Purification and the Aural Public Sphere in Latin America."
22 M. Sakakeeny, "New Orleans Music as a Circulatory System," *Black Music Research Journal* 31, no. 2 (2011): 291–325.
23 Among others, these include the infamous Mushroom record store which opened in 1969 founded by the Student Liberation Front as a method of protest against high-cost on-campus bookstores, Euclid Records which opened in 1983, the Louisiana Music Factory which opened in 1992, and Domino Sound Record Shack which opened after Katrina, owned by Seattle-born Matt Knowles.
24 J. Davis, "For the Records: How African American Consumers and Music Retailers Created Commercial Public Space in the 1960s and 1970s South," *Southern Cultures* 17, no. 4 (2011): 71–90.
25 Ibid., 73.
26 Ibid., 75.
27 Miller, *Bounce*.
28 *Interview with Sess 4-5* (2012), [Moving Image] Dir. Holly Hobbs, USA: Tulane University Digital Library. In this interview, Sess discusses how, amid the changing culture when it comes to digital and new media, his record store has endured because "bounce pays the bills."
29 Davis, "For the Records."
30 Record Store Day, "Peaches Records and Tapes," *Record Store Day*, April 19, 2007. https://recordstoreday.com/CustomPage/614 (accessed January 1, 2022). Here, the phrase "'bout it, 'bout it" not only refers to the slang of being true or authentic but also alludes to the song by New Orleans-based rapper Master P that shares the same name.
31 Davis, "For the Records."
32 Cooper, "Bouncin' Straight Out the Dirty Dirty."
33 S. LeMenestrel and J. Henry, "Sing Us Back Home: Music, Place, and the Production of Locality in Post-Katrina New Orleans," *Popular Music and Society* 33, no. 2 (2010): 179–202.
34 Davis, "For the Records." It might be worth noting that Nuthin But Fire's commitment to serving only a local audience such to the extent that they don't sell online. Whereas Peaches Records and Tapes has a robust online store, selling not only records but an extensive catalog of merch, Nuthin But Fire doesn't even have a reliably accessible website.
35 S. Floyd, *The Power of Black Music* (Oxford: Oxford University Press, 1995).

36 P. Nora, "Between Memory and History: Les Lieux de Mémoire," *Representations* 26, no. Spring (1989): 7–24.
37 K. Reeves, "Sites of Memory," in *History, Memory and Public Life: The Past in the Present*, ed. A. Maerker (London: Routledge, 2018), 65–79.
38 Ibid., 67.
39 V. Erlmann, "How Beautiful Is Small? Music, Globalization and the Aesthetics of the Local," *Yearbook for Traditional Music* 30, no. 1 (1998): 12–21.
40 Cooper, "Bouncin' Straight Out the Dirty Dirty."
41 S. LeMenestrel and J. Henry, "Sing Us Back Home: Music, Place, and the Production of Locality in Post-Katrina New Orleans," *Popular Music and Society* 33, no. 2 (2010): 179–202.

2

Firecorner

The Importance of the Reggae Record Shops in Black London and the Cultural Confluence of West Indian Music

Kenny Monrose

> *Music is our witness and our ally. The "beat" is the confession which recognizes changes and conquers time. Then, history becomes a garment we can wear, and share, and not a cloak in which to hide; and time becomes a friend.*
>
> —James Baldwin

Introduction

This chapter will highlight how reggae record shops in Black London vividly disrupted racial dichotomies, gender biases, and class position by presenting the voices of the few remaining reggae record shop owners in London and discusses how these holdings developed from humble beginnings to become successful distribution outlets that introduced reggae music to international audiences. The emergence of outlets to purchase reggae music in Britain since the arrival of emigrants on board vessels such as the *Ormonde* and the *Almazora* in 1947, and more famously the *Empire Windrush* in 1948, is silent within academic deliberation surrounding the import of Black-birthed music in postwar Britain. These holdings provided a number of unique and workable functions for maligned Black communities, who often were impelled to subsist within subterranean urban enclaves. First, they acted as a symbolic locations and meeting places for Black communities. Second, they were regarded as a safe licit space, which although marked blues dances and shubeens as illicit, allowed for West Indians, albeit temporarily, to free themselves from the shackles of racial discrimination. Moreover, as the development of reggae music grew in Britain from the late 1960s into the 1970s, new subcultural groups such as punks, skins, and suede heads began to incubate and attracted white youth to the anticapitalist and protest informative sentiment imbedded within the genre of reggae. These recordings, which were often absent from mainstream airplay, spurred a thirst for the acquisition of imported vinyl singles from Jamaica to become an integral part of British working-class youth culture. As a result, reggae record shops became a crucible for Black and non-Black engagement and to become acquainted and appreciative of West Indian culture via a mutual love of reggae. It is fair to conclude that the record shop did more to bring disparate communities together than politics, educational institutions, sport participation, or religious instruction were ever able to.

West Indian Arrival

For the newly arrived West Indians, mid-twentieth-century life in Britain was challenging. This was not only due to the climate, isolation compelled by migration, and economic paucity endured but also due to an ever-present atmosphere of racism, accentuated by the specter of official and unofficial color bars and racial lines that were an innate part of an everyday British existence.

> In Guyana we lived in an envelope of English culture and already had an unstinting view of its institutions and how the culture worked. But when the West Indians came to England from the 1950s onwards it was like going to the moon. You couldn't get a sense of different streets because they all looked the same. The houses we came to live in were Victorian or Edwardian multi-occupied apartment blocks and looked like one immense building connected together with so many people coming in and out of them like ants. And so, it took a long while to develop a frame of mind where you could distinguish one building completed another. It was London and everything looked monochrome and monotone.[1]

Although postwar British governments for a short spell adopted a relaxed attitude to new commonwealth migration and settlement, Black immigrants had always been seen as a concern since their arrival in 1948.

> Almost from the moment the first stages of arrival of black workers in the UK, they were perceived both within and outside government as a problem.[2]

West Indian migrants like the Irish before them performed lowly occupations for which the English were not suited either physically or temperamentally. Blacks were constructed into bearers of ascribed characteristics which were a threat to British society. These tropes were manipulated and diffused within the Powellite portend of an impending racial civil war said to be the inescapable consequence of Black settlement.[3]

> Antagonism towards minorities often revolves around particular stereotypes. Jews in the East End have been portrayed as illegal or dishonest traders; the Irish were

variously seen as lazy and as undercutting general wage levels: the Maltese were sexual predators and so on.[4]

However, a sizable number of the West Indians held little intention of remaining in Britain outside of their well-rehearsed five-year plan and remained stoic in their resolve to make good and to exercise their rights as British citizens. The acquired schema of social conduct within Anglophone dominions of the West Indies was dictated by the instep of the British Empire. *Britishness* for West Indians meant ideals tied to duplicitous epitomes of affability, decency, and lawful obedience. These idealistic aspirations laid an infrastructure upon which a formation of an imposing colonial-inspired cultural identity was developed and maintained. Therefore, any emergent trait of idleness, frivolity, or criminality was fervidly frowned upon and highlighted the fact that the majority of settlers to Britain were often skilled and of good character. In fact, it is instructive to keep in mind that Britain was privileged to host some of the most respectable, gifted, and talented citizens from their respected islands.

> The average Jamaican who came on the SS Windrush on 24 May (from Kingston) was not destitute. The destitute man did not have £28.10 (the passage cost). One or two might be unemployed but they were from a family background of support. So, they were above average as far as income was concerned.[5]

West Indian arrivals were patriotic and entered Britain possessing highly flavored Victorian and Edwardian morals and standards that were firmly grounded within a synergy of colonialism and of course seeped religiosity. This was reflected in music of the West Indies that made its foremost appearance in Britain upon the arrival of the *Empire Windrush* to Tilbury Dock in 1948, with calypso king Lord Kitchener performing "London Is the Place for Me" on a live BBC news report at the dockside. A few famed artists and assorted musicians from genres ranging from blues, gospel, jazz, and more noticeably calypso disembarked the vessel. For example, as well as Lord Kitchener, famed calypsonian's Lord Beginner, Lord Woodbine,[6] and blues singer and actress Mona Baptiste together arrived in "London" from Trinidad on June 22.

The Blues Dance and Shubeen

> *Shubeen Saturday night and Sunday morning,*
> *Could appear anywhere without a warning,*
> *Music like dirt, woman to match every man,*
> *If you did go a All nations what a bam bam.*
> *(Friday night Jamboree Bigga Monroe, 2017)*

Music venues where West Indians could congregate and enjoy music were sparse. The dancehalls of 1950s Britain did not especially cater to Black music or patrons.

Therefore, West Indians formed their own alternative space to enjoy music which manifested in the form of so-called blues parties.[7] The blues party, basement parties or simply referred to as "the blues" or shubeens, were an integral part of Black British social life from the late 1950s and crucial to the understanding of the development of the reggae record shop and the functions they served. In its original guise, the blues was a makeshift informal gathering located within a selected domestic setting, typically at the extremity of the working week away from the confines and rigors of Black life in Britain and was crucial in promoting cultural solidarity and collectiveness. The blues originally surrounded the legendary Blaupunkt "blue spot" radiogram, which was a central feature of the West Indian home.[8] Stuart Hall states:

> The radiogram occupied a space like a religious object. It was not just a tool, but tremendously important because it was a way of bringing back home into the new place through Caribbean music and when black urban music begins to be produced here in England it is used to play that. Though it was sanctified like religious pictures or family photos in the front room, it was a subversive machine because it was carrying a different message: a message about the past, about memory, about home, about a new generation, about making a life in this rather inhospitable cultural climate.[9]

Over time blues parties expanded and became increasingly popular, eventually becoming an embedded feature of West Indian life. Wherever a Black community was located one could find a blues.

> All roads led to the blues dance. It was our thing. It was safe. It was an extension of our front room, and the vibes was always nice.[10]

The subterranean quality of the blues was exploited to maximum benefit by those who facilitated them by virtue of their entrepreneurial nuance and skill and were of course turned into profitable business ventures. The spaces liberated for the blues were often undersized, airless, and cramped packed with bodies fused together moving in tandem to music.

> Patrons and organizers alike have a "ram dance" conception of space, considering that no space is ever too small to stage an event, as more persons there are bursting at the seams of the venue, the more successful the event is.[11]

An entrance fee was charged, and food and alcohol were sold, and over time the radiogram was replaced by the more powerful imposing and more dynamic Sound System, the formal term for a travelling PA system and its owner, who would set up a high-powered amplifier, turntables, and speakers in order to create a wall of sound. This segues into the vital role that Sound System played during this time, as it was a significant cultural possession for West Indian people as it was regarded as one of the very few pieces of cultural inheritance that they held ownership and control over.

The Import of Sound System Culture

Sound System is an important factor in the development of reggae record shops in Black London. A greater insight of this can be gained by positioning a condensed timeline of its emergence in Jamaica and eventual transatlantic diffusion in Britain.

> The first Sound really to take off was Tom the great Sebastian; run by a downtown Hardware proprietor named Tom Wong whose skilled selector was a man named Duke Vin. Based in a premise on Pink Lane off Orange Street, the outfit was in operation in the late Forties, and Vin kept in ruling the dancehalls at the start of the Fifties through the steady stream of records sent to him from America.[12]

Postwar industrial expansion, political shifts, and economic uncertainties were not solely witnessed in Britain but had a devastating impact on her overseas territories. In Jamaica, for example, manufacturing sectors forced workers from depleted rural communities to resettle in more economically buoyant towns. This produced several socially significant alterations that impacted on the intonation, cadence, and sound of musical tastes and preferences. Moving from small simple villages to the larger complex towns exposed individuals to a new wave of melodies and arrangements. For instance, locally based mento and quadrille folk-based compositions were steadily being replaced by the sound of *progressive* and *urbane* American rhythm and blues introduced to the island by returning ex-service personnel and migrants via Rediffusion broadcast.[13] These songs were often appropriated by orchestra bands within dancehalls or transmitted to the island. Steadily, however, bands proved to be too costly to event promoters and were replaced by Sound System as explained by record producer Bunny "Striker" Lee:

> Y'see, after the orchestra play all an hour, dem stop fi a break, an' dem eat off all the curry goat, an' drink off all the liquor. So the promoter never mek no profit—dat did prove too expensive fi the dance promoter. Dem alone eat a pot of goat! So when the sound come now, the sound no tek no break. When these few Sound Systems come, it was something different.[14]

Shortly after replacing the orchestra bands in the dancehalls, the popularity of the Sound System in Jamaica grew to unimaginable proportions. Thanks to pioneering soundmen, such as Clement "Coxsone" Dodd (Downbeat), Arthur Reid (Treasure Isle), and Vincent Edwards (Giant), and original Sound Systems, such as Tom Wong's *Tom the Great Sebastian, Sons Junior, Count Nicholas, Goodies, and Waldron*, a trend was set that has proven to be perennially etched in the history of reggae music.[15]

> The fledgling Sound System culture of urban Jamaica was transplanted into Britain during the 1950s and on his arrival Peckings began to supply records to Duke Vin of Ladbroke Grove, the first Sound System in this country.[16]

Vincent George Forbes aka Duke Vin the Tickler arrived in Britain as a stowaway from Jamaica in 1951. Vin had previously acted as a selector for Tom Wong's *Tom the Great Sebastian* Sound System in Jamaica.[17] Once settled in London, Forbes was disappointed at the lack of venues where West Indian music was played, so went about setting up a Sound System in 1954 for the enjoyment of his countrymen and fellow West Indians.

> (B)ecause there were only two Sound Systems back in the beginning, in fifty-six or fifty-seven, they were always in demand. So much that you couldn't book Duke Vin and Count Suckle.[18]

As mentioned, the radiogram used for blues parties lacked the power and fidelity that Forbes was accustomed to, therefore being acquainted with "running sound" sourced an amplifier and loudspeakers equipped to produce the sound similar to that he left behind in Jamaica. Along with Count Suckle, George Pryce (Peckings), and Lennie Fry, records were sourced from places, such as the Rainbow record store in Harlem (New York), to play songs from artists such as Louie Jordan, Professor Longhair, and Fats Domino and would set in place a trajectory that would eventually lead to the advent of the walking bass better known as ska in the late 1950s, rock steady in 1966, and eventually reggae in 1968. In time, prominent exponents of Jamaican music, such as the Winston Groovy, Prince Buster, Laurel Aiken, and Derrick Morgan, sojourned to Britain to further the momentum of the popularity of reggae that not only provided an uplift for Black people but also garnered appeal from British subcultures such as Skinheads and Suede heads, unlike Teddy Boys who terrorized Black people.

> As a West Indian, you could not walk on your own in certain places or at certain times. You had to walk in threes and fours. The "teddy boys" went around with bicycle chains. When they saw three or four Jamaicans together, they would not attack us in a group.[19]

A new urbane attire developed from the mod's captured the attention of working-class youth. This is seen as the garb associated with the Skin/Suede movement, which relied heavily on tones of apparel befitting of hard industrial labor such as donkey jackets, monkey boots, Dr. Martens boots preferably tanned in cherry (black was associated with policemen), Aertex shirts, and Harrington "Peyton Place" jackets. Membership to these subcultures was solidly developed on the basis of music. Unlike other subcultures at the time, such as the aforesaid Teddy Boys, race mattered not. This can clearly be seen via the genre of boss reggae[20] and Bluebeat,[21] which had a strong white following in the UK during the 1970s.

> Ska came in like a rush. I love my jazz, but when Ska came along it lick clean out my head. . . . Because it was something from our own Jamaica and that meant so much if you were living in England . . . it took the people back home a while to realize there was a market over here.[22]

The Sound System held many functions. One of its primary purposes was to act within similar capacity to that of a radio station and to showcase and promote new recordings. Many Sound Systems built their reputation on being adept in obtaining exclusive music. These exclusive records, known as *slates or wax*, were often test pressings featuring well-known or otherwise obscure artists vocalizing over unmastered transcription discs known as *dub plates* and were played sparingly and in moderation as they wore out and diminished in time if overused. For a record store owner (some of whom owned Sound Systems), regular attendance to Sound System dances was a must in order to gauge the reaction of patrons to any new music that was played so any demand could be met.

> I was the first sound anywhere in the world to play Police and Thieves by Junior Murvin from Island records on a seven 7-inch (dub) plate. We used to play all of Islands promotional records six months before they come out. We use to also play Virgin promotions as well because we were getting them straight from the producers to play over the weekend then go back to them on the Monday morning and tell them the response. The first time (Bob Marley & the Wailers) Rastaman Vibration LP was heard anywhere was when we played it at Allardyce Hall early 1976. That was the biggest dances ever to be held in Brixton-tell everybody Supertone tell you that.[23]

While it is important to keep in mind the focus on music that Sound System dances granted, it is equally essential to remember the confidence that Sound System provided within the development of Black cultural identity in Britain.[24] For example, Sound Systems established a platform from which the uncompromising position of Rastafari that was grounded within reggae music could be vociferously heard, and central to the philosophy of Rasta were messages of hope for dislocated and displaced Black people. This can be seen by the sheer number of Sounds adopting Rastafari titles and African handles such as *Moa Ambassa, Entebbe*, and of course *Jah Shaka*.[25]

Those outside of the dance space had very little understanding of the entrepreneurial and industrial nature of running a Sound System. Playing sound was a labor-intensive enterprise which for some was a full-time profession.

> I had a good job as a machine operator for Walls Ice Cream and at the weekend used to play at blues parties and things like that, but you know coming from Jamaica I wanted to get back into music so in 1969 I decided to leave my job and bought a Sound System.[26]

Men often took care of the heavy work associated with setting up and transporting and running the sound. However, the role of women was equally important, despite not obtaining the exposure it deserves. Women were often the hidden hands within the administration of running the sound and were at the helm of the organizational duties.

> Men took care of the music, whilst women's role involvement was confined to the more "private" activities associated with domestic responsibilities such as food, drink and fanatical management. In other words, it was confined to those very activities which were essential to the success of the dance.[27]

As Britain moved into the 1980s, race relations became increasingly dismal. Legalization imposed to outlaw racial discrimination had failed miserably, and unemployment among Black youth escalated to unprecedented heights. Inevitably, the relationship between the police and Black communities had crystallized into one of distrust, discord, and conflict.

> It was a ticking time bomb for a generation of kids born [in] the UK. They were supposed to do better than their mums and dads and were about to put themselves on the white-collar market having been advised at school that they should view themselves as British, and at home that once that had an education door would swing smoothly open.[28]

While this decade was symbolized by social unrest and political friction, it was during this period the impact of reggae in Britain reached its zenith. The creative and the artistic capabilities of the post-Windrush generation began to transform the manner in which the aesthetic of Blackness was seen in Britain. Moreover, the themes within reggae at this time spoke to the antiracist movements that gained prominence and attracted both Black and white subcultures and countercultures to merge. A good example of this was the confluence between the punk rock movement with its antiestablishment stance and the anticapitalist stance of reggae. Soon major record labels became aware of this union and seized the opportunity to capitalize on its emerging marketability. Equipped with global distribution networks, the majors rapidly developed subsidiaries of labels to attract fresh audiences to Jamaican music for the first time.

> Honest Jon's a record shop not far from the Roxy and on the fringes of Portobello Road—a favored punk hangout—become thing of a Mecca as it began selling punk and Reggae single side by side.[29]

Richard Branson's Virgin developed the frontline label in 1978 which pioneered in recording uncompromised reggae music from heavyweights such as Gregory Isaacs, I-Roy, the Mighty Diamonds, and Prince Fari. This built upon the success of creative geniuses such as Dennis Bovell and Lloyd Coxsone who had introduced lovers' rock to the UK in 1975 as an authentic British form of reggae music. Lovers' rock not only gave strength to the majors such as Virgin and Island but also prompted the rebirth of *skinhead reggae* in Britain with the emergence of two tone which, like lovers' rock, became an example—British artists performing West Indian-birthed music for a mainstream British audience.

> Reggae is what we did—it was the heartbeat of the Black community, but the message had an impact on other communities around us.[30]

Resistance, Resilience, and Record Shops

> The reliance on recorded music takes on even greater significance when it is appreciated that for much of the post-war period, Britain, unlike the US and the Caribbean, lacked both a domestic capacity to produce black musics and any independent means for their distribution. At this stage the BBC was not interested in including African and Caribbean music in their programs. When "pop" charts began to be compiled black shops and products were structurally excluded from the operations which generated them.[31]

Public places where Black people felt safe to gather and congregate were often labeled as "frontlines" and characterized as locations for illicit activity such as drug dealing, illegal gambling, and prostitution. This resulted in Black-owned businesses and even places of worship being subjected to police raids on so-called *coon hunting* expeditions and blanket stop-and-search operations.

> Locations where unemployed youths—often black youths—congregate: where the sale of and purchase of drugs, the exchange of stolen property and illegal drinking and gaming is not unknown. The youths regard these symbolic locations as their territory, police are viewed as intruders, the symbolic authority, they equate with the criminal rookies of Dickensian London.[32]

However, the creative instincts and a resilience of Black people in Britain matched their determination to resist the racism they encountered at the hands of the state. Such strength, in tandem with the entrepreneurial spirit imbedded within West Indian culture, is one of the primary motivations behind the growth and popularity of reggae retail holdings in London. Selling reggae records provided a service to the community and a secure route for economic stability for those who had the right connections.

> Black Britain prized records as the primary resource for its emergent culture and the discs were overwhelmingly imported or licensed from abroad.[33]

In Britain the movement from Sound System ownership into record vending echoed the well-worn pathways set by store owners in Jamaica, who, as mentioned earlier, sourced records independently, then built Sound Systems before segueing into retail, and as we have seen, many went on to build recording studios and record labels. Jerry Anderson of Hawkeye in Harlesden, one of London's oldest and most successful reggae record shops, states:

> We started shipping records from Jamaica to England in 1974 and went on to be the supplying hub for all the major Reggae shops around the country, from all those in London to Don Christie and Summit and those shops in Birmingham.

We opened the shop [in] 1977 and nearly 50 years later we're still here and very important part of the community.[34]

Opening the record shop was not to be my only investment from the ever-growing income that I was accumulating from running the Sound System. Another of my dreams was to launch a successful record label producing my own rhythm tracks and voicing local and Jamaican artists.[35]

(B)efore long the record shop was running things in Brixton. Some of our rivals such as Desmond Hip City on Atlantic Road; Larry Lawrence's shop, Ethnic Fight on Coldharbour Lane; and Junior Booth's shop JB'S on Acre lace, were all put under pressure by performance. When it came to getting the latest tunes, there was only one shop in control and all other Sound Systems and deejays came to our shop.[36]

The record shop, like barbers or hairdressers, eateries, and places of worship, was an essential fixture of an established Black community. These spaces not only acted as business ventures but held important communal functions. Barber shops, for instance, were secure meeting places for Black men to assemble. The same is true of hair salons which acted as a protected space to celebrate Blackness and Black beauty for Black women.

Likewise, reggae record shops possessed manifold functions and were arguably the center of the community.

When I first came to the UK from Canada in the 80s, I didn't know anyone. I didn't have any family here and didn't know where to go to meet other Black people. I found my way to Body Music in Tottenham and spent most of my time just hanging out there liming and getting to know people.[37]

As the popularity of reggae music grew, so did the communal functions of the record shop, by serving to the needs not only of the Black community but also for those outside it. It became an unapologetically Black space where Blackness could be celebrated but was also a space accessible to all regardless of race or ethnicity. While exclusion zones based on race did not exist within record stores they were not vividly disrupted any gender biases either, as Black and white women often not only patronized them but would on occasion be the administrators in such establishments.

Regal Records in Hackney was the place to be in the 90s. Nearly all the soundmen in London used to pass through there at one time or another and more time it was June, Eddie Regal's wife, that was playing the tunes and dealing good with the people them and trust me she knew her music.[38]

Commentators such as Jones (1988) and Jachimiak (2021) suggest that white people often felt daunted and intimidated entering reggae record shops.

> Independent Reggae record shops, with the majority of customers coming from the Afro-Caribbean community and being quite often situated amid quite intimidating ghettoized urban settings, were far from welcoming for any white fans of Jamaican music.[39]

> Reggae record shops with their predominately black clientele, often located in inner-city areas, could present obstacles for white record buyers. As Jon (Jones' respondent) pointed out, for white people unfamiliar with this environment, the process of entering a reggae shop for the first time could be an uncomfortable experience.[40]

Such statements disregard the core mechanics and collective nature of reggae record stores and arguably speak to white male fragility of some individuals in the presence of Blackness, within what these observers consider as proscription, as several white people did not share these experiences. Christine and Kerry mention:

> I used to go there (Daddy Kools) on my lunch break. It was down a little alley off of Oxford Street and always had Black guys standing outside. I never felt intimidated going there. Yeah, they'd look at me but never said anything bad to me. Being a white girl in there didn't bother me, I liked the music, and it was the only place I could get the records I wanted.[41]

> The local HMV was boring. Me and my Black girlfriends used to go to Davis Record arcade in Upton Lane on a Saturday afternoon to get our records. It was like going to a gig. I met Sugar Minott in there once—that wouldn't happen in Our Price would it![42]

In London, if white people chose not to enter the shops it was most certainly not due to any form of color bar or race-related restrictions. In fact, reggae record stores were significantly more welcoming than a number of white-dominated leisure (or business) spaces such as pubs, music venues, and football grounds that openly excluded nonwhites. It is also important to remember that several prominent reggae record shops in London as early as the 1950s had white proprietors. For example, Rita and Benny Isen (later known as King), a Jewish couple based in Stoke Newington, North London, were pivotal in the development of retailing and the distribution of Jamaican music via R&B records from 1959, as well as Keith Stone proprietor of Daddy Kool's and John MacGilivray and Chris Lane of Dub Vendor Records, and many more.

> Dub vendor (founded by British reggae fans John MacGilivray and Chris Lane in Clapham junction in 1976), had emerged to meet the demand for new Jamaican music.[43]

Conclusion

Reggae record shops acted as resting place, and a place of solace or comfort and succor, for Black community members in London. They provided a space where Black identities and cultural practices could be openly expressed and also provided an unrestrained visual example of the aesthetics of Blackness to the aural backdrop of reggae and other forms of afro-descendant music. Moreover, they were regarded as the hub of the community which was evidenced by the hive of activity that circled them. At the heart of them lay a fierce entrepreneurial spirit that prevailed despite the many social obstacles beset by many of the Black and nonwhite owners who simply loved and appreciated the music. Alas, these holdings are slowly becoming obsolete with only literally one handful of reggae record shops left in London. This is in part due to the advent of the internet with its new technology of downloading, file sharing, and streaming services, and in part to the closure of vinyl pressing plants in Jamaica. Buying records seems an evocative reminiscence from a bygone era and remains kept alive only by avid vinyl collectors and the ever-decreasing circle of soundmen. Reggae record shops broke barriers and went some way to encourage new meanings of chaotic and complex issues such as identities and race and developed new artistic pathways and trajectories for ensuing generations to follow. The study of new commonwealth migration and its accompanying sounds, from dockside to dancehalls to Sound Systems to stores, allows us to gain a privileged insight into a slice of undocumented British culture. They helped bridge the political divide between Black and white Britons so that a union of cultures could emerge.

Notes

1 Phillips, from M. McMillan, *Front Room: Migrant Aesthetics in the Home* (London: Black Dog Publications, 2009), 25.
2 J. Solomos, *Black Youth, Racism and the State: The Politics of Ideology and Policy*, 30 (Cambridge: Cambridge University Press, 1988), 30.
3 John Enoch Powell was a conservative MP from 1950 to 1974, as well as an Ulster Unionist. He was best known for his "Rivers of Blood" speech in 1968, which focused on the immigration of Blacks into Britain. This, however, was not the central theme of the speech as was widely reported. The thesis of Powell's speech was a directed critique of the Labor government's introduction of the Race Relations Act, which deemed it illegal to refuse housing, employment, and public service to individuals on the basis of color, race, or ethnicity. While the act was subsequently passed and Powell had his parliamentary position revoked, his eloquent speeches afforded him widespread white working-class support from dockers and meat packers. However, in the eyes of many, particularly those in Black communities, his legacy is one of bigotry and his name synonymous with white racism.
4 G. Dench, K. Gavron, and M. Young, *The New East End: Kinship, Race & Conflict* (London: Profile Books, 2006), 169.

5 M. Phillips and T. Phillips, *Windrush: The Irresistible Rise of Multi-Racial Britain* (London: Harper Collins, 1998), 59.
6 Lord Woodbine settled in Liverpool and became the early manager of the Beatles.
7 A makeshift domestic space used for music and leisure where West Indians in Britain would congregate. Often also defined as a "shubeen," which was deemed illegal because it had not procured a license to sell alcohol.
8 A top-of-the-line imported radiogram from Germany.
9 Hall, S. from McMillan, *Front Room*, 36.
10 K. Monrose, *Illegal Entrepreneurship, Organized Crime and Social Control: Essays in Honor of Professor Dick Hobbs* (London: Springer, 2016), 78.
11 S. Niaah, *Dancehall: From Slave Ship to Ghetto* (Ottawa, Ontario: University of Ottawa Press, 2010), 60.
12 D. Katz, *People Funny Boy: The Genius of Lee "Scratch" Perry* (London: Omnibus Press, 2006), 16.
13 A popular broadcasting platform in British West Indian territories especially Barbados and Jamaica.
14 S. Barrow and P. Dalton, *The Rough Guide to Reggae: The Definitive Guide to Jamaican Music from Ska through Basement* (London: Penguin, 2004), 11.
15 An individual who is part of a Sound System collective closely associated with a sound. They could be an owner, security, engineer, selector, a toaster (MC/Deejay), a box boy (one who transports speakers), or operator (one responsible for how the system sounds).
16 P. Gilroy, *There Aint No Black in the Union Jack: The Cultural Politics of Race and Nation* (London: Hutchinson Education, 1987), 164.
17 A person who chooses records and plays them on a Sound System, a Jamaican term for disc jockey.
18 L. Bradley, *Bass Culture: When Reggae Was King* (London: Penguin, 2000), 116.
19 Niaah, *Dancehall*, xvii.
20 Boss reggae aka skinhead reggae in 1969.
21 Bluebeat was a UK record label that released ska and R&B music founded in 1960 by Emil Shalit's Melodisc Records. It has become a generic term to describe all styles of early Jamaican popular music.
22 Daddy Vego Peoples Sounds Records cited in Bradley, *Bass Culture*, 142.
23 Wally B, Supertone Records Brixton (London), personal communication with author.
24 L. Back, "Coughing Up Fire: Sound Systems in South-East London," New Formations, Summer, No. 5, 1988, 141–52.
25 A messianic religio-political movement upholding the belief that Haile Selasie-Negus Tafari Makomen (His Imperial Majesty the First of Ethiopia) is divinely appointed as the human manifestation of Almighty God. In addition, Rastafari suppose that Selaisie's presence heralded the call for Black redemption and repatriation of all displaced Black people globally back to Africa.
26 Wally B, personal communication with author.
27 C. Williams, "We Are a Natural Part of Many Different Struggles: Black Women Organizing," 154, in *Inside Babylon: The Caribbean Diaspora in Britain*, ed. W. James and C. Harris (London: Verso, 1993), 153–63.
28 Bradley, *Bass Culture*, 255.

29 Ibid., 449.
30 Dennis Bovell, personal communication with author.
31 Gilroy, *There Aint No Black in the Union Jack*, 165.
32 J. Rose, *A Climate of Fear: The Murder of PC Blakelock and the Case of the Tottenham Three* (London: Bloomsbury, 1992), 65.
33 Gilroy, *There Aint No Black in the Union Jack*, 165
34 Jerry, Hawkeye Records Harlesden, personal communication with author.
35 M. Gordon, *From One Extreme to Another: From 1960s Brixton Life & Sound System Business to 1980s Jamaican Underworld & Life Transformation* (London: Tam Re House, 2013), 52.
36 Ibid., 61.
37 Quincy Jam, personal communication with author.
38 Shortman, personal communication with author.
39 P. Jachimiak, "Curious Roots & Crafts: Record Shops and Record Labels amid the British Reggae Diaspora," in *Narratives from Beyond the UK Reggae Bassline: The System Is Sound* (London: Palgrave, 2021), 218.
40 S. Jones, *Black Culture, White Youth: The Reggae Tradition from JA to UK* (Birmingham: Bassline Books, 1988), 160.
41 Kerry & Christine, personal communication with the author.
42 Kerry & Christine, personal communication with the author.
43 C. Melville, *It's a London Thing: How Rare Groove, Acid House, and Jungle Remapped the City* (Manchester: Manchester University Press, 2019), 57.

3

Journey of a Girl in a Plaid Skirt and Knee Socks

Holly Gleason

Record Revolution, Coventry Rd., Cleveland Hts., Ohio

A day-glow Elvis Costello overwhelms a massive picture window. Record Revolution epicenters the new wave and punk coming out of the UK and New York in Album Rock mecca Cleveland, Ohio. Yes, the Dead Boys had Cleveland roots; Chrissie Hynde was an Akron expat; Pere Ubu, Tin Huey, and Devo were exploring the existential edges of contemporary alt-music, but Cleveland remained flyover.

For a junior high school girl in a plaid A-line skirt and forest green knee socks, it's not a friendly environment on one's best day. But it was the *only* place to get *New Musical Express* or *Melody Maker*. Or find the records being written about. If you want to know, you got to go—and I would. Psychically armoring for the bald-faced derision, the scoffing, and condescension, the music was far more important than the textbook reaction to the clichéd notions—goodie-goodie or slut?—a girl in a school uniform elicits.

Whip-thin, pasty-faced clerks, sometimes with razor blades dangling from earlobes, lobbed disdain like gobbed spit. Thankfully, disdain doesn't get caught in your hair.

Until the day it happened. "Kid," barked the sunk-eyed guy behind the counter. "KID..."

Surely, he wasn't talking *to me*. The store was empty. I looked around.

Turning to meet his eye, he cocked his chin up and tilted his head back twice, indicating I should approach. Scared. I didn't want to buy drugs, and I really didn't want to get insulted at close range. I walked over slowly...

"You're the kid that likes Elvis Costello, right?" I did. I was. How'd he know?

Then again, how many kids in school uniforms congregated at the headshop, sometimes hippie outpost, absolutely punk rock record store sprawled across *three* full storefronts?

"Mmmmmmhmmmmmm."

"Yeah, you're gonna like this." He made a noise like clearing his throat and sucking snot into his head. Still scared, as he reached under the counter, I was also transfixed. What horror was he going to bestow upon me? A joke? Something so dreadful, the mocking would be clear?

He pulled out a record with a dark-haired girl, *a girl* around my age, wearing an oversized biker jacket. She stared into the camera, not come hither nor—as Chrissie Hynde said in "Precious" from the Pretenders' self-titled debut—"Fuck Off." Hair a little fly-away in the wind, she was looking into the camera as if she was looking at me.

Fool Around. Rachel Sweet.

I took it in.

"She covers 'Stranger in the House'. Does a good job . . ."

"Really?"

Sweet was like me; she liked Elvis Costello.

"An import. It's more expensive, but I think you'll like it."

I had twenty dollars in my knee sock. Christmas money from Nana. He held it out, let me look as he kept talking, "She does some oldies, stuff your parents' ll know. And she sings about Firestone High School in Akron . . ."

"She does? *Really?*"

We were having an actual conversation.

"She's from there. Probably your age."

BOOM! The insult drops. The record was—I believe—$12.99, but I was curious. I mean, she was almost from Cleveland.

Maybe, even more importantly, this record, this Rachel Sweet record, came in, and someone at Record Revolution thought about *me*. They didn't know my name; they barely noticed, only snarled while I shopped, but they'd remembered.

Record stores, as evidenced in Nick Hornby's *High Fidelity*, especially John Cusak's movie version, were their own nuclear family. Odd balls, obsessives, hip guys, kids who had to work somewhere, they all integrated into a collective that gave record stores their personality. But to have *them* notice *you*? That was meaningful, suggesting you might yet "belong" in this world of infinite but very specific cool.

And each store, along my journey through life, had a distinct vibe, a specific purpose. Almost like faith-healing, you recognize the gift from an early age—and not everybody gets the mojo.

Record stores were everything to a certain kind of kid. Bins and bins of people most of us would never be, who'd capture our imaginations and speak our burning, even yearning desires. Rage, loneliness, doubt, sexual identity, failure to fit in, inability to cope, or just something to bounce around to. What we couldn't tell our parents, teachers, friends, enemies, we could hear on these black vinyl discs—amazed someone else knew what we felt.

Beyond the sometimes outright hostility from clerks, who in the late 1970s seemed to live and die for the music, each store was its own kind of temple. Each seemed different in their manifest destiny, but all were a portal to somewhere you wanted to be.

As Penny Lane, the groupie heroine, tells the baby rock critic William Miller early in former *Rolling Stone* critic/Oscar winner Cameron Crowe's memoir-on-film *Almost Famous*:

"Never take it serious.

If you never take it serious, you always have fun.

If you always have fun, you never get hurt.
And if you ever get lonely,
you just go to the record store and visit all of your friends."

Truer words were never spoken.

Whether you would ever meet the faces on the covers face-to-face—David Bowie, Joni Mitchell, James Brown, Mott the Hoople—or not, you already knew them. They understood. And in that, they helped you understand, too, and not feel so alone.

John Wade, Shaker Square, Van Aken Rd., Cleveland, Ohio. Early/Mid-1970s

When my mother would get her hair done—meaning wash, set, and curled, popped under an actual hair dryer—I would stare at prehistoric movie magazines until I wanted to cry. Tedium is the enemy of all children. Though well-behaved, the weekly ritual was vexing. But every week, to Leo's we would go; tucked in a sticky vinyl chair, I would endure.

My incessant questions, begging and cajoling, would become annoying. I would be told to "sit down and read," the words falling like an axe chop-chopping wood. Silently, I would.

After enough weeks, my mother changed her game. When the whining "I've already read all these magazines" started, she'd reach into her purse, take out a five-dollar bill, and send me to Burrows Books, through a little walkway and down the wraparound sidewalk five stores and one restaurant away.

It was understood I could buy anything I wanted with the money, but I had to explain my decision. Back then, you could get two good paperbacks—or a paperback and a magazine. I'd spend more time deciding what I wanted than I would sitting in the stuffy shop that smelled of perm solution, Aquanet and peroxide dye. Until the day . . .

After the fancy wine store, there was a very basic storefront, usually filled with classical records and scores of sheet music in the window. Occasionally, jazz titles or "vocalists" like Dean Martin, Lena Horne, or Ella Fitzgerald might pop up. What made me walk through the door, hearing the tingle of bells announcing a customer, I do not know. But when I saw the pegboard on the walls and albums held by brackets for easy viewing, my heart raced.

Young people with long hair, buckskin jackets, psychedelic letters were on many of the covers. Looking like my wicked cool babysitters, they were—I was sure—people my wicked cool babysitters would know, maybe even own.

Five, six years old, I felt a rush of pure joy. Why? To this day, I have no idea. Our house had Louis Armstrong albums, Peggy Lee's *Is That All There Is*, the ironic satirist/songwriter Tom Lehrer, and Frank Sinatra. This wasn't that.

"Can I help you?" a man with a short-sleeved button-up shirt and tie asked.

"I don't know," I replied. Clearly overwhelmed, I was doing six kinds of calculus. My mother had not told me I *had* to buy a book . . . There were so many choices, how would I know . . . I was so short, I couldn't reach the bins filled with records, nor would I be able to take anything off the wall. "How much are they?"

I believe $3.99. Perhaps $4.99, and the man made an allowance for tax. I don't remember.

I loved *The Partridge Family*, especially Susan Dey with her long straight hair and non-impressed way of facing the family band's career snafus. Was telling an old man I liked them saying "I'm a little kid and don't know anything"?

And there was the album with the lady and the cat. The impossibly cool Mrs. Morton, one of my mother's best friends, had that record. She'd told me Carole King was a songwriter but an even better artist. And there was the cat.

Partridge Family? Carole King? Carole King? Partridge Family? A horrible choice. I knew *The Partridge Family Songbook* from television but Carole King? *Tapestry*? How could I tell? Finally, I pointed at the record I wanted. I couldn't remember the woman's name, but I mentioned the cat. He brought a copy of the album down. "Carole King," he said with a half-smile. "That's a good choice."

Reassured, I played it off like I knew. "Thank you."

He took my money. The cash register went PING! The record went in a brown paper bag, the top folded over the top edge. He may've given me change. I don't remember. I'd just bought a record! It was like a drug hitting my veins with total rapture. I full smiled.

Once out of the store, I steeled myself for interrogation. She hadn't said it *had* to be a book; she didn't insist I had to spend the $5 at Burrows. She hadn't. She absolutely hadn't.

My purchase was met with a nonplussed reaction. It wasn't she didn't care; she didn't understand. It made no sense. "Why would you do that?"

She wasn't yelling. She wasn't marching me back down there to get the money back. She wasn't even hissing or insulting. She turned back to Dennis, the hairdresser, asking for a little more lift on top and maybe some curl on the one side.

Bullet dodged, I waited to get home to truly understand what I'd done. Anticipation, excitement, curiosity. Peaking in the bag, I could read see the girl with the frizzy hair, the needlepoint canvas she was working, the cat and the sunshine pouring in through the lead-paned window with thin drapes hanging down.

Whatever it sounded like, I thought I might want to be this person. Her feet were bare. She wore blue jeans and a top that looked soft. She seemed so at ease. Yes, I think I'd like to be this lady.

Trying to not act too excited, I took my treasure to my room. Running my fingernail along the opening, the plastic gave way and the record came out in my hands. Shiny, black, I stared. Whatever it sounded like, it was *mine*.

What it sounded like, though, was a revelation. Pianos, acoustic guitars, drums that moved the songs along. It wasn't lush like my parents' records, or jazzy. It was . . . organic? True? Like the music I might hear at school or folk mass. And King's voice—a warm alto that almost sounded like sweet/salty caramel—was familiar and friendly.

But the songs? I didn't know *why* the earth would move or *what* a natural woman might be, but I understood feeling far away from where you were supposed to be, the idea of something being broken and the returning to a place you belong. Even more, her promise that I had a friend, the idea that—never sure whether I was a pretty child—I could be as beautiful as I feel?

YES! She was like the world's smartest babysitter or an upper-school girl willing to share the secrets. She sang "Where you lead, I will follow"—I did; ultimately *Rhymes & Reasons, Wraparound Joy,* later *Thoroughbred.*

No one had to know I'd found a secret decoder ring for life. Buy the right records. Listen to the right songs. Figure out the golden needles in the lacquer haystack; no matter what was going on at home, somehow you wouldn't be so alone.

Never care too much, then people can't take it away from you. Don't let the parents realize just how addicted to listening to *Dreamboat Annie* or *Horses* or local songwriter Alex Bevan's *Springboard* I was, how seeking the next "fix" I'd become. They should have known. Records became my primary hustle. Long before I could drive, I was figuring how to *get* to the record store, whether suggesting a meal close by, seeing if any upper-school kids would take me to Coventry with them or just whining, *whining* I *needed* a new record!

A drug of choice doled out in 45 rpms with the big holes in the middle—OR 12" 33 1/3 rpms that held full encyclopedias of awesome. Eavesdropping to figure out what I might want, or need, reading Jane Scott in *The Plain Dealer* to sense who I would like *or* who I could learn from.

Record Theater, Mayfield Rd. and S.O.M. Center Rd., Mayfield Hts., Ohio

As eleven turned to twelve, I realized how different record stores were. Record Theater at the Golden Gate "strip" mall was a big brightly lit room with rows of bins punctuated by stand-ups of suggestions—end caps—at the front of each. The music loud but not too loud. Posters everywhere for *every* thing: Stevie Wonder, Foghat, Judy Collins, John Denver, Richard Pryor, Todd Rundgren, Fleetwood Mac with the knacker-balls hanging between Mick Fleetwood's legs causing me to titter. Emmylou Harris with the peacock feathers in her hair made me buy *Quarter Moon in a Ten Cent Town* with no idea what it was; same for Valerie Carter's hat and *Stone's Throw.*

Record Theater was welcoming. My supercool grandmother could hang out, telling the clerks she "got what you're into: making your own scene." They dug the pixie-haired lady in the mohair coat, bringing her grandchild to search bin-to-bin before feeding her TGI Friday's potato skins then returning her to the always fighting parents.

Grandma wanted to know about Jackson Browne, *Endless Summer,* and Rickie Lee Jones ("she sounds be-bop to me!"). She'd talk about Benny Goodman and Bing Crosby before he was the Christmas TV special guy.

Nothing said love quite like a yellow bag with the Record Theater's yellow old-school Victrola logo. Even if—occasionally—attached to a lecture. My parents were separated; currying favor, my father had volunteered to get me a record. "The Marshall Chapman," I requested, having heard the most brilliant radio spot. Standing outside his Mustang when I emerged from school, I could see the bag behind his back. I noticed the cigar puffing but didn't read his annoyance.

"Uh, pro," he began. "Can you, uhm, tell me what your fascination is with this . . . I believe . . . woman?"

"Yes," I replied forthrightly. "They say she's the female Mick Jagger."

Nodding, he echoed, "Jagger, huh? You know the *name* of this record?"

Crossing the fingers of the hand that was carrying my book bag, I sheepishly shook my head. If you don't say it, it's not a lie, right?

"You have no idea?" he asked more directly.

"Uhm, no," I gulped, remembering the day in third grade he'd torn up my *Catcher in the Rye* because of the swear words.

"NO *idea?*"

I shook my head.

"Well, I don't know what kind of trash *this* is, but it's called *Jaded Virgin*."

Withdrawing the album from the bag, he held it out. "Do you even *know* what that means?"

Chapman was wearing an opera scarf, a guitar low-slung, tilting to one side in loose-fitting jeans. I grinned. My heart raced. This was going to be cool.

"No, Daddy, I don't. I think she's kind of rock & roll, and she sings about how sad and lonely it is on the road. Kinda like going to golf tournaments."

It shut him up. It shut him down.

The record swerved from lean rock to country to a little reggae. The yowling "Why Can't I Be Like Other Girls" was a reckless anthem of independence. I wasn't telling anyone about my private manifesto, but I blessed the people at Record Theater for bringing that record in, for giving my teen rebellion a theme song.

If you could send your dad to Record Theater, it was mostly because it was just a suburban store with cooler posters. The skinny kids with the feathered hair might not be as menacing as the guys at Record Revolution, but they knew contempt. They sold local bands like Deadly Earnest & the Honky Tonk Heroes, Buckeye Biscuit, and the Euclid Beach Band's albums; they even carried Wild Horses' staccato breakdown 45 "Funky Poodle." Just don't . . . for the love of God . . . ask about country music. Ever.

Yes, there was a fairly robust section . . . in the back . . . by the eight tracks behind plexiglass that were dropped onto a little belt that took them to the front . . . But the country section, like porn in video rental places, was not to be discussed. Private, embarrassing. Don't.

Asking for Waylon Jennings—having heard Deadly Earnest play "Ain't Living Long Like This"—I was march-of-shamed to the back, dropped in front of the Js. Jennings's section was more than two inches thick; I couldn't remember the song's name. The clerk didn't care, he enjoyed watching me sweat.

"I don't *listen* to that," he rebuked. "You're gonna rot your brain. You're not some hillbilly. This makes no sense. Maybe some Bad Company?"

Pursing my lips, I flipped through the jackets, cheeks blazing. Backwards and forwards. Jennings had a high-topped pompadour in some of the pictures, which made me uneasy. Stealing a look, hoping for reassurance, I saw scorn. I doubled down.

Dreamin' My Dreams—with a tight shot of a man in a white shirt and leather vest—seemed hopeful. Also, a little dangerous.

I had no idea, and I didn't get the song I was after. But that voice? That bottom string of those guitars, the way the tremolo went through you?

I had no idea what outlaws were—or why they mattered. This guy seemed like a rogue; like someone in a rock band who also played real country music would dig. Screw that clerk! I could be right, too.

Sometimes negative navigation could be as formidable a resource as the person who knows what to suggest. It hardens one's resolve; teaches the value of the well-deployed "Oh, yeah." Knowing you could be wrong, your will to *know* by your own ears, establishing owning your own mind.

Dreaming My Dreams might not have the song, but there was swagger. That deep, dark voice held a kind of brio I didn't experience in my nice Midwestern town; dangerous, different than the thrashing rage-on bands at Record Revolution. Two lessons, maybe three.

Be your own compass. Know the door opened for the wrong reasons may still yield real delight. Be open to more music than what you already love.

Record Theater—like Northern Ohio's Camelot at the real walk-around inside malls—was a chain. Covered in posters, the occasional stand-up version of some artist who was obviously a priority, they were homogenized to present no real threat. But they also offered none of the music industry's flash/pop/wow.

That arrived when Peaches opened a massive high-ceilinged, wood-paneled, concrete-floored outpost by the Randall Park Mall, then the largest shopping center in the world. Giant album cover mock-ups—of Bob Marley, Linda Ronstadt, Hall & Oates—adorned the entire perimeter of the building like a rock and roll tiara; a marquis "premiering" new records or in-store appearances stood over the entrance like a movie theater. Inside, an electric buzz, islands of records recommended, aisle-after-aisle of record bins and more of those hand-painted renderings of the art overwhelmed customers.

The music was loud, the sheer sonic and visual ballast leaving you invigorated. Like Vegas to a gambler, it was a record buyer's cocaine: you entered at your own risk or left your money at home. For hours, you could get lost, found, thrilled, and surprised.

This felt like big-time show business, whatever *that* was.

But they made sure when someone like local folk icon Alex Bevan released his second album, *Grand River Lullabye*, to do it up like a major release. He came in, played for a few hundred people who showed up, then waited in line to have their records signed. What was so sexy and immediate was tendered to one of our own. Suddenly, a local hero was a rock star. Even if only for the day, the excitement was palpable. Looking around Peaches, feeling the buzz in the air, people talking to each other at an accelerated pace about this "Skinny Little Boy from Cleveland, Ohio . . . who came to chase your women and drink your beer" delivered a message far more potent than the

"To Holly, With Love & Light" Alex Bevan scrawled on my album. The message was clear: it can happen to you. It can happen to someone *like* you.

Spec's, Town Center Mall, Glades Rd., Boca Raton, Florida

Landing at a prep school in South Florida—for two seasons of golf, trying to get "recruited" for a scholarship—I was freed from my plaid skirt but away from everyone I'd ever known. Taking care of a father who was having health issues, WSHE's steady diet of Molly Hatchet and Journey left me hollow and adrift.

Once again, a local record store saved my life. Spec's, founded in 1948 in Coral Gables, Florida, by Martin Spector was a small South Florida chain with an outpost at Boca Raton's Town Center Mall. With horrible fluorescent lighting and white record bins, it was so sterile, so housewifely, it made my heart sad.

But it was the only escape in an upper-middle-class suburb that offered solace, so my ninety minutes of refuge each day—lunch, plus an open period—afforded me the space to look at *Buckingham Nicks*, the pre-Fleetwood Mac Lindsey Buckingham/Stevie Nicks collaboration, and wonder how one knows when it's time to move on to bigger dreams or to consider Jimmy Buffett's legacy as a poet who could write about the characters I was encountering—snow bird tourists, senior citizens who'd lost things that mattered, drug smugglers, and tropical beach bums—in a way that was so much more than "Margaritaville" with its "lost shaker of salt."

Spec's was quiet, the opposite of Peaches and Record Revolution. But it gave me the ability to drill down into the credits, to walk to the counter and ask questions. The hired-for-the-bougie-suburb staff found me cooler than the white wine-swilling mommies, so they were always willing to try to figure it out.

Once, racing to pick up the leader of Pure Prairie League in Fort Lauderdale after a random meeting at Disney World's "Grad Night," the store manager sliced open the band's latest album to tell me the singer's name was "Vince Gill." Knowing there was no way the hotel would put a call through without the name, he helped a fellow music lover out—then ran a key partway across one side, saying, "Oh, it was damaged," because he knew I didn't have the money to pay for it.

And in their cutout bins, there was treasure untold. *Townes Van Zandt: Live at the Old Quarter* brought a starkness to singer/songwriters that stopped me in my tracks. How could I know those songs, sung like ghostly fingers scratching at a cold November windowpane, would bring me into a world of "Nashville's Credibility Scare of the Late 80s" with a map to its greatness. Not because I went seeking it but because for $3.99, I took a chance on a double-record set—and learned about songwriting austerity in my game of music lover roulette.

Mostly, though, Spec's was a place for an out-of-place teenager to daydream. To think about concerts I'd been to, artists I wanted to see. It taught me to recognize even in the less than ideal, the less "awesome," you could find genius. It was a matter of looking, seeking, once again being open.

Not every cutout was great or special. You took a lot of shots, some didn't connect. But that was half the fun of it, too. See who the writers were, or musicians, or guests—maybe that leads where you want to go.

Heck, James Taylor's *Walking Man* was in the bin. I didn't even have to turn it over. Just go where the cast-offs are and pay attention.

Open Books + Records, Ft. Lauderdale; Y&T Records, Coconut Grove, Florida; Spec's, S. Dixie Hwy, Coral Gables, Florida; Peaches, Ft. Lauderdale, Florida

In Miami, the indie stores—Open Books + Records and later Y&T Records—returned me to record store as gathering place. Not everyone loved every band you did, but if you could make the grade—and being a young freelancer for *The Miami Herald* as I was going through college, my ticket was punched.

Coral Gables' Spec's on Dixie Highway was more technicolor, a bit louder. Sometimes people recording at Criteria Studios, who didn't want the pressure of being seen at Peaches in Ft. Lauderdale and having folks know they were in town, would show up. You'd come around a corner, and there was Rob Halford from Judas Priest. How unlikely, and yet, even rock stars sometimes want to flip through the bins, see what their peers are doing—or find a record they need for inspiration.

More exciting was the idea that college radio was igniting a movement; the indie record stores were the glue for labels like Twin/Tone and IRS or majors like SIRE. When the Ramones in the first flush of their success came through, they decamped at Y&T; Henry Rollins did an actual poetry reading at Open.

Inspired by the DIY spirit, both record stores released albums. *The Land That Time Forgot* on Open Records showed the world South Florida was more than old people in plastic shoes and drug dealers as they compiled the best of the punk scene for a fifteen-track power blast featuring the Eat, the Bobs, the Front, the Spanish Dogs, and the Vulgar Boatmen; Open's *Live at the Button* LP from rock quarry worker Charlie Pickett & the Eggs received glowing reviews from *NME*, *Melody Maker*, and *Trouser Press*.

Y&T upped those stakes significantly, as Mary Karlzen was scooped up by Atlantic Records in New York and the Mavericks created a Nashville-bidding war. Later, For Squirrels would sign with Epic Records, before a tragic van wreck ended the bands' career. Owner Rich Ulloa recognized the talent; he gave the artists a way to get their music turned into records *on* his indie Y&T Records.

Why not? Why the hell not?

Hanging around record stores long enough, the audacity of dreams descends. Yes, you can reflect or alter your mood. You can climb into your records and manage your feels. You can discover truths you didn't know you needed. Play your cards right and someone will hand you the record that changes your life.

It's a secret club that knows where the cool shows are, the band to watch, even the inside gossip. If you get the joke, the possibilities are endless. When someone presses a record into your hand, tells you something—especially in those pre-internet days when records were sold to the stores on the phone or in-person—going on with a band that you love (or loathe), you feel attached to something you care about.

But lean in a little closer, especially back in the day, you see the folks running the stores making a difference for musicians they loved. When waiting on a major release was everything... when seeing an Alex Bevan or a Charlie Pickett bringing the heat... when hearing the story of poor Patti Smith falling off the stage in Florida from the guy who sold you *Horses*... that bond matters. In a world where nothing is sacred, it's holy.

When a band would see their album cover on the outside wall of Peaches or Tower Records, especially the one on the Sunset Strip, no words can capture that. The whooping or speechlessness is a body slam of "This is really happening."

And to that community, you can count on these stores to stage epic Record Store Day events, live in-store plays, food drives, or pandemic drive-bys (you call, they run the music you need to the car). The refuge of record stores really doesn't change.

Grimey's; 12th Ave So, Nashville, Tennessee; now E. Trinity Lane, E. Nashville, Tennessee

Michael Grimes—aka Grimey, co-owner of Nashville's Grimey's New & Pre-Loved Music—is a recovering bassist from several prominent local bands and a stint with Bare, Jr. Taking the notion of record store as gathering spot for music people, he opened the Basement underneath his store. One year, Metallica—rocked by Grimey's fan-forward approach—played the Basement in advance of their Bonnaroo set.

See it, touch it, believe it, do it. Record stores are good like that.

Even all these years later, in a new town or out of sorts, normalcy exists by talking about records across a counter from someone you don't even know. You can learn a lot about a city that way, not just the demographics but the heart and what the locals value. Twenty minutes of small talk can tell you almost everything... and that's never stopped.

Harder to find record stores, perhaps. But, just as when I was a kid in a plaid skirt, when you made the effort to seek the record store out, there's always something that makes it more than worth the effort. As a full-grown adult, I sometimes get the derisive scoffing reserved for "lame middle-aged people"; but that's okay, I grew up on that.

Now tell me where the Aaron Lee Tasjan and Valerie June records are, please. And hey, that's a great T-shirt! Have you got one in a medium girl cut?

4

The Cult of the Record Bar

Stephen Shearon

At its height in the 1980s, Record Bar was a force in the American music industry: the second-largest record-store chain in the United States, with at least 180 stores in 30 states. Two decades later, in August 2006, when the Record Bar brand was long gone, Bob Prout, a former Record Bar employee, posted on his personal blog a short essay, "The Cult of the Record Bar." Prout's post was triggered by Record Bar memorabilia he had found in his attic as he prepared for an upcoming move. The essay consisted of only a few paragraphs and a couple images, but it generated an outpouring of heartfelt sentiment from commenters. And the comments kept coming as Record Bar alumni (mostly former employees and a smattering of patrons) discovered it in the succeeding weeks, months, and years. A lot of people who had interacted with Record Bar once upon a time had missed the company, describing it as a great place to work, with great bosses, and the Record Bar era as one before innocence was lost—"when all that mattered was the next shipment of new releases."[1] A few commenters also mentioned Barrie and Arlene Bergman—in particular, Barrie—as the persons most responsible for the company's special culture: one of the reasons for all the positive comments.

Prout began his essay:

> For a number of years during the late 1970s and early 1980s I was a member of the Cult of the Record Bar. If you were a member too, you know what I mean. Most people thought of Record Bar as a business engaged in selling records and tapes and a varied assortment of music-related items . . . but in reality Record Bar was a cult. Once you were a member, you were a member for life. Right now there are people in their 40s and 50s who are leading seemingly normal lives, but hidden undeneath [sic] a (theoretical) mantle of respectability is a Record Bar employee.[2]

One of the images Prout included in his post was a caricature of "Uncle Barrie" (Barrie Bergman, then the company president and CEO) as the "Head Counselor" of the "Record Bar Summer Camp '81," in which he admonishes the "Boys and Girls" (mostly store managers and central office personnel) to remember the camp motto: "'Clean mind, Clean body, take your pick.'"[3] The reaction to Prout's post was strong enough that at the urging of several commenters, he started a private social media group on

Facebook named "Cult of the Record Bar." With over 2,100 members it thrives to this day, providing a virtual gathering place where those who care about all things Record Bar can reminisce and engage.[4]

Record Bar was a successful corporation. But its more lasting impact is the result of the culture its management cultivated: a culture built around family, youth, a commitment to fun, and strong business values.

Record Bar

Although Record Bar was probably best known as a large chain, its roots lay in three small, independent record businesses established in the Piedmont region of North Carolina in the years during and after the Second World War, businesses owned and operated by two families related by marriage and ethnicity. During its heyday, its executives served as leaders of the American music industry, but it was started by merchants more familiar with selling groceries than music.

Harry and Bertha Keyser Bergman, grocers in the city of Durham, began working in the record business in the 1940s when they acquired a jukebox company, Carolina Music Service, in nearby Burlington.[5] Shortly thereafter, in the late 1940s, Bertha's brother, Paul Keyser (1916–2001), newly discharged from the US Army after serving in the Second World War, settled in Durham and established a small, independent record store downtown.[6] This he named Record Bar, the first store so called. In 1950, the Bergmans followed Paul's lead and added to their Burlington jukebox business a retail record store, Musicland.[7] By the mid-1950s, both the Bergman children, Barrie (1942–2020) and Lane (born 1946), were assisting their parents and their Uncle Paul in the three businesses, simultaneously learning the retail record business from the ground up. The companies were restructured in 1960 when Harry and Bertha bought Paul's store. Paul moved his family to Jacksonville, Florida, where he eventually opened at least four Record Bar stores, independent of the Bergman businesses.[8] In Durham, meanwhile, the Bergmans, their children, their children's families, and an increasing number of employees and investors built their company into a multimillion-dollar corporation. In 1989, after approximately forty years as a viable independent business, Record Bar was sold to a European corporation, Super Club N.V. Super Club retained Record Bar's identity, but in 1993 sold it to Blockbuster Video, which then converted its several music-retail operations to a single company brand: Blockbuster Music. At that point, Record Bar's history as an independent entity ended.[9]

But Record Bar was more than its business history. For many of its employees and patrons it was a cultural oasis, a sort of Camelot—or, as Bob Prout asserts, a cult. In particular it was a place where those who embraced the music and cultures associated with the social upheavals of the post–Second World War era could engage with other members of their tribe. In piedmont North Carolina, that status was reinforced by its proximity to universities and colleges—Duke and North Carolina Central Universities (Durham), the University of North Carolina (Chapel Hill), North Carolina State University (Raleigh), and others—which were, at the time, important centers of youth culture and musical activity. Record Bar's significance as a cultural icon and social locus is clear from the comments one finds today on blogs and social media groups dedicated to the company's memory and community and in online obituaries of persons associated with the company.[10] In addition, Record Bar stood as a product and representative of the Jewish mercantile tradition in the overwhelmingly Protestant American South of the Civil Rights Era.

Family

Perhaps the first clue to understanding the Record Bar culture is that, at its core, Record Bar was a family business. A *Jewish* family business. It was established in a time and place marked by legal racial segregation, occasional unrest over civil rights, a conservative political climate, and the strong remnants of the Confederacy and the Temperance movement—all in a profoundly Protestant society. It cannot have been easy being Jewish in that environment.[11]

The Bergman family was part of a Jewish community in Durham centered around Beth El Synagogue, an Orthodox, later Conservative, congregation. The Beth El community consisted primarily of German, Eastern European, and Russian Jewish immigrants who had begun settling in the area in the 1870s. Initially, most of them worked in the new tobacco factories. But over the course of the late nineteenth and early twentieth century many in the Jewish community opened family businesses that catered to the larger Durham community, making their livings as merchants.[12] Although neither Harry nor Bertha were Durham natives, their work lives there fit this family-oriented, mercantile model.

The history of Record Bar—especially in its first two decades—reflects this orientation: the importance of family and of making business decisions based, to a significant degree, on the needs of family. Moreover, as the company grew and the Bergmans ran out of family to run it, they had to hire, and place their trust in, persons who were not family members. In so doing, they extended their sense of family to their employees. Barrie Bergman, the family member given control of the company in the late 1960s and the person most closely identified with it in the public's mind, seems (1) to have fully assimilated those family values and (2) to have had a special ability to help employees feel like they were part of the Record Bar family. Those employees responded in kind. Two decades after the company was sold (and after the development of a second successful business: Bare Escentuals, Inc.), Barrie expressed those values clearly in his book, *Nice Guys Finish First: How to Succeed in Business and Life*, and reiterated them in succeeding interviews.[13]

Family considerations were important to the Bergmans and Keysers when making business decisions dating back to at least 1933 when Harry Bergman (1914–91) first moved to Durham. The son of Russian immigrant fruit merchants in Jacksonville, Florida, Harry, in the early 1930s, was pursuing an undergraduate degree at the University of Florida in Gainesville. Because of the challenging economy (the Great Depression was in full swing) the family business went under, so in 1933 they moved to Durham, which had a healthier economy due to the vitality of the tobacco industry. The family opened a grocery store there and, according to Harry, "made a living from Day One."[14] Harry believed strongly in the importance of education and had hoped to complete his degree at Duke University, but the family could not afford it.[15] He therefore sacrificed his desire to complete his education in order to help his family survive.

Harry met Bertha Keyser (1913–79), a native of Washington, DC, in 1939 at the wedding of his brother Mitchell. They married in 1941 and made their home in Durham, where Bertha joined the Bergman grocery business.[16] To that business, of course, was soon added Carolina Music Service (the jukebox business), Paul Keyser's Record Bar in downtown Durham, and Musicland, the Bergmans' retail record store in Burlington. On September 24, 1960, Paul sold his Record Bar store to Harry and Bertha and shortly thereafter moved his family to Jacksonville, Florida. Family was the reason for this move, according to Lane Bergman Golden. One of Paul's three children had special educational needs. The public school system in Jacksonville met those needs; the school system in Durham did not. With the needs of this one child in mind, then, Keyser moved his family to Florida and continued in the family mercantile tradition by opening a Record Bar store there.[17] As for the Bergmans, one wonders if, aside from helping Paul, they too were perhaps motivated by family needs. At the time of the sale, son Barrie had just turned eighteen and was ready to begin his undergraduate studies at Duke University. And daughter Lane was only a few years behind Barrie; they would also have to finance her education. Perhaps the Bergmans needed additional income to pay for Barrie's and Lane's educations. According to Lane, her parents made it clear to Barrie that they could not afford to pay for him to live on campus, so expenses were an issue.[18] One thing is clear: soon after the purchase, they moved the Record Bar store down the block to Main Street, across from the old Durham County Court House—a location with more traffic probably—and in the process significantly enlarged the store's floor space. This became the company's primary location.[19]

The next major developments in the Bergman family business appear to have had two goals: growing the company and providing income to their two children and their children's families—not necessarily in that order. In 1963 the Bergmans opened a second Record Bar store in Chapel Hill, the site of the University of North Carolina. Given the large student population there and the dominance of that demographic among the record-buying public, it seems to have made business sense.[20] But perhaps just as important was Barrie's impending marriage to Arlene Macklin (1945–2016) and the newlyweds' need for income. Barrie and Arlene had met at a bar mitzvah when he was seventeen and she was fourteen[21] (the two often told the story of how Barrie wooed Arlene with a copy of Elvis Presley's recording, "G.I. Blues."). Arlene was a Chapel Hill native, graduating from Chapel Hill High School in 1963, and her parents owned Harry's Restaurant and Delicatessen, "a Chapel Hill landmark and favorite of many Carolina students from 1939 to 1972." At 173 East Franklin Street, Harry's was right across from the campus and almost within sight of the new Record Bar at 108 Henderson Street.[22] Record Bar #3, as it came to be known, opened in June 1963.[23] The Bergmans stocked the new store by closing Musicland and moving that store's inventory to the new location. Barrie, who was on the cusp of turning twenty-one at the time, and his fiancée Arlene, then eighteen years old, were made its managers. They married two months later, in August.

Similar motivations were evident four years later when the Bergmans opened their third Record Bar store. Whereas the Durham and Chapel Hill stores had been established in free-standing locations, this new store was opened in North Hills Mall, a new, enclosed shopping center in nearby Raleigh, the state capital. As with the Chapel Hill store, the Bergmans had family to manage it. Daughter Lane had earned her undergraduate degree at the University of Florida and, in 1966, had married her college sweetheart, Bill Golden (1944–2010). The Bergmans gave the couple the opportunity to join the family business for a probationary year, to see if they liked it. They did. So, in 1967, when the company opened the Record Bar in North Hills Mall, Bill and Lane were made the store managers. By all accounts the store was a roaring success from opening day, convincing the Bergmans that the company's future lay in placing retail stores in shopping malls—and that did indeed become the company's focus. As with Barrie and Arlene, daughter Lane and husband Bill were given incomes and the prospect of financially rewarding careers—close to family!

Barrie's Ambitions

Barrie Bergman's love of music was clear, according to Lane. When he was working, she says, especially early on, he was often wearing headphones, listening to music. But no matter how much Barrie loved music, he was more interested in the music *business*. In a supplement published by *Billboard* in December 1984 to celebrate Record Bar's twenty-fifth anniversary (in 1985), Barrie was quoted as saying,

> "I've been infatuated with the record business since I was a teen," . . . adding that he would read the trades "cover to cover" each week. "My heroes were marketing vice presidents," . . . "I know that's kind of bizarre, but that's the way I was."[24]

Most of the marketing guys, of course, were in New York and Los Angeles, and Barrie's dream was to go to New York and work for a record label. When Harry was ready to open the new Chapel Hill Record Bar, apparently he and Barrie talked. In Barrie's telling, his father wanted Barrie to stay at home and manage the new store, to work in the family business. Barrie in turn made it clear that he didn't want to manage a record store for the rest of his life, that there had to be more for him. So a compromise was reached. Barrie told his father that he was willing to give it five years, but that if he didn't see things happening in five years, he was going to New York. Harry apparently agreed, so Barrie and Arlene began managing the Chapel Hill Record Bar, and Barrie continued to learn business from Harry.[25] This episode reminds me of what Martin Davidson, a childhood friend of David Geffen, said in the documentary *Inventing David Geffen*: If you're a nice Jewish boy from Brooklyn, "you own your own store. You take over your father's business and you 'do better.'"[26] It appears, then, that Barrie decided, at his father's urging, to stay five years to see what would happen, to see if he could "do better." And indeed, five years later, in 1968, Harry and Bertha, then in their mid-fifties, handed over the reins of the company to Barrie, Arlene, Bill, and Lane, all in their twenties. And this was the beginning of Record Bar's explosive growth.

The Next Generation Takes Charge

In 1968 both Barrie and Bill Golden were made vice presidents of Record Bar. Under the new generation's leadership, Record Bar gradually ceased to be a mom-and-pop operation and became a major record-store chain. At the time, Record Bar consisted of a home office/warehouse and three stores: one each in Durham, Chapel Hill, and Raleigh. By the holiday season of late 1971, three years later, the company had fourteen stores located in North Carolina, South Carolina, Tennessee, and Georgia, with one geographic outlier in State College, Pennsylvania.[27] Beginning with the North Hills store (Record Bar #4), the stores were placed in malls, mostly enclosed malls, in urban areas and college towns. The 1970s was a period of significant growth for both the record industry and Record Bar. Twenty stores were opened in 1973 alone, and in 1974 the company began holding conventions for the employees. The first was held in downtown Atlanta, the business center of the American Southeast, the company's home region, but the second was in Los Angeles—far from the company's base of operations but right in one of the centers of the recording industry. It was also, in the minds of many who were associated with the company, the beginning of the golden era of Record Bar.

The Cult of the Record Bar

In the years that Barrie ran the company—that is, between 1968 and 1989—he more than anyone else seems to have set the tone for the company culture. Four cultural elements appear to have been most influential: youth; family values; a commitment to fun; and business values. Of the numerous cultural elements that he, Arlene, Bill, and Lane contributed that made the Record Bar environment special, certainly a part of it was youth. When Barrie and Lane's generation took over management of Record Bar, all of them were young, born between 1942 and 1946. And when they sold their shares of the business in the 1980s, they were only in their forties.[28] They came of age during the post–Second World War period and were generally the same age as the popular artists whose recordings they sold. Barrie and Arlene in particular seem to have identified culturally with many of the artists and music-industry figures they encountered and to have become quite close to some of them: Harry and Sandy Chapin; Lyle Lovett; and Phil Walden, cofounder of Capricorn Records, stand out. Nor did Barrie hide the fact, at least in later years, that he and Arlene smoked marijuana, prolifically at times and sometimes to their embarrassment. And as Ashkenazi Jews, they were less likely than their Protestant neighbors to grow up with a strong Temperance mindset. In other words, they drank and enjoyed alcoholic beverages.[29] It is also clear that Barrie especially loved music and knew it well. A visit to his blog, produced when he was in his seventies, shows that he ended each post with a relevant lyrical quote from a popular song. He seems to have had a deep knowledge of the repertoire. And he continued to attend live concerts and comment on them knowledgeably, long after he had left the business.

As for family values, I provide evidence above that many of the Bergmans' business decisions were driven by family considerations: forgoing a university education in order to help the family; helping a brother-in-law by buying his store; expanding the business so they could finance their children's educations; and opening new stores in order to provide their children's families with secure incomes and future prospects. Barrie at least seems to have internalized those values.[30] As the company grew and the Bergmans found it necessary to rely on employees who were not family members, they extended that sense of family to their employees.

But Barrie had absorbed, or been born with, even more basic values. Reading his writings, reading about him, listening to his recorded interviews, and most especially reading comments about him in the wake of his death, several qualities stand out: his basic humanity; his respect for, and respectful treatment of, others; his humor; his intelligence; his likeability; and his honesty. He seems to have been a genuinely likeable, kind, and generous man. And although I don't have lots of hard evidence to back this up, it seems that Arlene, Lane, and Bill brought similar values to the company.

In addition to these many humane qualities, Barrie in particular seems to have been committed to having fun, to creating a fun workplace. He states in his book, "Arlene and I were in it for the fun. It never occurred to us that we were going to get rich. We were just trying to make a living and have fun in the process."[31] And in a 2014 interview Barrie explains that, for him, the growth was about keeping it interesting.

> The reason I always wanted to open [new] stores—it wasn't about money, it's just cause it was so much fun. I loved the action. And I loved what was going on. It was a great business; it was a great business to work in; it was always fun to go to work. And because it was our company, we could build it in any way we wanted to. So we built a really ethical organization that was a good place to work. And that paid off, over and over and over.[32]

All that said, Barrie was also a businessman and not beyond making choices designed to benefit the business. That meant firing employees who were stealing or lying or not executing their duties effectively. That meant making money by growth, expansion. And finally it meant agreeing to sell his interest in the family business when he was offered what he considered to be an exceptionally good price.[33]

The Record Bar employees seem to have responded to this treatment. Based on their published comments, many of them, now decades later, and after working for other employers, look back on their time at Record Bar as some of the best years of their lives.[34] It was fun. It was financially rewarding and professionally satisfying.

They valued their interaction with the recorded music of the time and the customers who came into their stores to buy it. And they are now members of the cult of the Record Bar.

Notes

1. Unknown [Bob Prout], "The Cult of the Record Bar," *I Don't Pretend to Know: An Assemblage of Random Thoughts, Unresearched Opinions and Flagrant Hyperbole*, August 25, 2006. http://onomatopeyote.blogspot.com/2006/08/cult-of-record-bar.html (accessed September 22, 2021).
2. Ibid.
3. Ibid.
4. Prout initially named the group "Record Bar Survivors" but quickly changed it to match the title of his blog post.
5. Fred Goodman and Jim Bessman, "Record Bar: The 25 Year 'American Dream,'" "25 Years. A Tribute. Record Bar" ["A Billboard Advertising Supplement"], *Billboard*, December 8, 1984, RB3. For a slightly different explanation, see also Barrie Bergman's comments starting at 6:15 in University of California Television (UCTV), "Barrie Bergman Rock 'N Roll CEO," YouTube video, 58:56, October 8, 2014. https://www.youtube.com/watch?v=D66n0s31710&t=157s (accessed October 16, 2021).
6. Lane Bergman Golden, telephone conversation with the author, September 23, 2021.
7. Goodman and Bessman, "Record Bar," *Billboard*, RB3.
8. Ibid.
9. This statement applies only to the company Paul Keyser established in Durham in the late 1940s and sold to his sister and brother-in-law in 1960. Keyser later established stores of the same name in North Florida, but they were not part of the Durham company. Lane Bergman Golden, telephone conversations with the author, September 23, 2021, and April 11, 2022; Goodman and Bessman, "Record Bar," *Billboard*, RB3; and "Paul Keyser," 2021. https://www.findagrave.com/memorial/88145100/paul-keyser (accessed October 14, 2021). Another Record Bar, completely separate from those just mentioned, was owned and operated by Norman Hammel (1917–2002) in Belleville, Illinois. Roger Schlueter, "Answer Man: The Record Bar Was a Belleville Music Mecca," *Belleville News-Democrat*, September 1, 2016. https://www.bnd.com/living/liv-columns-blogs/answer-man/article99234697.html (accessed October 14, 2021); and "Norman John Hammel," 2014. https://www.findagrave.com/memorial/132119766/norman-john-hammel (accessed April 12, 2022). Interestingly, both Keyser and Hammel were US Army veterans of the Second World War.
10. [Prout], "The Cult of the Record Bar"; moonpie23, "R.I P. [*sic*] Barrie Bergman," 2020, *Duke Basketball Report*. https://forums.dukebasketballreport.com/forums/showthread.php?45693-R-I-P-Barrie-Bergman (accessed October 26, 2021); "Barrie Bergman: July 7, 1942–September 11, 2020," Hudson Funeral Home and Cremation Services. https://hudsonfuneralhome.com/obituary/barrie-bergman/?fbclid=IwAR1jjQGW3yYAzUsAuqvDBEdWUju5_ovyeVnfHQieNipgiRqIEp__3qIzX6Y (accessed May 9, 2022); "Barrie Bergman, 1942-2020," *The News & Observer*. https://www.legacy.com/us/obituaries/newsobserver/name/barrie-bergman-obituary?id=8403883

(accessed May 9, 2022); "Arlene M. Bergman: February 5, 1945–July 28, 2016," https://hudsonfuneralhome.com/obituary/arlene-m-bergman/ (accessed May 31, 2022); "Arlene M. Bergman," https://www.legacy.com/us/obituaries/newspress/name/arlene-bergman-obituary?id=10080119 (accessed May 31, 2022); "William Golden," https://www.legacy.com/us/obituaries/roanoke/name/william-golden-obituary?id=28028800 (accessed May 31, 2022); and "William Bill Golden," https://www.connerbowman.com/obituaries/William-Golden-34385/ (accessed May 31, 2022). The "Cult of the Record Bar" private Facebook group was also reviewed.

11 Leonard Rogoff, *Homelands: Southern Jewish Identity in Durham and Chapel Hill, North Carolina* (Tuscaloosa: The University of Alabama Press, 2001), 1–5, 150–244, 310–17; and Eli N. Evans, *The Provincials: A Personal History of Jews in the South* (New York: Atheneum, 1973), vii–xi, 310–26; Lane Bergman Golden, telephone conversation with author, September 23, 2021; the author's personal knowledge and observation.

12 Beth El Synagogue, "History," https://betheldurham.org/history/ (accessed October 16, 2021). The best book-length study of this community is Rogoff, *Homelands*. Eli Evans, whose father Mutt Evans was mayor of Durham from 1950 to 1962, provides insight into the Jewish community of Durham and Chapel Hill and its relations with the rest of the community in his idiosyncratic ethnographic study, *The Provincials*.

13 Barrie Bergman, *Nice Guys Finish First: How to Succeed in Business and Life* (2009).

14 Goodman and Bessman, "Record Bar," *Billboard*, RB3.

15 "Harry Bergman," https://durhamhebrewcemetery.org/details.php?id=497 (accessed May 31, 2022).

16 "Bertha Bergman," https://durhamhebrewcemetery.org/details.php?id=498 (accessed May 31, 2022). Such marriage stories are typical of those buried in Durham Hebrew Cemetery.

17 Lane Bergman Golden, telephone conversation with the author, September 23, 2021.

18 Ibid.

19 Open Durham, "201-203 East Main/Record Bar," https://www.opendurham.org/buildings/201-203-east-main-record-bar (accessed October 11, 2021) (note the comment on this page by Audrey Evans, dated December 31, 2013, in which she describes the store before it was moved to Main Street: that is, presumably when Paul Keyser owned it); and Lane Bergman Golden, telephone conversation with the author, September 23, 2021. The Wikipedia article on "Record Bar" states that Paul Keyser's store at the corner of East Parrish and North Church Streets had 800 square feet of floor space and that in 1962, after the Bergmans moved the store to the corner of North Church and East Main Streets, they expanded the floor space to 3,300 sq. ft. Although this sounds accurate, I have not been able to find a reliable source for this information.

20 Bergman, *Nice Guys Finish First*, v; Fred Goodman and Jim Bessman, "Highlighting Product through Store Design," "Record Bar: The 25 Year 'American Dream,'" "25 Years. A Tribute. Record Bar" ["A Billboard Advertising Supplement"], *Billboard*, December 8, 1984, RB14; and Goodman and Bessman, "Record Bar," *Billboard*, RB3.

21 "Arlene M. Bergman," https://www.legacy.com/us/obituaries/newsobserver/name/arlene-bergman-obituary?pid=180845327 (accessed May 31, 2022); and "Arlene M. Bergman," https://hudsonfuneralhome.com/obituary/arlene-m-bergman/ (accessed May 31, 2022). Based on their birthdays (and if this information is true), they met sometime between July 7, 1959, and February 4, 1960.

22 "Arlene M. Bergman," https://hudsonfuneralhome.com/obituary/arlene-m-bergman (accessed October 18, 2021).
23 Company employees referred to the Record Bar warehouse as #1; the Durham store on Main Street, #2; and the Chapel Hill store, #3.
24 Fred Goodman and Jim Bessman, "Barrie Bergman: Positioning Record Bar as Innovative Leader," "25 Years. A Tribute. Record Bar" ["A Billboard Advertising Supplement"], *Billboard*, December 8, 1984, RB4.
25 Barrie seems to have chafed under his father's management. In his 2014 interview Barrie said, "I really hated working for him. I worked for him for, for, for a long time. Until he finally got out of the business. And he was real critical. But he taught me a great deal." Barrie said this about his father when he was seventy-two years old, twenty-three years after his father had died and many years after he and Arlene had moved to California. ("Barrie Bergman Rock 'N Roll CEO," 5:30-6:15. See also Bergman, *Nice Guys Finish First*, vi.) If there was animosity between Barrie and Harry, it appears to have been mutual. Harry's obituary on the Beth El Synagogue website for the Durham Hebrew Cemetery names his parents, three brothers, a sister, his wife, and his daughter Lane. Nowhere does it mention a son ("Harry Bergman," https://durhamhebrewcemetery.org/details.php?id=497 (accessed May 28, 2022). Bertha's obituary is similar ("Bertha Bergman," https://durhamhebrewcemetery.org/details.php?id=498 (accessed May 28, 2022). *Note*: Lane Bergman Golden disputes any animosity between her parents and Barrie, stating that they were very, very proud of him (Lane Bergman Golden, telephone conversation with the author, June 14, 2022).
26 *American Masters: Inventing David Geffen*, DVD, directed by Susan Lacy, New York: Thirteen, 2012, beginning at 04:49.
27 Home of Pennsylvania State University, one of the largest universities in the United States.
28 Harry and Bertha had been in their mid-forties when they bought Keyser's Record Bar and were in their mid-fifties when they handed the management of the company over to their children's generation. Bill and Lane sold their shares to the private equity company General Atlantic in 1987. Barry and Arlene and their partners sold their shares in 1989 to Super Club N.V (Joseph B. Swain, telephone conversation with the author, December 31, 2021).
29 Bergman, *Nice Guys Finish First*, 79–80, 112. Having grown up as a Southern Baptist in that area, I clearly recall the strong social prohibition against alcohol. Social drinking simply was not part of polite society among many Protestants in that area at that time.
30 I limit this statement to Barrie only because I do not have enough evidence to attribute the same to Lane, Arlene, and Bill, but their association with the company certainly suggests that they shared those values.
31 Bergman, *Nice Guys Finish First*, 29.
32 "Barrie Bergman Rock 'N Roll CEO," starting at 6:50.
33 Bergman, *Nice Guys Finish First*, 162–3.
34 The best way I have found to get a sense of the man is to read his blog, secondarily to watch the 2014 interview of him at the University of California, Santa Barbara, and to read his book. (Barrie Bergman, "BarrieBergman," https://barriebergman.typepad.com/ (accessed May 31, 2022); "Barrie Bergman Rock 'N Roll CEO," and Bergman, *Nice Guys Finish First*.)

Magic in Here

Brisbane's Alternative Record Stores from the 1970s to the Digital Age

Ben Green

Independent record stores Rocking Horse Records and Skinny's Music both opened in Brisbane, Australia, in the mid-1970s and became hubs for the local music scene across generations. As vital sources for recordings and information from the world of alternative music, both stores endured police raids for allegedly obscene material through the repressive pre-1989 political era while employing musicians behind the counter and hosting hundreds of all-ages "in-store" shows. The place of these businesses in the cultural history of today's cosmopolitan Brisbane became clear in emotive journalism and popular sentiment when the twenty-first-century challenges of rising urban rent and declining physical sales of record albums threatened both terrestrial businesses. Skinny's closed in 2007, while Rocking Horse openly threatened closure in 2011. However, both endure in their own ways. Rocking Horse remains open in its central location and, in 2015, the former Skinny's management opened Sonic Sherpa in the inner Brisbane suburb of Stones Corner. Both occupy smaller premises than they did at their peak but, among a handful of new and old Brisbane record stores, they are living representations of more than four decades of local cultural memory. They now demonstrate strategies for maintaining the financial and cultural viability of a bricks-and-mortar music shop in the digital age. While both businesses engage in substantial online trading, their websites and social media emphasize the in-store attractions of atmosphere, conviviality, and collective music listening. Their physical and digital presences carry reminders of their historical and continuing relationships with local and international scenes, presenting each record store as a site of trans-local music culture and a way for consumers to connect with it. This chapter will explore the historical and contemporary presence of Rocking Horse, Skinny's, and Sonic Sherpa based on secondary sources and around twenty-five years of personal observations.

Imports and Alternatives in Pig City (1970s–1980s)

Rocking Horse Records opened in 1975 under a stairwell in the Rowes Arcade, a building in the central business district (CBD) that in true Brisbane style had undergone multiple uses and renovations since its late Victorian beginning. Shop founder Warwick Vere, having just moved from Sydney to work as a legal assistant, identified a gap in the market for an import record store.[1] As Vere told Brisbane music historian Andrew Stafford, magazines like *Rolling Stone* and *New Musical Express* ran reviews of "stuff that just wasn't available here locally," due to an "enormous undersupply by the Australian record companies of what should have been available."[2] Thus, Stafford suggests the imported magazines stocked by Rocking Horse had an even greater cultural impact than the records on the shelves, among the young musicians for whom the store quickly became a meeting place. Vere recalls the store being so busy on Saturday mornings that patrons had to wait their turn outside, yet before midday the arcade caretaker would begin pulling down the shutters as the city effectively closed for the weekend.[3] A couple of blocks away, toward the southern end of the CBD, Skinny's Music was established in 1977 in the Queen's Arcade. Founder Colin Rankin recalls it was "a tiny little store but the rent was only $35 a week, which was pretty good for a central city store." As to its remit, Rankin asserts there were very few "dedicated" record stores and those that existed "really concentrated on the mainstream and didn't cater for any of the alternative music." He places his first ever sale, a Joan Armatrading album, into that category while admitting it might not be considered as such these days.[4]

Skinny's and Rocking Horse now dominate local history to an extent that obscures their early peers. As early as September 1976, Discreet Import Records was advertised as "handling all latest major overseas releases plus the finest range of specialist music in Brisbane."[5] Discreet was also located in a city arcade. Brisbane musician Robert Forster, best known as a founder of the Go-Betweens, recalls:

> In Brisbane in the early to mid '70s, the only places offering any kind of interesting shopping experience were in the inner-city arcades. Elizabeth Arcade, in particular, had a string of shops that not only offered goods found nowhere else in town, but also an enclosed tunnel-like atmosphere that threw off the city's tropical heat and its equally boiling corrupted political and police culture.[6]

As well as the radical Red and Black Bookshop, exotic restaurants, and a hip hair salon, this tunnel housed Discreet "with its eclectic mix of import records." Brisbane academic and musician (and fellow Go-Between) John Willsteed corroborates the significance of both arcades and record stores:

> Like Warwick at Rocking Horse round in Adelaide Street, Phil and Mary [at Discreet] created homes for the bored . . . Lovely new imports, wrapped in plastic . . . Punk and new wave, prog and no wave. Jazz.[7]

Willsteed observes elsewhere that "Rocking Horse is well remembered because it still exists, but Discreet Records was a powerful force back then."[8] Other establishments included the inner-suburban Toowong Music Centre, advertising "special orders taken for customers"[9] and from which proprietor Damien Nelson cofounded the Able Label with the Go-Betweens in 1977, as well as Wizard Import Records on inner-city Albert Street, stocking "American and British albums, specializing in rock, country, blues."[10]

One reason why Rocking Horse and Skinny's hold an important place in Brisbane's cultural memory is their endurance through the repressive political climate of 1970s and 1980s Queensland, as mentioned by Forster in the previous quote. The state was governed by the Country/National Party from 1957 to 1989, led by premier Sir Joh Bjelke-Peterson from 1968 to 1987. Against a broader liberal turn in Australian politics and culture, Queensland became known as "the deep north" and Bjelke-Peterson its "hillbilly dictator"[11] based on the restriction of civil rights, heavy-handed police force, and high-level corruption (later judicially established). For Brisbane punks, whose mere appearance in public drew police attention and whose shows were typically raided by the notorious Special Branch, the state capital was "Pig City," so named in a 1983 single by the Parameters and later adopted as the title of a cultural history by journalist Andrew Stafford.[12] A crucial site of resistance, politically and culturally, was 4ZZZ, the state's first FM radio station founded in 1975 just months after Rocking Horse opened. With a community model based on public funding, listener subscriptions, and substantial volunteer labor, the station took an explicitly noncommercial and independent approach to news, information, and music.[13]

The early years of 4ZZZ were marked by a cultural transition between the hippie and punk eras,[14] with local tastemakers taking cues from the imported magazines stocked by the record stores (though Brisbane band the Saints pre-dated the UK punks with independently released single "(I'm) Stranded" in 1976).[15] The station's annual Hot 100 songs poll tells a story: the year 1976 was topped by the Beach Boys' "Good Vibrations," followed by the Rolling Stones, the Beatles, Cream and Led Zeppelin, while the 1980 top five featured Sex Pistols' "Anarchy in the UK" followed by Talking Heads, more Pistols, Dead Kennedys, and Devo.[16] Through the late 1970s Rocking Horse and Skinny's are mostly absent from the pages of 4ZZZ newsletter *Radio Times*, leaving advertising space to the competitors mentioned earlier, but in October 1983 the front-page bears pictorial advertisements for those two record stores only. An adjacent editorial, illustrating the concerns of the time, mentions recent live shows by Midnight Oil and the Dead Kennedys before calling for increased commitment to the causes of Aboriginal and Torres Strait Islander land rights, nuclear disarmament, and land conservation.[17]

While not as explicitly political as 4ZZZ, Rocking Horse and Skinny's were at least sanctuaries for cultural enemies of the state, and they were regarded as subversive enough to become targets themselves. Skinny's proprietor Colin Rankin recalls:

> I think it was after an election that was particularly tumultuous at the time that we did have the police close the store down one busy lunch hour. . . . They closed the doors and searched everybody and the premises, I got taken out the back and

was virtually strip searched. But I always took that to be a political statement more than anything.[18]

Rocking Horse, having relocated from the Rowes Arcade to larger and more central Adelaide Street premises, was visited on February 14, 1989, by an undercover officer seeking rude records for "a wild Valentine's Day party," before uniformed police raided the shop and brought charges for the exhibition and sale of obscene material. The offending items included 1960s instrumental "Do the Shag" by the Champs, Sonic Youth's *Master=Dik* EP, the Hard-Ons' *Dickcheese*; Guns n' Roses' high-charting *Appetite for Destruction*, Lydia Lunch records, and albums by the Dead Kennedys.[19] Dead Kennedys singer Jello Biafra empathized with Rocking Horse in a piece published in 4ZZZ's *Radio Times*, reflecting on his own encounters with Queensland police during a 1983 tour (noting "I feel safer on the streets of Harlem or Berlin") and observing ominous similarities with a religious-right censorship push in the United States.[20] The Dead Kennedys' song "Too Drunk to Fuck" was aired and analyzed in the ensuing court proceedings, before the eventual, begrudging dismissal of all charges by the presiding magistrate, who stated:

> there is a considerable market for these items and that is an interesting comment, but not unexpected, upon the taste of the general community . . .
>
> On the balance I come to the conclusion that merely selling these things, all of them, or having them for sale without public performance and without undue public display, is not something which is offensive to current community standards.[21]

The gap between state authorities and community standards was emphasized four months later, when the two-year Fitzgerald Inquiry delivered a far-reaching report on corruption, swiftly followed by electoral upheaval (the first Country/National Party loss in Queensland for over thirty years), systemic reform, and the criminal convictions of the police commissioner, four government ministers, and numerous police.[22] Pig City was changing.

In-Stores and Scene Infrastructure (1990s–2000s)

In the fresh air and sunlight of the 1990s, Brisbane's music scene grew from underground roots to national prominence, in conjunction with the global alternative rock boom. Local bands like Screamfeeder, Regurgitator, Custard, and Powderfinger (colloquially known as "the big four") became mainstays of a national scene, with an all-ages audience attending festivals like Livid and the Big Day Out, championed by the Australian Broadcasting Corporation's youth radio station Triple J and its televised counterparts, all-night video show *rage*, and Saturday morning variety show *Recovery*. Skinny's, like Rocking Horse, emerged from its arcade in 1986, only to dive into a new hideaway below the footpath on Elizabeth Street, a block or two away from the main shopping and business precincts. Simon Homer, who bought the business from Colin Rankin in 1997, appreciated the location:

> Elizabeth St was always on the outskirts, and we have used that to our advantage . . . That whole walking down the stairs into a dark, skanky rock venue adds to its credit. With our in-stores, (which) we have been doing as all age shows on Saturdays, (it) just adds to the mystique of walking down those stairs and seeing how it was old-school style.[23]

This quote describes my experience as a teenager in the late 1990s. By this time, Brisbane was well into a phase of political liberalism and cultural cosmopolitanism, though the music scene remained attentive to its "Pig City" past[24] and a rising, reactionary backlash exemplified by the One Nation Party formed in 1997 in Ipswich, less than an hour's drive west. Among a flourishing of new venues, events, and artists, Rocking Horse and Skinny's embodied the city's alternative cultural heritage, with a palpable sense of continuity to their function as key spaces of nonmainstream consumption and sociality.

Regularly, on a Saturday, my friends and I would take the hour-plus train to Brisbane from our home in the neighboring city of Gold Coast. After a round of the "op shops" (charity secondhand outlets) in search of the vintage clothes that could still be found at cheap prices, we would circle through the city's record stores, including secondhand hoards such as the vast Record Exchange (established in Brisbane in 1985). The excursion almost always culminated in Rocking Horse or Skinny's for the main event: an "in-store," live performance, usually by a local band promoting their new album, which they would sell and sign after the show. CD and T-shirt racks would be pushed aside, and the same bands we would see in teeming showgrounds and steaming big tops at festivals could be enjoyed up close from a cross-legged vantage point, recorded on a handheld cassette device and, if you bought a CD or pulled a poster off the wall to have signed, engaged with personally. International artists would usually only appear for a signing. Skinny's, starting with signings by major Australian acts like the Skyhooks and Cold Chisel in the 1970s and 1980s, ultimately hosted more than 500 bands.[25] In contrast to that carpeted subterranean world, Rocking Horse of the 1990s and early 2000s held its in-stores upstairs, on the wooden floorboards of a loft-like space with sunlight pouring in from windows overlooking the street. In both spaces, one's attention would drift over the accrued ephemera on every surface, like the Cramps poster signed "STAY SICK!" that loomed above the Rocking Horse stairwell for as long as I could remember. History was palpable and global culture seemed within reach through these portals.

Local musicians and scene figures were also found behind the counter. When cult icons the Apartments formed in 1978, guitarist Michael O'Connell worked at Rocking Horse and reportedly bonded with bandmates over record collections.[26] Twenty to thirty years later I encountered many familiar faces there. Evil Dick from Strutter, the Aampirellas, and HITS was ever present—I remember the gruff rocker's surprisingly enthusiastic response to the relatively gentle Yo La Tengo album I brought to the counter, based on a shattering festival set we had both seen—along with Tam Patton from electronic outfit Full Fathom Five; experimental musician Kahl Monticone; Tom Beaumont of garage rockers the Standing 8 Counts; Alex Gillies of sludge rockers No Anchor, whose gothic woodcuts have adorned various bands' works and who generously chased me up the street to ask if I was "the guy who always asks about Boredoms stuff because we've got some in"; Jimi Kritzler of noisy punks Death:Wolf, On/Oxx, and Slug Guts, who distributed zines over the counter before authoring a weighty book documenting the "ugly Australian underground"[27]; Harriet Hudson of Good God, Southern Comfort, and Miss Destiny, who wisely but futilely urged me to buy more secondhand Lydia Lunch LPs than the one I allowed myself; Damon Black of heavy psych-rockers Secret Birds and weekly appearances as an indie dancefloor DJ, and other half-known locals of various ages and styles. Skinny's had a similarly scene-connected crew, including Damon Cox of Intercooler (and more) and Kate Cooper of Iron On who met at the store and went on to international success as An Horse. Their bands released discs on Plus One Records, run by Skinny's proprietors Simon Homer and Denise Foley since 2004. Foley became Executive Officer of QMusic, Queensland's peak music industry body, in 2005 and served for nine years in which she oversaw the development of its BIGSOUND program into the southern hemisphere's biggest new music festival and conference.[28] Previously, Homer ran the stoner rock-focused label Rhythm Ace Records founded in 1999 with partners including Skinny's assistant manager Steve Bell, who as well as working in the store from 2000 to 2006 managed bands like Giants of Science and Iron On, hosted shows on 4ZZZ, and wrote for *Time Off* before assuming the role of editor in 2007. Since the 1980s, *Time Off* was one of Brisbane's inky "street press" magazines, alongside *Rave Magazine*, founded in 1991 and published by former Skinny's owner Colin Rankin, and the more club-oriented *Scene* established in 1993. These weekly publications, funded by advertising and free to collect from stacks inside the doorways of shops including Skinny's and Rocking Horse, each ran their own interviews with local and international artists, reviews of new releases and shows great and small, and most importantly the weekly gig guides. A visit to the record store could yield a week's reading material and social diary for an entire share house. The stores themselves were bottomless troves of information, with the passions and pedigrees of staff translating to recommendations, handwritten stickers and signs, and a nonstop curated soundtrack.

Closures and Cultural Memory (2000s–2010s)

In 2004, Rocking Horse moved around the corner to an even more prominent frontage on Albert Street. The new premises seemed to attract a broader clientele, inviting casual shoppers with a wide footpath entrance and popular sale items on the front racks while still harboring a deep collection including the specialist worlds of dance and hip-hop downstairs. During one lunchtime visit, I ran into a manager from a previous office job, a late middle-aged man who said with polite wonder, "What I like most is all the old ones they have, that you haven't seen in so long." This was the time of "fifty quid bloke," famously described by magazine publisher David Hepworth as "the guy we've all seen in Borders or HMV on a Friday afternoon, possibly after a drink or two, tie slightly undone, buying two CDs, a DVD and maybe a book—fifty quid's worth."[29] My friends and I, then in our twenties, evaluated albums by whether they were worth picking up as $10 classics or $40 imports—with these independent stores still the main means of access to titles without local distribution, although word was now spread through blogs and online forums in addition to magazines. Some in-stores were held at street level on Friday evenings, in addition to the weekend afternoon, downstairs performances. However, there seemed to be fewer of these than before.

The independents were by this time competing with multiple megastores. The massive HMV that had been in the central Queen Street Mall since the 1990s was joined in the early 2000s by Borders, a few doors down from Skinny's but rising over two levels joined by escalators, with listening stations providing samples on demand. Near Rocking Horse on Adelaide Street, JB Hi-Fi presented bright yellow racks of competitively priced CDs and DVDs, filling a level below the home electronics. Within a decade there were two more JB Hi-Fi stores in the central business district. If the CD era was in decline, the 2000s seem in retrospect like the last days of the Roman Empire, expanding recklessly before the proverbial tent collapsed. The HMV in Queen Street closed in 2010, the last in Australia to do so.[30] The Borders outlet shut in 2011, midway through a wave that took all others in the country that year.[31] Skinny's closed at the end of 2007 and was followed within months by Toombul Music, another independent store that had operated in the northern suburb of Nundah since 1981.[32]

Online music distribution was an obvious factor in this decline but not the only one. Quoted in a news report that emphasized the historic significance of Skinny's and its closure, Simon Homer explained that online platforms like eBay had actually provided new opportunities for secondhand sales, but that same market had been devalued by record labels dropping the price of their back catalog CDs. He observed, "having a label, we can see it from both sides, seeing digital sales go up and CD sales go down," and the silver lining of closing Skinny's was that Homer could now focus full-time on the Plus One label.[33] Skinny's held one final in-store featuring musicians who had worked in the shop and bands who were signed to the label: An Horse, Intercooler, Iron On, Little Lovers, and Mary Trembles.

When Rocking Horse announced its imminent closure a few years later in June 2011, the effects of the advancing digital revolution were accompanied by material and local concerns: the high rental cost of a central location, the slow return of city retail in the wake of a destructive "100-year" flood six months earlier, and competition from three surrounding JB Hi-Fi outlets. The announcement was publicized in a report by prominent music journalist Noel Mengel in the state's major newspaper, *The Courier Mail*, which had played a key role in exposing government corruption in the 1980s.[34] Warwick Vere said that despite searching for an alternative site for a year, no realistic options had emerged to justify the "giant leap of faith to sign a three-year lease in these conditions." Interestingly, Vere saw the challenge from online trading as a matter of perception—"everyone imagining they are getting a better deal online"—citing an increase in his own sales through Amazon even though the items cost 20 per cent more than they would over the counter.³⁵ The announcement inspired an outpouring of nostalgia and support on blogs and websites, which by this time had replaced the physical street press, as well as social media. More than 7,000 people clicked "attending" on a Facebook event in support of a stock-clearing, capital-raising sale,³⁶ while a range of local and ex-local music personalities offered statements on how much the store meant to them.³⁷ Former competitor Simon Homer commented from his new perspective as an independent label owner:

> One of the great things about a store like Rocking Horse is it stocks all kinds of local releases, albums, EPs, singles, seven-inch vinyl, whereas the major chain stores prefer albums. Thank God for digital music for indie bands and labels because if stores like Rocking Horse go, where do you go to get the physical product?
> You buy them from bands at gigs, but bands don't play seven nights a week.³⁸

Within a few weeks, Rocking Horse announced on their website that due to "pleasing 11th hour developments"—understood to involve an arrangement with the landlord in light of their long relationship and the real risk of an empty tenancy—they would continue trading for the foreseeable future.³⁹ Vere said he was touched by the outpouring of emotion while noting that it included "a lot of wailing and gnashing of teeth . . . by people we hadn't seen in ages."⁴⁰ The store's high standing in local cultural memory had been highlighted, leaving the question of whether this would translate to economic sustainability.

Renewal and Continuation (2010s–2020s)

The broadly parallel stories of Rocking Horse and Skinny's commenced new chapters in 2015. Rocking Horse celebrated its fortieth birthday with a consolidation: cutting floorspace by more than half to just the basement level, with a new stairwell providing street access (and video game retailer EB Games above), while the store's management grew to include long-term employee Tom Beaumont as well as Ric Trevaskes, who brought substantial secondhand stock following the closure of his local Egg Records branch. Trevaskes had originally worked at Rocking Horse as a teenager, going on to found Egg Records in 2000 with his brother-in-law Baz Scott, with flagship premises in the inner-western Sydney suburb of Newtown as well as a series of shorter-lived outlets in Brisbane city and the inner-southern bohemian enclave of West End.⁴¹ Further southeast, in the haltingly gentrifying inner suburb of Stones Corner, another partnership with deep roots was (re)formed when Skinny's veterans and former label comanagers Simon Homer and Steve Bell opened Sonic Sherpa. This new store, featuring racks donated by Rocking Horse, settled into one end of a row of single-story retail beside busy Old Cleveland Road, overlooking a small carpark shared with a drive-through bottle shop attached to the Stones Corner Hotel.

Bell thought Homer "was crazy for the longest time" but came around to the business proposal as the vinyl revival accelerated.[42] When Rocking Horse faced closure in 2011, Vere had speculated that "if we could turn every CD we had into a vinyl record then we would be right,"[43] and this is the direction he pursued, albeit gradually, along with the understandably delayed task of reducing staff.[44] Throughout the 2010s, the growth of vinyl sales was the main theme of media coverage of both Rocking Horse and Sonic Sherpa, often in conjunction with Record Store Day promotions.[45] Vinyl records became the focus of in-store displays, with an ever-expanding range of new releases and reissues overtaking the secondhand stock sourced from overseas (notably Japan) as well as local procurements, alongside new and secondhand CDs, shirts, and books. Bell observes that while Sonic Sherpa initially attracted an older crowd with connections to Skinny's, this has been joined over time by increasingly younger vinyl shoppers.[46] Both Rocking Horse and Sonic Sherpa are regular stallholders at the record fairs proliferating in the region, as well as operating festival "pop-ups" with scheduled signings by performing artists.

The businesses now maintain substantial digital shopfronts, but these retain connections to the physical stores. Both the Rocking Horse and Sonic Sherpa websites feature browsable and searchable catalogs and sales points, with a choice of collection or delivery. Their homepages highlight new arrivals, staff selections, and promotional blurbs, echoing the physical displays and cover stickers found in-store. Rocking Horse continues to trade through eBay and Amazon using the iconic red rocking horse logo. Both businesses run frequently updated pages on Facebook and Instagram, which are notably focused on their respective physical outlets. New stock, and whatever Sonic Sherpa deems "catch o' the day," is typically promoted through photographs of vinyl displays, with shop interiors, quirky objects (like a Santa-suited Homer Simpson mannequin in Rocking Horse) and occasionally staff or customers in frame, accompanied by invitations to "come in," "browse," and "pick up" copies. The celebrated histories of both businesses are also referenced; for example, "on this day" trivia posts by Sonic Sherpa about famous albums and artists might include anecdotes about times when Homer and "Belly" (as he is known by all) saw concerts and met musicians, as well as their personal tastes. The social media pages share pictures of artists themselves delivering records and merchandise, and Sonic Sherpa features a series of "Brisbane bands buying Brisbane bands," implicitly emphasizing the theme of mutual support within the local music scene—including the role of independent stores. Social media is also a key avenue through which both businesses promote in-store performances and signings.

As in the past, the connections between these independent record stores and the Brisbane music scene extend beyond the promotion and sale of locally produced records and merchandise. Rocking Horse continues to employ local musicians while Sonic Sherpa is now home to two record labels, Homer's Plus One as well as the newer Coolin' By Sound label run by Bell and his partner Michelle Padovan. Both stores are sponsors and subscriber discount outlets for 4ZZZ. In recent years Bell has coached the Brisbane Lines team in the Reclink Community Cup, an annual fundraising event featuring an amateur Australian Rules football game between the Lines, comprising local media (mostly 4ZZZ) personalities, and the musicians' team: the Rocking Horses. Scene connections are perhaps best exemplified in the continuing tradition of in-store performances at Rocking Horse and Sonic Sherpa. Sonic Sherpa in particular has become a regular stop for touring bands of modest fame, who complement standard gigs with an afternoon, all-ages set in the store. In-stores are also a common way for local artists to promote new releases. More spectacularly, on Record Store Day 2019, Sonic Sherpa set up a stage in the carpark in collaboration with the neighboring Stones Corner Hotel, for an afternoon and evening of live performances. A notable addition to this practice is West End's Jet Black Cat record store (established 2011), whose proprietor Shannon Logan organizes Saturday afternoon performances in a neighboring park.

Brisbane's in-store culture was severely restricted during the Covid-19 pandemic of 2020 and beyond. Public health directives limiting live music attendance to one person per two square meters effectively ruled out performances in the cozy confines of Rocking Horse and Sonic Sherpa. However, in contrast to the post-flood retail slump that so affected Rocking Horse in 2011, both businesses maintained visibility and trade through online avenues that are now normalized.[47] Anecdotally, people turned to record collecting in the same way they took up gardening, crafts, and musical instruments. It is significant that in doing so, despite the many options available online, patrons remained loyal to these local providers. Arguably, two years of "social distancing" reemphasized the social significance of Brisbane's independent record stores, both as places of physical interaction and as iconic cultural spaces, embedded in local and trans-local networks, rooted in cultural memory, and contributing to local identity.

Acknowledgments

"Magic in Here" is a song by the Go-Betweens and the first track on their "reunion" album *The Friends of Rachel Worth* (2000). Thanks to Warwick Vere of Rocking Horse Records and Steve Bell of Sonic Sherpa for assisting with queries during writing. Thanks to the book editors for encouragement and feedback.

Notes

1. Meredith McLean, "Warwick Vere: 40 years of Rocking Horse Records," *Embrace Brisbane*, 2015. https://embracebrisbane.com.au/warwick-vere-40-years-of-rocking-horse-records/ (accessed August 11, 2022).
2. Andrew Stafford, *Pig City: From the Saints to Savage Garden* (Brisbane: University of Queensland Press, 2004), 68.
3. Andrew Stafford, "The Stayer," *Notes from Pig City*, April 24, 2016. https://www.andrewstaffordblog.com/the-stayer/ (accessed August 11, 2022).
4. Scott Casey, "History Lost as Skinny's Shuts Its Doors," *Brisbane Times*, December 7, 2007. https://www.brisbanetimes.com.au/national/queensland/history-lost-as-skinnys-shuts-its-doors-20071207-ge9f3o.html (accessed August 11, 2022).

5   4ZZZ, *Radio Times* 1, no. 11 (1976): 4.
6   Robert Forster, "Hair Care," *The Monthly*, November 2008, 19.
7   John Willsteed, "It's Not the Heat, It's the Humidity" (PhD diss., School of Media, Entertainment and Creative Arts, Queensland University of Technology, 2017), 76.
8   Tony Moore, "When White Chairs Was the Centre of Brisbane's Counter-Cultural Revolution," *Canberra Times*, October 10, 2018. https://www.canberratimes.com.au/story/6002000/when-white-chairs-was-the-centre-of-brisbanes-counter-cultural-revolution/ (accessed August 11, 2022).
9   4ZZZ, *Radio Times* (1976): 8.
10 4ZZZ, *Radio Times* 2, no. 6 (1977): 4.
11 Evan Whitton, *The Hillbilly Dictator: Australia's Police State* (Crows Nest NSW: ABC Enterprises, 1989).
12 Stafford, *Pig City*; Ben Green, "Whose Riot? Collective Memory of an Iconic Event in a Local Music Scene," *Journal of Sociology* 55, no. 1 (2019): 144–60; Ian Rogers, "'You've Got to Go to Gigs to Get Gigs': Indie Musicians, Eclecticism and the Brisbane Scene," *Continuum* 22, no. 5 (2008): 639–49.
13 Heather Anderson, "The Institutionalization of Community Radio as a Social Movement Organization: 4ZZZ as a Radical Case Study," *Journal of Radio and Audio Media* 24, no. 2 (2017): 251–69.
14 Geoff Wood, "Weasel Goes the Pop: Counter-Culture Meets Punk Culture Head On," *Radio Times*, December 1985, 34–6.
15 Stafford, *Pig City*, 39–43.
16 4ZZZ, *Radio Times* (1985): 25.
17 4ZZZ, *Radio Times*, October 1983.
18 Casey, "History Lost."
19 Stafford, *Pig City*, 175.
20 Jello Biafra, "Jello Speaks," *Radio Times*, August 1989, 14.
21 Stafford, *Pig City*, 178–9.
22 Crime and Corruption Commission Queensland, "The Fitzgerald Inquiry," 2019. https://www.ccc.qld.gov.au/about-us/our-history/fitzgerald-inquiry (accessed August 11, 2022).
23 Casey, "History Lost."
24 Green, "Whose Riot?"
25 Casey, "History Lost."
26 Stafford, *Pig City*, 99.
27 Jimi Kritzler, *Noise in My Head: Voices from the Ugly Australian Underground* (Melbourne: Melbourne Books, 2014).
28 Scott Fitzsimons, "QMusic's Denise Foley to Remain in BIGSOUND Role after Stepping Down," *The Music*, December 11, 2014. https://themusic.com.au/news/qmusic-denise-foley-to-remain-in-bigsound-role-after-stepping-down/RWFWWVhbWl0/11-12-14/ (accessed August 11, 2022).
29 Tim de Lisle, "Melody Maker," *The Guardian*, March 1, 2004. https://www.theguardian.com/music/2004/mar/01/popandrock2 (accessed August 11, 2022).
30 Rohan Williams, "HMV Goes into Administration," *Scenstr*, January 15, 2013. https://scenestr.com.au/music/hmv-goes-into-administration (accessed August 11, 2022).
31 Chris Zappone, "End of Story: Borders to Shut Remaining Stores," *Sydney Morning Herald*, June 2, 2011. https://www.smh.com.au/business/small-business/end-of-story-borders-to-shut-remaining-stores-20110602-1fhsx.html (accessed August 11, 2022).

32. Shannon Molloy, "Net Downloads Killed the Music Store," *Brisbane Times*, June 15, 2008. https://www.brisbanetimes.com.au/national/queensland/net-downloads-killed-the-music-store-20080615-gea1rm.html (accessed August 11, 2022).
33. Casey, "History Lost."
34. Noel Mengel, "Record Store Faces Toughest Hurdle," *The Courier Mail*, June 28, 2011. https://www.couriermail.com.au/ipad/final-track-for-rocking-horse-records/news-story/5fa1a74fc2f543f9e665b85c6adecfa0 (accessed August 11, 2022).
35. Ibid.
36. John Ritchie, "Rocking Horse Records Will Remain Open," *Music Feeds*, July 20, 2011. https://musicfeeds.com.au/news/rocking-horse-records-will-remain-open/ (accessed August 11, 2022).
37. suz, "Brisbane's Music Community on How Much Rocking Horse Records Means to Them," *Pedestrian*, June 3, 2021. https://www.pedestrian.tv/music/brisbanes-music-community-weighs-in-on-the-closure-of-rocking-horse-records/ (accessed August 11, 2022).
38. Mengel, "Record Store."
39. Ritchie, "Rocking Horse."
40. "Rocking Horse Records Saved from Closure at 11th Hour by Mysterious Benefactor," *The Courier Mail*, July 25, 2011. https://www.couriermail.com.au/business/rocking-horse-records-saved-from-closure-at-11th-hour-by-mysterious-benefactor/news-story/0be13d52d7008617e62f80e7f1c5ab62 (accessed August 11, 2022).
41. Stafford, "The Stayer"; Raylene Bliss, "Sydney's Largest Vinyl Market Comes to Glebe as Sales Soar by 70 Per Cent," *Daily Telegraph*, April 11, 2018. https://www.dailytelegraph.com.au/newslocal/inner-west/sydneys-largest-vinyl-market-comes-to-glebe-as-sales-soar-by-70-per-cent/news-story/9db67d093647f4ebb6669021f3e2b0d6 (accessed August 11, 2022).
42. Daniel Johnson, "Back in the Groove: A Hole New Spin on Record Store Day," *InQueensland*, August 27, 2020. https://inqld.com.au/culture/2020/08/27/back-in-the-groove-a-hole-new-spin-on-record-store-day/ (accessed August 11, 2022).
43. Mengel, "Record Store."
44. Stafford, "The Stayer."
45. For example, Katherine Feeney, "Vinyl Spins Back into Fashion," *Sydney Morning Herald*, April 19, 2012. https://www.smh.com.au/entertainment/music/vinyl-spins-back-into-fashion-20120417-1x5ag.html (accessed August 11, 2022); Johnson, "Back in the Groove".
46. Johnson, "Back in the Groove."
47. Ibid.

6

High Fidelity across Twenty-Five Years

Record Shops, Taste, and Streaming

Jon Stratton

With all the psychedelic art on all the great sleeves
Much cooler and much cheaper than the CDs
And that's the way I caught the disease
I caught the disease for LPs

—"LPs" by Jeffrey Lewis and the Voltage

High Fidelity started out as a novel written by Nick Hornby and published in England in 1995. In 2000, a film was released based on the book. In 2020, the streaming service Hulu showed a television series of ten episodes based more on the film than the book. The novel's protagonist is Rob Fleming, the 35-year-old owner of Championship Vinyl, a London record shop. Rob's live-in girlfriend Laura walks out on him at the beginning of the novel, and he responds by examining his previous relationships. The book constructs an intimate connection between Rob and his working in a record shop based especially on his breakup with Charlie with whom he lived for two years while at college.

Rob employs two young men in the shop, Barry and Dick. Both are, even moreso than Rob, music nerds. Barry wants to play in a band, but it must be a band that plays the kind of music of which he approves. Dick is socially very awkward and, of the three, the one who most mediates life through his musical knowledge. However, to the surprise of Rob and Barry, Dick finds himself a girlfriend Anna whom, in characteristically male style, he tutors about "quality" music. To the horror of Dick, Rob, and Barry, Anna's favorite band is Simple Minds.

In the film Rob is now Rob Gordon and his record shop, rather than being in North London, is in Wicker Park, Chicago. His breakup with Charlie is still a key moment in the film but does not have the causal importance of the novel. The record store in the television series is situated in Crown Heights, Brooklyn. This Rob is Robyn Brooks, younger, African American female, and the two employees in the shop are an African

American woman named Cherise and a gay white male named Simon, who had dated Rob before coming out. As the series opens, we find Rob a year after her breakup with the Black Englishman Mac, about to start dating again. In a knowing change, Rob's previous most significant breakup was with Kat, her only female relationship we know about. Kat is the series equivalent of Charlie in the book and film. Her being female highlights the masculine connotation of Charlie's name.

The three versions of *High Fidelity* span twenty-five years of massive change in the music industry, from a time when vinyl's popularity was eventually supplanted by compact discs, which in turn were replaced in a commercial world increasingly dominated by downloading and streaming services. By 2020, record shops selling vinyl were something of an anachronism and those that still existed did so as specialist shops for vinyl collectors and, sometimes, also selling to customers who think that music sounds better in analogue reproduction or who value vinyl albums for their physical presence. Thus, while in the book and film the shops sell CDs (although looked down upon by the Robs and their employees), in the television series the shop has an illuminated sign inside saying, "No CDs." We also need to note the shift from England to America as the location for the film and television series. In England, the distinction between specialist record shops and those shops selling popular, chart records has been more well-defined than in America.

Specialist Record Shops

To understand the meaning of the record shop in Hornby's novel, we need to go further back in time than when the book is set. During the 1960s, the UK had two types of record shops. Chart and other popular records were mostly offered by shops selling electrical appliances and household goods, of which Woolworths was perhaps the most popular. In 1965, the company upgraded its store at Gallowtree Gate in Leicester becoming the first Woolworths to have a dedicated record department. This development spread across the over 1,000 Woolworths stores in the UK.[1] These Woolworths record departments were the antecedents of the high-street record shops and chains like Virgin. Virgin opened its first large shop in Oxford Street in 1971 and its first megastore in 1979.

In 1971 the Tape Revolution opened on Finchley Road. It was the beginning of what became the largest high-street chain. By 1976 there were six shops selling cassettes and eight tracks. In that year the shops were renamed Our Price Records and refocused on vinyl. By the first half of the 1980s, the chain was the second-largest record retailer after Woolworths. In 1988 the name was changed again to Our Price Music as the stores started selling CDs and finally to Our Price in 1993 when the chain numbered around 300 shops. W H Smith, a retail conglomerate, acquired Our Price in 1986 and a controlling interest in the Virgin Record shops in 1994.

At the same time, there were also specialist record shops catering to people who wanted to buy records, often imports, by less popular and less well-known artists. One of the first, and generally regarded as one of the best, was Dobell's in London's Charing

Cross Road. Chris Barber, the English trad jazz trombonist, remembers the shop in the late 1940s and 1950s:

> To my mind, Dobell's was the first record shop that was really special. Doug Dobell had a real understanding of how to do this. The other shops tended to be all the same, but Dobells really set out to get jazz and blues. It knew what music we wanted and how to get it.[2]

It is significant that Dobell was also a collector. Jazz records were his particular interest.

Dobell was the archetype of a particular kind of person who worked in specialist record shops: the knowledgeable assistant who often knew more than the customers and who did not follow the dictum that the customer is always right. Richard Williams, the English popular music critic, tells this story:

> A great record shop sometimes requires a form of negotiation that doesn't apply widely in the world of retail: the need for customers to prove themselves worthy of making a purchase. The first time I encountered this phenomenon was in Dobell's, when I bought the Ayler album [*My Name is Albert Ayler*] and was brusquely informed by a man with a beard that "the guy can't play". That was fifty odd years ago.[3]

Williams's point is that this behavior, which he regards as typical of some shop assistants who are often collectors, assumes that the person working in the shop is more knowledgeable than the customer. Their knowledge is meant to be translated into a taste preference which is used to guide the customer's purchase. What Barber identifies, the distinction between the shops selling popular records and the shops selling records for specialist tastes, became common throughout the 1960s.

In Hornby's novel, we know that Championship Vinyl is a specialist shop. Rob tells us that he sells "punk, blues, soul, and R&B, a bit of ska, some indie stuff, some sixties pop—everything for the serious record collector, as the ironically old-fashioned writing in the window says."[4] We know from the shop's name that its focus is on vinyl. However, it also stocks some CDs. One important scene in the book centers on an obscure album by the Sid James Experience. Rob tells us that Barry loves this album and, indeed, has two copies of his own. However, nobody buys the copy in the shop. Finally, it sells: "It gives us all strength: if someone can walk in and buy the Sid James Experience album, then surely anything good can happen at any time."[5] However, it is not to be. The next morning the man is back. He doesn't like the album and to Barry's distress exchanges it for a secondhand Madness CD. Hornby's point is that in Barry's opinion, the man clearly has no taste. He doesn't return the Sid James Experience album for another album that Barry, Dick, and Rob, would consider of quality. Rather he picks a Madness album, a popular, mainstream group, part of the ska revival of the late 1970s and 1980s who had many hit singles. Adding insult to this injury, the man buys the album on CD. In order to understand the disdain involved here we need to think about attitudes to compact discs.

CDs and the Online Marketplace

Compact discs went on sale in 1983. One watershed in their increasing popularity was the release of the Dire Straits album *Brothers in Arms* in 1985. As Kirk Miller explains, recorded digitally, the album was engineered to be heard on CD. It sounds "both pristine and intricate . . . and was made to be heard on the most advanced technology of the time."[6] *Brothers in Arms* sold more than a million units on CD when it was released. By 1988, sales of CDs overtook sales of vinyl in both the UK and America. In 1991, 3.8 million vinyl albums were sold in the UK and in 1995, the year *High Fidelity* was published, 1.41 million vinyl albums were sold.[7]

Among record collectors and those who worked in the specialist record shops (now called independent, or indie, shops to distinguish them from the high-street chains such as Our Price and the megastores like Virgin) there was a prejudice against CDs. The CD format was associated with popular music—chart hits and similar mainstream music—whereas people with taste bought vinyl. One rationale for this was the claim that digitization of music made it sound more clinical, that analogue recordings on vinyl gave a more authentic and warmer sound. In *Vinyl: The Analogue Record in the Digital Age*, Dominick Bartmanski and Ian Woodward argue: "Auditory warmth, richness and the much-vaunted high fidelity of the musical message account for vinyl's lasting and its air of 'holy script' for serious music aficionados."[8] A consequence of these claims was a segmentation in the market for music which entailed collectors, and the shops that catered for them, continuing to value vinyl.

This provides a way to understand just how lacking in taste the man was by exchanging the Sid James Experience vinyl album for a CD. Hornby's joke is actually at the expense of Barry who, taste arbiter that he is, thinks he has finally found someone else who likes this unknown album by a group which in reality doesn't exist (strangely there is a covers band with this name which started in 1991) and is named by Hornby after a mainstream old-school, English comedian known best for his roles in the popular *Carry On* film series full of dated kitsch sexual innuendo, only to discover that the man cannot explain why he chose the album and is happier with a CD by a popular group.

It is a little surprising that Rob's shop sells CDs, but the answer may lie in the format still being something of a novelty. In the film, there are rows of CDs against the wall. This suggests that this Rob is making some concessions to shifts in the dominant format. Though an independent shop, it sells mainstream and chart records as well as specialist, minority interest records. By the time of the series in 2020 this Rob's shop establishes its credentials as an indie shop for collectors with the "No CDs" sign. This Championship Vinyl is even more firmly located in a niche market than the one in the novel. At one point in the 2020 series, Rob comments: "Half the neighborhood think we're washed up relics, the other half think we're nostalgic hipsters. They're both kind of right."[9] We shall return to this comment later.

In the novel Rob starts his career in record shops by working in the Record and Tape Exchange in Camden.[10] The first Record and Tape Exchange opened on Goldhawk

Road in Shepherd's Bush in 1970. The Camden shop opened in 1982. There is here a chronology problem for the novel. Rob breaks up with Charlie in 1979 and then finds himself working at the Camden Record and Tape Exchange. In order to understand this sacrifice of reality for effect, whether it was intentional by Hornby or not, we need to gain insight into the role of the Record and Tape Exchange shops. These shops bought most secondhand albums brought into them and resold them at prices set to achieve rapid turnover. Leafing through the racks customers could find rare and much sought-after records among otherwise uninteresting stock. Cartwright offers this reminiscence from Gary Jeff who worked at the Goldhawk Road store in the late 1980s:

> Back then a lot of people were re-buying their album collections on CD, so dumping vinyl collections on Record & Tape. We got great collections in, often very cheap. . . . After eighteen months, I moved to the Camden store, where lots of rock musicians hung out.[11]

Camden, with its market, had become a subcultural scene in the late 1970s and this continued through the 1990s. The Camden Record and Tape Exchange had a reputation for selling intriguing albums from when it opened.

Jeff, who worked at the Goldhawk Road store in the late 1980s, recalls the extent to which indie record shops were an important link in the organization of popular music:

> Across the UK, more than any other nation state, record shops have often operated far beyond their quotidian retail function: musicians have met, formed bands in them and been informed by them; record labels have grown out of them; close friendships and intimate relationships have taken shape across the racks; . . . Most importantly, all manner of music has been engaged with and enjoyed.[12]

As Jeff bears out, this was especially true of the Camden Record and Tape Exchange. Having Rob work there before he established his own shop emphasizes the allure of record collecting. It helps us to appreciate that even Rob's out-of-the-way shop played a part in the organizational structure of popular music which was in some ways similar to the role played by shops that supplied the items for collectors such as stamps for stamp collectors. At the same time, specialist shops occupied an additional role in the circuit of a popular art form, providing the music and the knowledge for mostly young men and women who were intent on being popular music artists or who were just interested in music beyond the charts.

Taste, Pleasure, and Elitism

In the film *High Fidelity* (2000), Louis, who is African American, comments to Rob, Dick, and Barry: "You guys are snobs. Yeah, seriously, you're totally elitist. You feel like under-appreciated scholars so you shit on the people who know less than you."[13] In the novel, even Rob finds Barry's elitism extreme, describing him as a "snob obscurantist."[14] The negative aspect of the knowledgeable indie shop employee is that the knowledge can be used as a means of exclusion. Lee Ann Fullington makes this point in relation to gender: "Masculinity may be conflated with knowledge about, and a specific taste in, music, and this music often excludes female artists, which in turn excludes female and male fans of such artists."[15] Historically, women have not been expected to have the musical knowledge that male shop employees had. Fullington explains: "In this displaying of knowledge, if the woman is on the other side of the counter, she may be faced with a hostile customer who will not take her seriously, as she, a woman, must not know anything about such male preserves as rock, jazz, or certain strains of dance music."[16] In this context, we should remember the whiteness and heterosexual maleness of Rob, Dick, and Barry, and their customers, in both the novel and the film. In the television series, it is significant that the writers made the Barry figure both female and African American, and Simon queer. Louis makes his comment about Rob, Dick, and Barry's elitism as a racial outsider.[17]

The novel and the film include versions of the same scene which plays out the indie record shop employees' use of knowledge as elitism. A man enters the shop, he is, as the novel's Rob describes him, "middle-aged, from the sound of him, and certainly not hip in any way whatsoever."[18] He wants to buy a copy of Stevie Wonder's "I Just Called to Say I Love You" for his daughter's birthday. It is a popular song reaching number one in both the UK and US charts in 1984. Barry claims the shop has a copy and then refuses to sell it to him. When the man asks why, Barry says: "Because it's sentimental, tacky crap." In the film Barry adds, "Do you even know your daughter? There's no way she likes that song."[19] What this addition highlights is a generational divide not present in the novel and which works to associate record knowledge with hipness and youth.

Rob's shop doesn't have a copy of the single. For that the man should have gone to one of the high-street chain stores. However, given when "I Just Called to Say I Love You" was released, by 1995, eleven years later, it was unlikely to be available in the high-street stores. It certainly would have been unavailable in such stores in 2000. Also, old vinyl singles were being deleted as CDs became the mainstream format of choice. It's possible that the man was in a difficult situation, trying to buy an old single, sixteen years old by the time of the film, which was not stocked in the chain shops and, for reasons of taste, not stocked in the indie shops. The best way to get hold of the song would have been to buy the soundtrack from the film *The Woman in Red* in which it was used and for which Wonder is credited with the soundtrack. It would have been purchasable perhaps on vinyl in 1995 and certainly on CD in 2000.[20]

Discussion of the elitist knowledge of those who worked in the specialist shops bears directly on aesthetics and personal taste. The presumption was that there was a connection between the obscure and little known and a claim to quality. The obverse was the claim that the popular, chart hits and the artists who made them were lacking in quality. As Lauren Ziegler puts it:

[P]opular music and its many branches, be they pop, EDM, trap and so on, have been considered plebeian preferences, music for the untrained ear, the underdeveloped brain and just plain bad. Aphex Twin, Brian Eno and Talib Kweli are profound and significant, canonical essentials to any serious collector. David Guetta, Jason Derulo and Tyga on the other hand? Get out.[21]

We can remember the scene in the *High Fidelity* film where Barry presses a vinyl copy of Bob Dylan's *Blonde on Blonde* onto a customer while saying: "That is perverse. Do not tell anyone you don't own fucking *Blonde on Blonde*. There, it's alright now."[22] Louis suggests Rob, Dick, and Barry are like underappreciated scholars. The analogy is with the distinction between high culture and popular culture.

We may not like the elitism of the people, mostly male, who work in record shops but it used to be their knowledge, and their willingness to share it, which helped keep the traditions of popular music alive. Writing about *High Fidelity* from the point of view of 2020 and the world of digitization and streaming, Carl Wilson comments on the novel and film:

> Today, *High Fidelity* stands as an elegant parable about the way taste operated in between the upheavals of the 1960s and the digital transformation of the new millennium—how the trappings of subcultural cool could serve as a substitute for actual politics or ethics, and function primarily as a weapon of exclusion.[23]

As we have seen, this is only partly correct but for what came next it was the emphasis on elitism and exclusion that was important.

Shazam, Spotify, and Substituting Morality for Aesthetics

We must now examine the impact of the digitization of music on the record shop through the 2020 television series *High Fidelity*. In episode one, a man enters the shop and after listening to what is being played takes out his smartphone and Shazams it. Cherise asks him if she can help him and when he says no, he's good, she replies: "You do know that there's an actual person standing right here in front of you?" The customer responds: "I'm sorry, I just thought Shazam would be easier."[24] Cherise then constructs a hypothetical conversation with the man in which he asks what the track is and she tells him, providing further information and telling him where the album is located in the shop. In short, Cherise provides a version of the traditional indie record shop interaction between employee and customer which, as the man demonstrates, is now obsolete.

Spencer Kornhaber suggests that Cherise "isn't out to shame the Shazamer so much as connect with him."[25] Rather, the scene can be read as an enactment of the present meeting the past. As the man leaves, echoing the man in the novel and in the film who leaves after being abused by Barry, Cherise comments loudly and aggressively: "The problem with these kids is that the generation has completely fucked off."[26] Her point is not so much that the man doesn't want to communicate as that he isn't interested in Cherise's music knowledge because he can get all the information he needs from Shazam; and he doesn't buy the album. What the customer knows is that Cherise doesn't just want to chat about the track. Rather, she wants to tell the customer why she's playing it, why it's good. He doesn't want to know. The track has caught his attention. He likes it. He uses Shazam to find out what it is. Rather than buying it in the store he will go home and download it. In her exchange with him Cherise marks herself as part of the dying breed of record shop cognoscenti, replaced by the increasing availability of music, and information about it, on the web. Kornhaber elaborates this point:

> Even though some *High Fidelity*-style shops catering to vinyl collectors have survived the extinction of big-box retailers, streaming and downloads have chipped away at the super-listener's pretexts for arrogance: special knowledge (entire discographies are now explorable with a click), special access (few B-sides can hide from Google), and curatorial chops (algorithms can DJ your life). Cloistered listening has become more common, as Spotify and the omnipresent earbud turn an entire art-form into an on-demand, all-you-can-stream personal utility.[27]

Indie record shops have been replaced by digital recognition systems and music need no longer be bought on either vinyl or CD but can be downloaded or streamed on platforms like Spotify which began in 2006.

However, there is a problem with this way of finding out about and accessing music. Algorithms linked to streaming offer you music that, based on your established preferences, they are pretty certain you'll like. The streaming company may give you artists you've never heard, or indeed never heard of, but the music is a result of choices made by an algorithm that has learned what you like. The consequence is that you are never challenged. We might turn here to Roland Barthes's distinction between *plaisir* and *jouissance*. In the translation of Barthes's discussion:

> Text of pleasure: the text that contents, fills, grants euphoria; the text that comes from culture and does not break with it, is linked with a comfortable practice of reading. Text of bliss: the text that imposes a state of loss, the text that discomforts (perhaps to the point of a certain boredom), unsettles the reader's historical, cultural, psychological assumptions, the consistency of his tastes, values, memories, brings to a crisis his relation to language.[28]

When our music choices are made by an algorithm that knows what we like, we live in a world of *plaisir* never shocked by new and different music.

One of the informal roles of the assistant in the indie record shop was to suggest new music, different music that might give the listener pleasure through the barrier of displeasure. There is often a moment in the shop when the listener starts by disliking what is being played but, listening longer, realizes it's actually pretty good. In the film *High Fidelity*, Rob puts on "Dry the Rain" from *The Three EPs* by the Scottish group the Beta Band at the same time murmuring to Dick that he will sell five copies to the customers browsing in the shop.[29] Suggesting the retro tendency of the Hulu series, in episode six the same scene is enacted with Rob putting on Swamp Dogg's "Lonely" from *Love, Loss, and Auto-Tune*. In each scene, a customer turns to Rob and asks who the artist is, then says, in the film "It's good" and in the series, "It's great." Each Rob has introduced their customers to an artist they've never heard before. The retro quality of this scene in the television series is emphasized when we remember the man who Shazam's Lescop's "La foret." In the time before streaming the shop assistant might say: "Hey, you should hear this," suggesting that it's out of your comfort zone. Or they might say, "Hey, I think you might like this given you like . . . ," which is more like the way an algorithm works. It might still stretch your listening boundaries a little. This relates to what annoys Cherise. The Shazamer doesn't engage her in conversation so she doesn't get an opportunity to explain why he might like what he's listening to. Her hard-won knowledge and ability have become redundant in a world that privileges *plaisir*.

The effects of digitization have transformed aesthetics. Now the traditional gatekeepers, the indie record shop employees, can no longer keep their elitist position. Scarcity has been replaced by excess. This in turn relates to the development of a proliferation of niche genres. Nobody can be an expert on all the genres. Within each genre there are debates about aesthetic quality. At the time of Hornby's *High Fidelity* there were fewer genres and access to music was limited to the mass media, friends, including, for example, the mixtapes that almost always boys would make for girls to "improve" their music knowledge and broaden their aesthetic understanding, and record shop employees. These gatekeepers were able to be both knowledgeable and elitist, though many were knowledgeable and willing to give advice. English literature criticism in the 1950s and 1960s was dominated by the ideas of F. R. Leavis. Leavis considered that one role of the critic was to bring out of great literary texts their moral complexity. Indeed, for Leavis, moral complexity was a key feature in what made a text great literature. In the world of digitalized music Leavisian aesthetics have been stripped back to morality.

In episode two of the television series, a middle-class white woman comes into Rob's shop asking to buy Michael Jackson's *Off the Wall* album for her boyfriend. This scene reworks the novel's scene in which the man asks to buy Stevie Wonder's "I Just Called to Say I Love You" for his daughter. Now, however, this random customer is anachronistic. Why does she want to buy a vinyl album for her boyfriend when she can download it for him or, if he is old-fashioned enough to like physical media, she could buy a CD for him online? Perhaps the boyfriend is somebody who prefers the sound of analogue. We never find out. This anachronism leads into a shift in the employees' deployment of their knowledge from aesthetics to morality. There follows a discussion in the store about the ethics of selling an album by an accused pedophile. Cherise, playing the part taken in the novel and film by Barry, is loudly against the sale. She calls Jackson a pedophile, the women interjects "allegedly" as Jackson was never convicted.

The debate is not about the aesthetic quality of Jackson's hugely popular album, as it would have been in Hornby's *High Fidelity*. As the claims to taste weaken with the increased access to diverse music other discussions related to morality take their place. One consequence is the connection with Leavisite literary criticism becomes overt. Rob asks Cherise why she continues to listen to Kanye West given his fondness for Trump and his tenuous grasp of history. Cherise replies: "Are you fucking serious?

Having shitty politics and a second grade understanding of American history is a tiny bit different from being a goddam child molester" (2020). The record shop assistant has become a moral rather than an aesthetic arbiter.

Discussing the Shazamer incident, Kornhaber argues: "The much-discussed 'death of the snob' in the Internet era explains part of the shift on display."[30] Now the indie shop assistant's musical knowledge is put to different use. In this new world where everybody has access to huge amounts of music, and to information about it, Rob puts her knowledge to use making sophisticated moral judgments. There is a flattening of the hierarchy of musical knowledge. Kornhaber describes the older attitude as snobbery. It was an elitism born of a knowledge that enabled the employee, who sold the music, to discriminate between insiders, who had, or aspired to have, the knowledge, and outsiders, those without the knowledge and who often did not even care about the knowledge. In this flattened world, Rob is repositioned as a nerd rather than an elitist. We should remember here Rob's remark about half the neighborhood thinking of them as nostalgic hipsters. Straw remarks that "Hipness and nerdishness both begin with the mastery of a symbolic field; what the latter lacks is a controlled economy of revelation, a sense of when and how things are to be spoken of."[31] While all three Robs might like to think of themselves as hipsters they are more nerdish in their general sense of social inadequacy, something disguised for the earlier Robs by the link between their elitism and their accumulated knowledge. Television series Rob feels able to sell the woman the Michael Jackson album because she knows that Quincy Jones did the horn charts and, from her perspective, she's selling Quincy Jones's amazing work, not Jackson's. This may sound casuistic, but Rob is making a moral judgment based on her detailed, we might say nerdish, knowledge of the album.

Conclusion

The three iterations of *High Fidelity* traverse fundamental shifts in the positioning of the record shop. In the mid-1990s the record shop still functioned as a place to buy music unavailable in the high-street chains which focused on chart music. These chains had become increasingly pervasive since the 1970s. At the same time, the customers of the specialty, or indie, shops were collectors looking not only for particular items of music but for rare examples of the release of that music, first releases, import copies, test pressings, and so forth. By 2000, when CDs had become the dominant medium on which to buy music, a debate was taking place about whether vinyl, and analogue, was a better medium for listening to music than CDs and digital. At this time, indie shops continued to stock vinyl, sometimes, as in the case of the shop in the 2020 version of *High Fidelity*, not stocking CDs at all. At that time, vinyl was the preferred medium for collectors and CDs were viewed as the medium for music bought and listened to by those solely interested in chart music, that is, most music consumers.

With the escalating importance of downloading and streaming as ways of accessing music, the high-street chains sold off their shops, went bankrupt, or moved into selling other media such as DVDs and computer games. They rebranded themselves as selling entertainment rather than just music. At the same time, many of the indie record shops closed as streaming sites specializing in alternative music such as Bandcamp, started in 2008, took much of their clientele. With a certain nostalgia, and a deliberate echo of Rob in the television series explaining that half the neighborhood think of the people who work in that *Championship Vinyl* as nostalgic hipsters, Cartwright sums up the change,

> [F]or over a century the record shop was an essential part of a city's fabric, seducing shoppers with beautiful noise and acting as chief muse to latent creative energies. Then, as we entered the twenty-first century, the record shop began to appear redundant, a leftover from an analogue age. In a world where you can now have all the music you want on your phone, who needs to enter a bricks and mortar structure to access the sounds you desire? Not, it seems, many people.[32]

Cartwright estimates that across the UK the number of record shops has fallen from around 1,000 in 2000 to less than half that in 2017.[33]

However, since 2006 there has been a vinyl revival, and this has impacted record shops. In 2020, 27.5 million vinyl albums were sold in the United States, which represents a thirtyfold increase since 2006. In the UK vinyl sales started rising a year later in 2007. In 2020, 4.8 million vinyl albums were sold, the highest number for thirty years. While some of these sales in both countries would have been online, many sales would have been through the remaining record shops. According to the June 2021 article in the UK newspaper *The Independent*: "Vinyl record sales surged through the pandemic as music lovers fattened their collections, and audio cassettes began a comeback as well, keeping business spinning at record stores."[34] These shops, previously surviving selling to those with specialist interests and collectors, have now found themselves with a new clientele, people wanting to buy vinyl albums by artists often considered mainstream. In the United States: "By Billboard's tracking, the list of the top 15 most-selling vinyl LPs year-to-date in 2021 includes albums from Taylor Swift, Olivia Rodrigo, and Harry Styles—as well as Michael Jackson's *Thriller*, Fleetwood Mac's *Rumours*, and Queen's greatest hits."[35] Before the release of Adele's *30* in 2021, Sony ordered half a million vinyl copies. In this context, the woman trying to buy a vinyl copy of Michael Jackson's *Off the Wall* in television Rob's shop was not so strange after all.

Ideally, indie record shop employees used their specialized knowledge to inform and advise their customers. It was, essentially, a tutelary role. At the same time, it could rapidly morph into being exclusivist and elitist, as it does in all three *High Fidelity*s. What this chapter has also argued is that there was a moral aspect to the knowledge about music. This becomes clear in the television series but is also present in Barry's attack on the man who wants to buy the Stevie Wonder single for his daughter. The premise of the Robs and their employees is that good music is often music that is difficult to listen to, that how to listen to it often must be learned, but that the reward is an ability to think critically about popular music and gain a greater pleasure in listening to it.

Notes

1. Garth Cartwright, *Going for a Song: A Chronicle of the UK Record Shop* (London: Flood, 2018), 33.
2. Richard Williams, "A Record Shop Life," *The Blue Moment*, April 13, 2018. https://thebluemoment.com/2018/04/13/a-record-shop-life/.
3. Nick Hornby, *High Fidelity* (London: Penguin, 1995), 30.
4. Ibid., 118.
5. Kirk Miller, "35 Years Ago, Dire Straits Physically Changed How We Listen to Music," *Inside Look*, May 2020. https://www.insidehook.com/article/music/dire-straits-brothers-in-arms-compact-disc.
6. Peter Tschmuck, "The UK Recorded Music Market in a Long-Term Perspective, 1975–2016," *Music Business Research*, July 30, 2017. https://musicbusinessresearch.wordpress.com/2017/07/30/the-uk-recorded-music-market-in-a-long-term-perspective-1975-2016/.
7. Dominick Bartmanski and Ian Woodward, *Vinyl: The Analogue Record in the Digital Age* (New York: Routledge, 2015), 166.
8. *High Fidelity* (2020), [TV programme] Hulu.
9. Hornby, *High Fidelity*, 24.
10. Cartwright, *Going for a Song*, 161.
11. Ibid., 12.
12. *High Fidelity* (2000), [Film] Dir. Steven Frears, USA: Touchstone Pictures.
13. Hornby, *High Fidelity*, 114.
14. Lee Ann Fullington, "Sneaking into the Boys' Club: Gender and the Independent Record Shop," in *Practising Popular Music: International Association for the Study of Popular Music Twelfth Biennial International Conference Proceedings*, ed. A. Gyde and G. Stahl (Montreal: IASPM, 2003), 304–5.
15. Ibid., 302–3.
16. It is worth noting here that the American film *Empire Records*, mentioned earlier, also had a white cast of employees and customers.
17. Hornby, *High Fidelity*, 42.
18. Ibid., 43.
19. *High Fidelity* (2000), [Film] Dir. Steven Frears.
20. Lauren Zeigler, "Highbrow vs Lowbrow and the Problem with Musical Elitism," *Howl and Echoes*, July 6, 2016. https://howlandechoes.com/2016/07/highbrow-lowbrow-musical-elitism.
21. *High Fidelity* (2000), [Film] Dir. Steven Frears.
22. Carl Wilson, "Can High Fidelity Survive the End of Taste?" *Slate*, February 25, 2020. https://slate.com/culture/2020/02/high-fidelity-hulu-taste.html.
23. *High Fidelity* (2020), [TV programme].
24. Spencer Kornhaber, "The New Rules of Music Snobbery," *The Atlantic*, March 2020. https://www.theatlantic.com/magazine/archive/2020/03/high-fidelity-hulu/605539.
25. Wilson, "Can High Fidelity Survive the End of Taste?," argues similarly that Cherise "wants to use music not to assert superiority and distance but to forge human connections."
26. *High Fidelity* (2020), [TV programme].
27. Kornhaber, "The New Rules of Music Snobbery."

28 Roland Barthes, *The Pleasure of the Text*, trans. R. Miller (New York: Hill and Wang, 1975), 14.
29 *High Fidelity* (2020), [TV programme].
30 Kornhaber, "The New Rules of Music Snobbery."
31 William Straw, "Sizing Up Record Collections: Gender and Connoisseurship in Rock Music Culture," in *Sexing the Groove: Popular Music and Gender*, ed. S. Whiteley (New York: Routledge, 1997), 9.
32 Cartwright, *Going for a Song*, 12.
33 Ibid.
34 *The Independent*, June 8, 2021. https://www.independent.co.uk/news/vinyl-records-surge-during-pandemic-keeping-sales-spinning-maine-americans-portland-joshua-tree-national-park-sales-b1861996.html.
35 Emily Stewart, "The Supply Chain Crisis, Explained by Adele," *Vox*, November 22, 2021. https://www.vox.com/the-goods/2021/11/22/22797290/adele-vinyl-record-supply-chain-delays.

7

Reflections from the Girls behind the Counter

Lee Ann Fullington

In the late 1990s, when I was in my early twenties, I worked in a large independent record shop. I was one of only four women on a staff of twenty to twenty-five full and part-time employees. I was one of two women at the counter and the other two women worked part-time in the stock room, processing used CDs. Most times, four or five employees staffed the counter. The owner, the general manager, the counter manager, the buyer who dealt with new stock, the buyer who dealt with used CDs, the classical buyer, the LP appraiser, and all the rest of the staff were men. The shop was a bit of a boys' club, and this could be good, this could be awful, or, given the day, it could just be.

Over the years I worked at the counter, I met a few regular customers who were women and from my vantage point, it was men, men, men throughout the shop, with only a smattering of women. I always wondered why this was. Did women not listen to records? Did women not buy CDs? That seems absurd. Could women just not be bothered with the shops? With digging through dirty records? Putting up with dudes looking over their shoulders? I'm not sure there's a definitive answer to any of this. However, this conundrum haunted me after I left the shop. I moved on to a different job and then shortly thereafter, postgraduate study for an MPhil in popular music studies, where I attempted to understand the culture of these shops. I met several women owners and staff members during my research. This chapter draws on my own experiences behind the counter and interviews I did between 2002 and 2004 with owners, staff, and customers, and all names used are pseudonyms. In my nonscientifically drawn convenience sample of shops that were easy to get to on public transit, or simply shops I knew of from my own travels digging for records, or word of mouth, men always outnumbered women. When I asked interviewees about their customers, each participant, be they owner or a staff, told me that based on their own experiences, it's mostly men. For instance, Melissa, who managed a shop in a busy city shopping district neighborhood, joked that she often thought about keeping statistics:

> "It burns me out being here so much and dealing with all the indie rock weenie boys, coz it's so many more guys that come in here. I said I was gonna mark down one day, like the boy to girl ratio."[1]

As a fellow woman who worked in an indie record shop, it may be that the women I interviewed opened up to me due to our common bond as girls in the proverbial boys' club. Some of our experiences were joyful, some frustrating, and some downright awful and scary.

The Good

Working in an independent record shop in my early twenties in the late 1990s was a formative experience. I'd already been quite obsessed with music and had trawled the racks of CDs, new and used, and the huge back room full of low-priced used CDs throughout my undergraduate days, amassing a decent collection of Britpop, goth, hardcore, alternative, post-punk, indie. I loved guitars, walls of sound, jangly pop, and false nostalgia for bygone eras. I put an application in for part-time work when the sign went up, and though I didn't get the position the first time around, someone else left and the manager called me to offer me a job a few weeks later. As a customer, it hadn't exactly occurred to me that there didn't seem to be women working there, but I had noticed it was always men who rang me up. When I started working behind the counter, I found out the person they'd hired before me was another woman, so now there were two of us in this sea of dudes.

Usually, it was just me or the other full-time woman staff member at the counter, with several of the guys. The counter was long, about 20 feet or so, and it was raised up on a platform, so we literally looked down on the customers. The owner said this was so we could watch out for shoplifters, but really it just put us all heads and shoulders above the people we were waiting on. Or, for the two of us girls behind the counter, it put their eyes at an angle to gaze up at our chests.

This job was dirty. You had to squat or kneel to retrieve CDs from the bottom drawers, and you wore out the knees of your jeans. Same with putting the "Dollar LPs" in the cardboard boxes under the racks of the desirable LPs. Pricing records? Dirty hands. Putting records into the polyvinyl sleeves? More dirty hands. If you didn't have mold or dust allergies prior to working here, you were definitely on the way to developing them as the air in the shop was always swirling with motes and spores.

I'd already had a fairly eclectic love of music, and through working in the shop I developed a much broader set of tastes. We (theoretically) took turns picking CDs or LPs to play over the shop's stereo, and I learned so much from my coworkers. I picked up a love of free jazz and reggae, underground hip-hop and garage rock, a deeper appreciation of classic rock (Saturdays meant deep cuts of Jimi Hendrix or Led Zeppelin for a few hours midday, a time-honored tradition at the shop), and I started buying records in earnest, converted to a love of vinyl by immersion. I had a voracious appetite, an employee discount, and we all had first dibs (and a lot of healthy competition) for used LPs and CDs as they came in. In hindsight, these golden days of sleeving, pricing, filing, and bullshitting at the counter were some of the best times I'd had and left indelible prints on my personality and taste in music. Being on display behind the counter or out on the floor filing records, I ended up developing a thicker skin and a hardness from dealing with all the men, both customers and staff members. Rudeness was rampant waiting on men, and I often wearied of having coworkers groan when I put on the Smiths or some other "girlie music" or missed my turn at the stereo because one of the guys jumped the queue. It was expected that I defer because I was simply just a gal and the dudes needed to hear whatever it was they were test driving on the turntable. So even though I loved some of my colleagues dearly and spent a lot of time with them outside of work, when it came down to perceptions of serious music fandom, I wasn't even on their radar.

My record collection may not be considered large by some standards, but I do have quite a few finds in it and so many LPs from my days at the shop. (I probably have at least a thousand records and singles and quite a few LPs and 7"s that would fetch a decent price but I'm not selling!) Having access to so much good stuff coming in and having the record appraisers putting things aside for us before they got priced based on them knowing our tastes and want lists were glorious. So, though my tastes may not have been taken as seriously by the guys at the counter, the buyers certainly did keep my preferences in mind. For example, our main LP appraiser knew I had a thing for Manchester bands and when a collection came in that was heavy with the Fall, Durutti Column, A Certain Ratio, and of course the Smiths, these LPs and 12" singles didn't make it out to the floor but straight to my personal "hold" shelf in our break room. I bought these hundred or so records over a few months as we could keep stock in our hold shelves and buy things over time. (And yes, every so often, the owner purged our individual shelves if we took too long to pay off our holds. There are more than a few records I think I have, and then when I look for them on my shelves now, I remember I didn't get around to buying it and I kick my past self for letting these slip through my hourly waged fingers.) I am grateful to have serendipitously landed that part-time job that quickly became full-time and fed my hunger for new sounds. As is evident by now, my reasons for working the shop were myriad. However, I was hired not for my musical expertise but for my retail experience and good-natured demeanor. Though it may not have been obvious, our general manager did like to hire women, but we didn't stick around as long as some of the guys who ended up staying for decades.

As my interviews reveal, some record shop owners and managers hire women to work behind the counter to encourage women customers to come in. As Matthew, a shop owner, stated, it "breaks the ice" for women and makes for a more inviting atmosphere and less of a "boys club."[2] Woman customers did indeed seem to respond to this, as Vicki, a regular customer of Matthew's shop, commented, "I do like being helped more by a woman than a man."[3] Furthermore, Lizzie, a record shopper I encountered in my interviews, explained:

> If some women are working at a place, women customers are going to feel, even subconsciously, more like they belong in that place. Also, a woman clerk will put on whatever music she likes during her shift and sometimes she might put on music that, for whatever mysterious reason, appeals to chicks more than guys, so the women shopping might get into it and enjoy the atmosphere more because of it.[4]

Women customers may try to chat with or interact with other women customers as a kind of bonding or acknowledgment that they're rarities in such shops. Sarah likes to say hello to another woman customer if she sees one because they're often the only women in the shop. Finding another woman with the same interest in music and record shopping is sometimes rare, as the clientele in smaller shops is so often male. Interacting with a woman staff member is also important to some women record shoppers and may sometimes lead to a connection that goes beyond the shop. Vicki explained that she had become friendly with a woman staff member at a local shop because they both were into the same band. As she recounted:

> "Once I was going to go see a certain band, so I was buying their newest album. The girl at the counter got so excited that I bought it we decided to meet up at the show and now every time I want to see a show and have no one to go with, I ask her."[5]

Melissa, the manager of Matthew's shop, liked interacting with younger women shoppers like Vicki across the counter. She explained that when she was younger and just starting to get into serious record and CD shopping, she knew that staff members in some shops would not pay her any mind, but the ones who did made her feel so happy and accepted. Now that she was on the other side of the counter, she tried to do the same thing.

"You know what I love? I love all the young girls that come in and I know they come in and they're like, 'Oh the redhead cool girl!' So, I try to be extra nice to them [and] make them feel comfortable. I love when they come in and they're like, 'OK, I got the Bikini Kill, now what else should I get?' That is my favorite thing. It's such a good feeling . . . it makes them so happy."[6]

Thus, Melissa recommended music to younger girls without them feeling patronized. She enjoyed sharing their enthusiasm when they connect with music, and by treating the women customers well and making them feel comfortable, her warmth will keep them coming back to the shop.

Helena also tried to keep the shop she co-owned and managed, in her words, a "female friendly space."[7] From her own personal experience record shopping, she felt "instantly on edge" if she went into a shop with only men behind the counter. She did not like feeling like she did not belong, and the men behind the counter sometimes gave the impression that she was not welcome in their "boys club." Freya also noted this problem as a female record shopper:

> "When, you, as a woman, walk into a new shop where you never have been in before the men there are looking very unfriendly so as they would say: 'Oh god, a woman—what is she doing here? Couldn't she go buy clothes?'"[8]

Such condescension is off-putting and sometimes would manifest as outright hostility to women customers. The flip side also happens, whereby male customers mistreat (or worse) female staff.

The Bad (Behavior)

One of my tasks was pricing used classical CDs. The classical manager complimented my being the only one who paid attention to where to put the price tag. The left side of the spine was where he preferred us to sticker CDs, as the customers would read them with the spine facing out, as they were shelved in cubby holes on the wall. If this spot on the spine had pertinent information on it (e.g., the conductor's name), we were supposed to put the sticker anywhere on the spine that was blank or at least didn't cover anything important. Apparently, all the other counter staff just slapped the tags on the usual spot but, due to my being more meticulous, this became my "special" job. Classical CD customers had some intense men within their numbers who would hover at the counter to try to see what I was pricing and snatch things out of the long boxes, sometimes even out of my hand, as I was cocking the price sticker gun. Would they have done this to one of the guys? Maybe, maybe not. A side effect of having this responsibility was that the customers trying to trawl the boxes as I priced would sometimes ask questions about composers and I'd have to admit to having no expertise in classical music, that my skill was simply correctly using the pricing gun. Some were gracious when I told them; others, however, were frustrated and wondered what the hell I was doing working at a record shop if I couldn't answer simple questions about obscure versions of classical works.

Soon after I began working at the shop, the general manager and I had a talk about what it was like to be a girl at the counter. He said from his times managing chain record stores and now at this independent shop, women at the counter were sitting ducks and he hated that we caught unwanted attention from men and had to put up with being flirted with, accosted for dates, or told we didn't know anything about music and shouldn't be here. However, this is where being an independent shop shines—he told me we didn't have a customer service policy and if someone was being awful to me, I could walk away or curse them out. It was up to me whatever I felt like doing in that situation. And indeed, this happened on a fairly regular basis. Thankfully, the other guys at the counter had my back and would step in and complete the transaction. Some would tell me to hide in the back when certain terrible customers were lurking in the shop so I wouldn't have to deal with them. There was a protectiveness there, a closing of the counter staff ranks. No matter how awful any of us were to these guys, they always bought the records and kept coming back.

Many owners and managers pointed out that they would always support their staff in disrespectful situations, especially women, as we were harassed much more often. For female staff, our comfort and safety are a concern, for we are subject to unwanted male gazes, unwelcome attention, and are targets for robbery. In a smaller shop, we may not be able to leave the counter and must endure these myriad manners of harassment.

As did I, Melissa has also dealt with harassment and threats to her personal safety. She complained that it was always male "weirdo" customers who hassled her, and she sometimes had to resort to being rude and nasty to protect herself. For the most part, Melissa was very friendly toward customers (which some men took as her flirting with them), but in some cases she was mean to male customers because "they took away your right, so you take away theirs."⁹ When she felt threatened, she did not act with the expected retail politeness, which *is* a customer's right in her opinion, but they have infringed on her right to be comfortable in the shop, so they've forfeited this expected kindness.

One of the creepiest regular customers at the shop I worked at was a man in his sixties, who collected jazz records. He'd come to the counter to ask a question, and if it was me or the other woman who worked at the counter, he'd bark at us to "get him" and point to any of the men who happened to be at the counter as he didn't think we could answer any questions about music, especially jazz. Fine with me, as the less I had to interact with him, the better. However, when it was time to check out, he would slink up to the counter, place his pile of records onto the glass, and because it always seemed like we were the lucky ones at the counter at the time, either myself or the other woman would ask if he was ready to check out and he'd look up and genuflect and then say "yes, mother" in some strange tone of voice as if he were an injured child seeking solace.

Melissa also dealt with one customer who always stared at her breasts, making her extremely uncomfortable and thoroughly objectified. When she told him to stop, he said he was "just reading your t-shirt."¹⁰ Also, when the weather is hot, she noted that she got more sexually charged comments from some men customers:

> Like in the summer, the air conditioner barely works, and in here it's so hot. I just wear a slip, you know? And of course, I'm gonna get more comments that day, you know? It's like, 'Dude, its 94 degrees in here. I'm really hot. I need . . .' I totally consider it my turf here and I should be absolutely comfortable here. That's why I just resent the, I resent the regulars who are weirdos.¹¹

The Ugly

Where I used to work, the car park for the employees was about a ten-minute walk from the shop, and at closing time (9:00 p.m.) it was dark, the street wasn't that well-lit, and the car park was in the shadows. I rarely had to walk to the lot alone as the few of us who were on the closing staff went together. On several occasions, men would wait for me (I was the only girl who closed) and then try to walk me to my car or try to get me to go for a drink. They knew when we closed, and they would lurk near the door. I never felt safe leaving the shop alone at night.

One time, a tall, stocky man I had waited on numerous times in the past told me that I was beautiful and that he'd like to slip me in his pocket and take me home and have his way with me. And that he'd be waiting for me in the parking lot after the shop closed. And sure enough, when I left and walked out with a coworker, he was indeed in the shadows. Thankfully, he didn't follow us, but he did turn up in the shop numerous times after that incident and I never waited on him again, as one of my coworkers would step in and handle the transaction.

Male customers have physically assaulted Melissa. She shuddered, "the one guy hit me once, because he doesn't know how to deal with women."[12] Another asked her out on a date, and when she declined, he hit her on the back. Often, when these customers come in, if she was not alone at the counter, she would stay in the back office until they left. The shop had a small staff, so it was usually Melissa and one other person, as it was only her and the owner working full-time there. Her coworkers, all male, were very supportive of her and she noted that "Sometimes I don't even have to say it. Like [my coworker], who's here on Sundays, knows when the drunk guy is here, he hops on the register and I pull the CDs and try to keep out of focus."[13] Furthermore, Matthew commented about safety that "we've had to throw people out, tell them not to come back. It scares me sometimes."[14] However, though they may throw these customers out, they may come back when the owner isn't in and try to push their luck.

Melissa has also had to deal with teenage boys threatening to rape her. Though they may just be being rude and obnoxious and making a disgusting joke, a remark like that will put a woman member of staff ill at ease, for the reality is that something horrific could happen.

> We used to have our basement open, and it was a horrible idea... We had our used stuff down there and our vinyl... it was like really dangerous down there. I had these kids, these four kids be like, "We're all alone down here with you. We could rape you." I chased them up the steps and was like, "Get the fuck out!"[15]

The shop has since closed the basement to customers to try and prevent additional menacing encounters.

While Melissa's experiences may seem extreme, they are representative of some of the problems female staff face. Male customers may see women staff as easy to take advantage of, easy to steal from because as Melissa said, "sometimes when I'm alone here, I can tell people come in and think, 'oh, we can rip this girl off,' because I'm little and I look so friendly."[16] However, Melissa calls herself the "tough one" in the shop; she "will kick people out when they need to be kicked out."[17] Pamela, a shop owner, also told me a story about how she, though small in stature, confronted a shoplifter. "[I] just kind of snapped and this kid was like 6 feet tall and I'm [was] dragging him out of the store by the back of his hair."[18]

In terms of being authoritative and upset that our agency can get taken away, Melissa notes that "It's just funny, though, that I'm the only girl, but I'm definitely the toughest person. And I'm here so much that I consider it my turf, like my territory. And that's what makes me angry that sometimes I want to hide in the back. I should be comfortable here."[19]

A related problem is that a woman's authority can easily be undermined. As Pamela further explained:

> I'm really small—I'm 5'1" and 95 pounds and I'm 35 and most people would probably think I was in my mid 20s. I think a lot of that just has to do with my size, because people tend to think that when you are small, you're younger...

Occasionally—and this is a good example—I had one guy walk in, he asked one of my employees something and they didn't know and they came in the back to get me and as I was walking out of the back he goes, "THAT'S THE OWNER?!?!" and I just stopped right where I was and I said, "I don't care what you're looking for, or anything, but we don't have it." And with that I turned around and went back into the back! Why should I have to take that in my shop? And he goes, "C'mon, c'mon, c'mon, I didn't mean it that way!" and my employee was like, "I think she's done with you."[20]

It's a common experience that as women we've all dealt with harassment and needing to "go hide in the back." Our male coworkers are supportive of us for the most part and will step in to take care of situations that arise. Eddie, the general manager of a large shop, spoke to me about "male customers who have taken it a bit too far with female staff."[21] At this shop, they ask the offensive customer to leave—"Get lost creep!"—and if he comes back again on a later visit, one of the male staff members will wait on him. Eddie explained, "We're a team, and if somebody didn't want to serve somebody else, then, OK, don't."[22] He told me about a harassment situation:

I remember, probably about 4 or 5 years ago, there was a guy who used to come in once a week and buy classical stuff of all sorts. And he always used to try and make sure he was served by Nora, who worked with us at the time. But the conversation was just a list of snotty double entendres, and she got really tired of it, she wasn't upset about it, she's just like, "This guy is a creep. I'm not going to bother dealing with this." So, we're just like, "OK, just go somewhere and pretend to be busy and I'll deal with it." He doesn't come in anymore.[23]

Thus, it is more important for the members of staff to feel safe and comfortable in their working environment than to make a sale. As these sorts of incidents are not happening on a constant basis, losing a few sales to harassment does not impact the record shop as much as losing a member of staff and having to train someone new to take her place.

The Run-Out Groove

What Melissa said is the bottom line—as women working in the shop, we want to be comfortable and safe. The record shop is our territory, our turf, and our home as well. Though women may not own, staff, or shop in indie record shops as much as men do, we should not feel alienated, threatened, or unwelcome. It's about agency, wanting to not be harassed at the counter, followed to the car park, or be continually objectified by an unwelcome, unsolicited male gaze. We were given the permission to be awful to badly behaved customers and this was a way to reclaim that agency, to reclaim ourselves from the gaze. We wanted to work in these shops to be around music and be around other people who are into music. For all the moments of joy—turning someone on to a new band or an album they've never heard, meeting someone who has the same obscure taste, or finally finding that elusive LP—there are also times that make you wonder why you do it at all. In the 1990s and early 2000s, record shops were very much places dominated by men, as Melissa colorfully summed up:

> It's like a boys' club everywhere. Because it's so many more guys that come in than girls. And I don't know why. And also, it's just the people that stay longer and will hang out here coz they love being in music stores, they're all dudes. Yeah. Drives me fuckin' nuts. It's like, you know, its soooo cool. I have to clean up homeless men's piss on the floor and get hit on. It's not THAT cool.[24]

I agree wholeheartedly—it could be exhausting and frustrating and even terrifying at times working with so many men and waiting on so many men. But those glimmers of light with meeting great people, finding amazing bands, and being paid to do so did help a lot. Though I can't say I loved every minute of working at the shop, the good outweighed the bad. And the hardness I developed through working there has served me well in life—a strange bonus to all those LPs I acquired.

Notes

1. Melissa, Interview by author, May 2, 2002.
2. Matthew, Interview by author, May 2, 2002.
3. Vicki, Email message to author, July 21, 2002.
4. Lizzie, Email message to author, March 17, 2002.
5. Vicki, Email.
6. Melissa, Interview.
7. Helena, Interview by author, August 6, 2002.
8. Freya, Email message to author, July 6, 2002.
9. Melissa, Interview.
10. Melissa, Interview.
11. Melissa, Interview.
12. Melissa, Interview.
13. Melissa, Interview.
14. Matthew, Interview.
15. Melissa, Interview.
16. Melissa, Interview.
17. Melissa, Interview.
18. Pamela, Interview by author, September 29, 2002.
19. Melissa, Interview.
20. Pamela, Interview.
21. Eddie, Interview by author, September 14, 2002.
22. Eddie, Interview.
23. Eddie, Interview.
24. Melissa, Interview.

Part II

Cultural Geography of Record Stores

8

"Ways of Living"

Touristification, Gentrification, and Curatorship in Spanish and Portuguese Record Stores[1]

Fernán del Val

The beginning of the twenty-first century witnessed the closure of many record shops due to the different crises affecting the record industry. Recently, however, different authors have observed and explained how independent record stores have proliferated again in cities around the world. The causes of this resurgence in a context of economic recession are diverse.[2] On the one hand, some authors argue that independent record stores have become spaces of curatorship and expert knowledge on music and records. Contemporary cultural consumption is based on dynamics which focus on the experience of the user. Record stores no longer sell objects but experiences, experiences which are based on rummaging through the shop's shelves or on exchanging knowledge with the workers. Record stores follow a pattern of "gourmetization" of culture. Just as someone goes to a gourmet shop in search of a fine wine or cheese, these stores offer the same experience but with music (in particular vinyl).

These shops also offer a specialized catalogue which goes beyond what bigger companies, such as Fnac in Spain and Portugal, can offer. It should also be pointed out that different changes have taken place in how record stores sell their products. The introduction of online sales, via the stores' own websites or platforms such as Discogs and eBay, has enabled stores to extend their business to fans and consumers around the world. Furthermore, online trade and social networks such as Facebook and Instagram are extremely useful in promoting shops and their activities. Another fundamental element for the survival of these stores is the huge growth in vinyl sales in countries such as the United States,[3] Sweden,[4] Portugal,[5] and Spain.[6] Indeed, vinyl is becoming increasingly appreciated and sought after due to its sound qualities, covers, and art as well as its value as an object.

These arguments have served to initiate research on record stores in Portugal and Spain over the course of a two-year period (2017–19). In this time, twenty-seven in-depth interviews with the owners and workers of record stores were carried out, along with many ethnographic studies and observations in the shops. During the fieldwork stage, some of the ideas presented earlier were corroborated (e.g., the importance of online sales), whereas other arguments were not confirmed. For example, the curatorship practices described by Hracs and Hansson were not observed to be essential in the legitimation of these stores. Neither did the interviews corroborate this issue. On the contrary, during the conversations the idea emerged of record stores as spaces for collective learning in which workers and fans share knowledge and mutual learning, an idea previously put forward by Gracon.[7] Therefore, record stores are not spaces in which, thanks to their subcultural capital, workers advise and teach buyers about what to listen to and what not to listen to. Some of those interviewed stated that buyers and fans have many tools at their disposal to know what new albums are released or how to estimate the value of a secondhand record. Some workers even recognized that, thanks to the buyers, they found out about new albums, new genres, or scenes.

Another significant question which emerged during the course of these conversations and ethnographies, which has not received much prior attention, concerns the importance of tourism and gentrification processes in the record store boom. Both Porto and Barcelona, the places in which the ethnographic studies were carried out, are cities which have undergone "touristification" and which have been the subject of gentrification processes since the 1990s. Although Barcelona is a large city with a population of 1.6 million inhabitants, compared to a mere 215,000 in Porto (increasing to 1.7 million in its metropolitan area),[8] certain similarities can be detected in the processes of urban redevelopment of the two cities. According to Degen and García[9] and Gusman et al., following the 1980s and 1990s, urban reorganization and gentrification processes took place in these cities, displacing low-income social groups from the urban centers and replacing them with a population with a higher level of economic and cultural capital.[10] In both cities, these processes were linked with sporting and cultural events (the Barcelona Olympic Games in 1992 and the UEFA European Football Championship in Portugal in 2004), which made it possible to modernize the urban centers of the cities, improving transport systems and infrastructures. Following the economic crisis of 2008, the arrival of tourists was encouraged via low-cost travel companies and the implementation of alternative models of accommodation such as Airbnb. These processes have, obviously, generated a certain degree of tension and ill-feeling. In Barcelona, the gentrification and touristification of the city have led to protests from residents of central neighborhoods due to the increase in rental prices, the replacement of local shops and businesses by franchises, and the massification of public spaces.[11] In the case of Porto, gentrification goes a step further with the appearance of processes of touristification and "studentification," in which the normal dynamics of gentrification are carried out not by local agents but rather by the international middle and upper classes, as Carvalho et al. explained.[12]

Regarding record stores and their connections with these processes of gentrification, Hendricks highlights the fact that vinyl and record stores have become distinctive elements of the hipster culture, in the same way as other businesses such as barber shops, cupcake shops, and burger restaurants. In the opinion of Hracs and Jansson, record stores located in modern and gentrified neighborhoods, such as Gracia in Barcelona and Rua Miguel Bombarda in Porto, enjoy a better reputation and greater credibility and authenticity compared to others in different locations. In this regard, those interviewed explained that the increase in tourism had been fundamental for the sustainability and economic viability of their shops.

Therefore, in this chapter, the connections between touristification and gentrification processes and record stores will be analyzed. In addition, based on the ethnography carried out, an analysis will be made of the types of record stores in these areas and their main characteristics and those of their workers. The discourse of those interviewed in relation to tourism, the location of the shops, and the vocation and lack of job security in their trade will also be analyzed.

An Analysis of Record Stores in Barcelona and Porto

In order to identify the location and to carry out a census of record stores, Google Maps was used, along with the VinylHub application of the Discogs website. Data obtained from the websites and social networks of the shops themselves have also been used, along with information collected during the interviews and from the ethnography. Using this information, two types of tables have been created. The first, focused on the stores, includes data on the year the shop opened, the type of material on sale (new or secondhand), the formats on sale, the main genres in which they specialize, whether the store sells any other type of products or performs any other functions (distribution, record label) and whether the store is located in a gentrified area. The second tables analyzes the profiles of those interviewed and includes data on their sex, age, and level of education, as well as the roles they play in the musical scenes of the surrounding area.

In analyzing these data, it can be seen that in the case of Porto, of the eleven shops described, more than half (six) have been opened since 2010 in the context of an economic recession. Another two stores opened in the decade 2000–2010 and three in the 1990s.

As far as the type of material is concerned, two shops are specialized in new music, three in secondhand products, and six sell both new and secondhand records. As regards music genres, five shops focus on general interest (mainly pop rock) and six are specialized. All of the stores sell vinyl, while five do not sell CDs. A different type of shop can be identified from the others and can be defined as hybrid: shops whose main function is not selling records or are not only record stores. This is the case of Black Mamba, a cafeteria and record store, and Circus Network, an art gallery and record store. Furthermore, it can be observed that record sales are shared with other functions in several of the stores (record distribution, concert ticket sales, merchandising sales, etc.). In terms of location, nine of the eleven shops are in central and gentrified neighborhoods (Table 1).

In the case of Barcelona, out of the thirty-three stores identified, more than half (eighteen, 55 percent) were set up between 2010 and 2019. Four shops were opened between 2000 and 2009, six in the 1990s, two in the 1980s, and three in the 1960s. Therefore, the resurgence of this type of business is clear. However, in comparison with the case of Porto, it can be observed that there is a notable number of stores which have been in existence for almost fifty years.

Table 1 Record Stores in Porto

Store	Opened	Type of material	Formats	Genres	Other functions	Gentrified neighborhood
Black Mamba	2014	New	Vinyl, cassette	Hardcore, punk	Cafeteria	Yes
Bunker Store	2014	New and secondhand	Vinyl, CD, and cassette	Metal	Merchandising, ticket sales	No
Circus Network	2012	New and secondhand	Vinyl	Electronic, funk, soul	Record label, art gallery, co-working	Yes
Discos do Baú	2017	Secondhand	Vinyl	General interest		Yes
Louie Louie	2004	New and secondhand	Vinyl, CD	General interest	Merchandising	Yes
Materia Prima	1999	New and secondhand	Vinyl, CD, and cassette	Electronic, avant-guard	Magazines	Yes
Muzak	1999	Secondhand	Vinyl, CD	General interest		Yes
Piranha CD	1995	New and secondhand	CD, Vinyl	Metal, post-punk, dark		No
Porto Calling	2012	New and secondhand	Vinyl	General interest		Yes
Tubitek	2014	New	Vinyl, CD	General interest	Record distributor	Yes
Vinyl disc	2008	Secondhand	Vinyl	General interest		Yes

Table 2 Record Stores in Barcelona

Stores	Opened	Type of material	Format	Genre	Other functions	Gentrified neighborhood
BC Store	2011	New and secondhand	Vinyl, CD, cassette	Hardcore, punk	Record label	Yes
Blue Sounds	2009	New	Vinyl, CD	Blues, jazz		No
Crokan's Mutant Store	2019	Secondhand	Vinyl, CD	Electronic		Yes
Curtis Audiophile	2019	New	Vinyl	Soul, funk	Bar	No
Daily Records	1994	New and secondhand	Vinyl, CD	Punk, power pop	Record label	Yes
Decibel	2014	New and secondhand	Vinyl	General interest		Yes
Disco 100	1978	New	Vinyl, CD	General interest		Yes
Discos Edison	1979	Secondhand	Vinyl, CD	General interest	Clothing, books	Yes
Discos Impacto	1991	Secondhand	Vinyl, CD	General interest		Yes
Discos Paradiso	2010	New and secondhand	Vinyl	Electronic, house, soul, funk, pop, etc.		Yes
Discos Revolver	1991	New and secondhand	Vinyl, CD	General interest	Merchandising	Yes
Discos Tesla	1991	Secondhand	Vinyl	General interest	Merchandising	Yes
El Genio Equivocado	2018	New and secondhand	Vinyl, CD	Indie, pop	Record label, merchandising, books	Yes
Jazz Messengers	1980	New	CD, vinyl	Blues, jazz	Distributor	No
Kebra Disc	1983	New and secondhand	Vinyl, CD, cassette	Punk, hardcore, rock		Yes
King Atupali	2009	?	Vinyl, CD	Reggae	Merchandising	Yes
La Spaziale	2019	?	Vinyl	Electronic		Yes
Libertine Records	2019	New and secondhand	Vinyl	Electronic		Yes
Lostracks Records	2010	New and secondhand	Vinyl	Electronic, techno		Yes
Mondosinfonola	2015	Secondhand	Vinyl	General interest	Jukebox shop	No
Music World	1991	Secondhand	Vinyl, CD	General interest		No
Naima Records	2017	Secondhand	Vinyl, CD, cassette	General interest		Yes
Nut Records	2016	New	Vinyl	House, techno		Yes
Pentagram Music Store	2009	New and secondhand	CD, Vinyl	Metal	Merchandising	Yes
Revolver Records	2000	New and secondhand	Vinyl, CD	General interest		Yes

(Continued)

Table 2 (Continued)

Stores	Opened	Type of material	Format	Genre	Other functions	Gentrified neighborhood
Rhythm Control Barcelona	2016	New and secondhand	Vinyl	Electronic, techno		Yes
Subwax Barcelona	2011	New	Vinyl, CD	Electronic	Record label, distributor	Yes
Surco	1974	New and secondhand	Vinyl, CD	General interest		Yes
Tvinyl	2016	New and secondhand	Vinyl	Electronic	Merchandising	Yes
Ultra-Local Records	2012	New and secondhand	Vinyl, CD, cassette	Indie, pop		Yes
Vinyl Vintage	2013	Secondhand	Vinyl, cassette	General interest	Record players	Yes
Vinilarium	2017	Secondhand	Vinyl	Electronic		Yes
Wah Wah	1992	New and secondhand	Vinyl, CD	General interest	Record label	Yes

Taking into consideration the genres on sale, thirteen stores can be considered to be general interest record stores. The rest (twenty shops) are specialized in specific genres, particularly electronic music (nine stores), which is a genre of great relevance in Barcelona. As far as the type of material on sale is concerned, six shops only sell new records, nine sell secondhand products, and sixteen sell both types. As for formats, vinyl is shown to be fundamental and is sold in all of the stores analyzed. CDs are maintained as a sales format in the vast majority of shops, although there are eight stores which only sell vinyl.

The existence of hybrid stores can also be observed in the case of Barcelona. This is the case of Curtis Audiophile (cafeteria and record store), Discos Tesla (T-shirt and merchandising shop and record store), King Atupali (clothing and record shop), Mondosinfonola (jukebox and record store), and Vinyl Vintage (record player and record shop). It is striking that several shops also perform other functions, such as record label (five shops), merchandising sales, and record distribution. The sustainability of these stores is related to the diversification of the business. As far as location is concerned, it is worthy of note that only four out of the thirty-three stores studied are in ungentrified neighborhoods (Table 2).

This sample of record store workers provides a social and educational profile similar to that put forward in other research.[13] As can be observed, all of the owners and workers are male with an average age of forty-eight. Eleven of those interviewed are university graduates. Furthermore, seven of the interviewees perform other functions in the musical scenes of their city, for example as programmers, DJs, in fanzines, record labels, or on radio programs (Table 3).

In the case of Barcelona, those interviewed had a similar profile to that observed in Porto. With the notable exception of one woman, all of the other employees were male, with an average age of forty-five. As was the case in Porto, more than half of the interviewees were university graduates and nine of the twelve people interviewed were found to occupy positions and occupations connected to musical scenes (Table 4).

Table 3 Record Store Workers in Porto

Interviewee	Position	Sex	Age	Level of education	Other occupations in the music scene
A.	Employee	Male	43	Vocational training	DJ/Concert promotor
A.	Owner	Male	30	University graduate	DJ/Record label
M.	Owner	Male	45	Vocational training	DJ
M.	Owner	Male	47	University graduate	
F.	Owner	Male	56	University graduate	DJ
P.	Owner	Male	53	Vocational training	Fanzines/Radio/Festival program
R.	Owner	Male	50	University graduate	
J.	Employee	Male	53	University graduate	
M.	Owner	Male	52	University graduate	Fanzine/Record label/Radio
N.	Owner	Male	66	University graduate	
P.	Owner	Male	36	University graduate	Record label/Concert promotor

Table 4 Record Store Workers in Barcelona

Interviewee	Position	Sex	Age	Level of education	Other occupation in the music scene
G.	Owner	Male	49	University graduate	Music journalist/Management/DJ
M.	Owner	Male	45	University graduate	DJ
D.	Owner	Male	44	University graduate	
E.	Owner	Female	38	University graduate	
D.	Employee	Male	47	University graduate	
J.	Owner	Male	52	Baccalaureate	Record label
P.	Owner	Male	38	Vocational training	Musician/Fanzines/Distributor/Record label
E.	Owner	Male	36	Vocational training	Cassette label
J.	Owner	Male	60	Vocational training	DJ/Record company
J.	Employee	Male	41	University graduate	Music journalist/DJ/Radio
G.	Employee	Male	36	University graduate	DJ
D.	Owner	Male	52	Vocational training	Manager/Record label/Radio
J.	Owner	Male	53	Vocational training	

Tourism, Gentrification, and Job Insecurity

Via the information gathered in the interviews with the record store workers, it is possible to look in more depth at some of the aspects mentioned: the relationship between tourism and record stores, the importance of the location of the stores, and issues concerning vocation and job insecurity in this sector.

Record Stores and Tourism

As has been mentioned earlier, many of the interviewees stressed that the growth of tourism has been a key factor in keeping the stores afloat during the economic crisis:

> Tourism has been a great help for Porto. Five years ago, the situation was much worse. Tourism saved the day. I don't know if the shop would be open if it weren't for tourism. (R., 50, Porto)

> Why have so many record stores sprung up? Because we saw that vinyl was successful and that tourists were willing to pay high prices. (N., 66, Porto)

> The crisis of 2008 was very hard; the downturn was drastic. It led to a great crisis for us. Between tourism and the people who never stopped buying records, we got by. Here, tourists are very important clients. In this location, yes. Because lots of them pass by. We survive thanks to them. (J., 60, Barcelona)

One aspect clearly perceived in the interviews is that, as previously mentioned, tourists are a source of significant income for these stores. However, those interviewees whose shops are not so touristic see tourists as having a lack of knowledge and are of the opinion that they bring down the cultural level of the shop. They perceive tourists as consumers without criteria, who only buy records of no artistic or cultural value. These ideas can be linked to discussions regarding authenticity in popular forms of music. For these workers, who are critical of tourism and mass mainstream consumption, an authentic record store should have its own character, a distinctive line, and a solid collection of records (records with artistic value and hidden gems). They argue that classic rock albums can be found anywhere.

But what do tourists want? In the case of Porto, different interviewees highlight the quest for something different, for example, Portuguese editions of international records, which are rarities for collectors. Due to the country's colonial relationships with Brazil, Angola, and Cape Verde, some collectors seek African and Brazilian music in these stores. Another sought-after genre is fado and records of this style can be found in the majority of the shops analyzed, independently of their specialization. However, in the Barcelona stores flamenco can only be found in some mainstream shops. Some interviewees agree on the fact that tourists buy much more than Spanish customers, spending more money and buying more records in each visit.

The Location of the Stores

As far as the location of the stores is concerned, in the interviews it could be noted that the "city center," understood in different ways, is the desired location in most cases. The problem found by shop owners is the price of rent, which prevents them from being located in the center, where customers and tourists pass by more frequently.

One sought-after location, mentioned by Hracs and Jansson, is gentrified streets or neighborhoods. A good example of this is Calle Tallers, in Barcelona. This street has a long tradition of record stores and musical instrument shops. Although many of them have disappeared, seven (Discos Impacto, Revolver Records, Discos Revolver, Discos Tesla, Daily Records, Pentagram Music, Kebra Disc) can be located between Calle Tallers and Calle Sitges, a side street off Tallers. Also, in this network of streets located between the end of Raval and the beginning of Las Ramblas, many different shops of vintage clothing, cannabis-based products, sunglasses stores, burger restaurants, student bars, trainer stores, and tattoo parlours can be found.

In the case of Porto, Rua Miguel Bombarda, where the record stores Muzak and Materia Prima can be found, with Discos do Baú located nearby, is also a street characterized by the presence of shops linked to art (secondhand objects, art galleries), as well as murals, barber shops, and a shopping center (CC Bombarda) with alternative clothes, food, and beauty shops.

Record Stores as Places for Shared Knowledge

In the discourse of those interviewed, the idea emerges of a relationship formed with the customers based on sharing knowledge. This can be used to learn; customers have a lot of knowledge and much can be learned from chatting with them. On the other hand, there may be a financial interest: customers can identify emerging groups or trends which are not well known.

> Contact with customers is of great importance to me. I love talking about music, customers learn from me and I learn from them. My shop is different, it has a very social interactive side. Almost every afternoon, somebody comes not necessarily to buy records. The shop acts as a meeting point. (M., 47, Porto)

In the more specialized shops, dedicated to metal, punk, electronic music, or jazz, it can be perceived that they are meeting points for a particular scene. They are part of a community, of a network of relationships built around that scene. For fans, they are places to meet, talk, and form friendships and relationships which extend beyond music. Also, it is perceived in the ethnography that this phenomenon only arises in men. No women were observed participating in these meetings or conversations.

> when I set up the shop, I did so as a cultural centre. The customer asked you whether you knew such and such an artist, there was an exchange of information, you talked about the concerts you had been to. The guy who listens to Julio Iglesias has the same sensations as I do with Kiss. Emotionally, they are similar reactions. (D., 52, Barcelona)

> One thing which has been lost is the figure of the opinion leader, because customers often arrive and you recommend things that they don't know and then they meet and tell each other, "Wow! I hadn't heard that before", it creates the atmosphere of a club and everyone is enriched. They are things which nourish. (D., 47, Barcelona)

Vocation and Job Insecurity

One of the topics which emerged from the interviews is that of the vocation of working in a record store. Regarding this matter, two perceptions can be identified: those who see it as a pleasant job but one which could be exchanged for another and those who see it as their purpose in life. The argument of "being your own boss" came up in several interviews as something extremely positive, albeit linked to another issue: the sacrifices necessary to make a success of the shop.

> It's not a problem for me to get up early. The shop normally opens at 10 but I'm there at 8. I could do other things but I come to do what I like, to defend what I have to defend. (M., 47, Porto)

This is a recurring theme in artistic worlds. It is understood that the passion for art balances out the small economic profits. There is a discourse about sacrificing certain aspects of your life (economic, material, family) to exercise one's passion.

> you understand that the investment was worth it because life is not just your bank account, we have emotional capital that we earn . . . although the shop has

occupied many hours that I could have spent with my family, I spent that time well because it helped me to grow, to be a better person. (P., 53, Porto)

However, at the same time, the topic of job insecurity appeared in several conversations. Workers in record stores do not normally earn much money, working conditions are not good, and there is no way of knowing how sales will evolve. Some of those interviewed recognized that they hardly get by with the shop, that they have to complement it with other jobs, or admitted that they cannot pay their employees as much as they would like.

I have mixed feelings. Doing what you like is great, going there, being relaxed, being in control of your day and your life, it's your business, you're not tied to a boss who gets on your nerves. But when you can see that things aren't going well, sadness comes over you. I'm here wasting my hours, wasting my time, you feel like you're bringing records, putting on concerts for people who don't respond, who don't buy anything, you're fighting for nothing, you feel like a sucker! (P., 38, Barcelona)

Conclusions

To conclude, several aspects must be highlighted. First of all, the fact that more than half of the stores analyzed have opened within the last decade demonstrates the importance that these shops have in the consumption of music today. Furthermore, we believe that, in this revival, the role played by tourists as consumers with greater purchasing power than national consumers is of great significance, extending beyond the explanations given by other authors concerning record store curatorship. The location of the shops in gentrified areas is also relevant without ignoring, of course, the significant increase in sales of vinyl.

One aspect which has not received much attention to date is the new hybrid format being implemented in shops. Stores seek to extend beyond their classical settings and offer more elements to their customers (the consumption of experiences). An ambivalent attitude has been demonstrated on the part of record store workers regarding tourists. On the one hand, they have been a lifeline, while on the other, they are perceived to be mass consumers, without informed criteria when choosing their purchases.

As far as the sociological characterization of the workers is concerned, the data collected here are similar to those of other prior studies: masculinized sector, middle-aged, further studies on the whole, and significant subcultural capital, thanks to other roles fulfilled in musical scenes. The profile of the customers is also clearly masculinized. As has been explained, in some shops, particularly those linked to specific musical scenes, the establishment of social relationships which extend beyond commercial transactions has been observed, in which workers and fans meet and share their knowledge of music and other topics, along the lines highlighted by Gracon.[14]

It must be stressed that these patterns only arise between men, with women being excluded from this interaction.

Finally, the double perception held by the workers of their trade should also be stressed. On the one hand, they have an emotional connection with their work and a high degree of job satisfaction, but on the other hand, there is a lack of job security in many regards and their future survival may be complicated.

Notes

1 This research has been carried out, thanks to postdoctoral funding from the Fundação para a Ciência e a Tecnologia, reference SFRH/BPD/121274/2016.
2 See P. Guerra, "Alta Fidelidade: um roteiro com paragens pelas lojas de discos independentes em Portugal na última década (1998-2010)," *Sociologia, Revista da Faculdade de Letras da Universidade do Porto* XXI (2011): 23–48; J. M. Hendricks, "Curating Value in Changing Markets: Independent Record Stores and the Vinyl Record Revival," *Sociological Perspectives* 59, no. 2 (2015): 479–97; B. J. Hracs and J. Jansson, "Death by Streaming or Vinyl Revival? Exploring the Spatial Dynamics and Value-Creating Strategies of Independent Record Shops in Stockholm," *Journal of Consumer Culture*, first published December 6, 2017. https://doi.org/10.1177/1469540517745703.
3 Hendricks, "Curating Value in Changing Markets."
4 Hracs and Jansson, "Death by Streaming or Vinyl Revival?"
5 Guerra, "Alta Fidelidade," 23–48.
6 F. Val Ripollés, "Por qué seguimos escuchando vinilos? Repaso a algunas investigaciones recientes," *Telos* 110 (2019): 78–85.
7 D. Gracon, "Exiled Records and Over-the-Counterculture: A Cultural Political Economic Analysis of the Independent Record Store," dissertation presented to the School of Journalism and Communication and the Graduate School of the University of Oregon (Oregon: University of Oregon Libraries, 2010).
8 See I. Gusman, P. Chamusca, J. Fernandes, and J. Pinto, "Culture and Tourism in Porto City Centre: Conflicts and (Im)Possible Solutions," *Sustainability* 11, no. 20 (2019): 5701. MDPI AG. Retrieved from http://dx.doi.org/10.3390/su11205701.
9 M. Degen and M. García, "The Transformation of the 'Barcelona Model': An Analysis of Culture, Urban Regeneration and Governance," *International Journal of Urban and Regional Research* 36 (2012): 1022–38.
10 J. Sequera, *Gentrificación. Capitalismo cool, turismo y control del espacio urbano* (Madrid: Catarata, 2020).
11 According to A. Arias, "Turisme i gentrificació: apunts des de Barcelona," *Papers: Regió Metropolitana de Barcelona: Territori, estratègies, planejament* 60 (2018): 130–9.
12 L. Carvalho, P. Chamusca, J. Fernandes, and J. Pinto, "Gentrification in Porto: Floating City Users and Internationally-Driven Urban Change," *Urban Geography*, 2019, doi:10.1080/02723638.2019.1585139.
13 Guerra, "Alta Fidelidade," 23–48.
14 Gracon, "Exiled Records and Over-the-Counterculture."

9

Living Popular Music in "High Fidelity"
Portugal's Independent Record Stores, 1998–2020[1]

Paula Guerra

From Lovers to Other Lovers: Independent Record Stores on the Road to "High Fidelity"

This chapter explores the importance of independent record stores in Portugal. It highlights their relevance as catalysts for specific scenes and sociabilities and examines how they can be understood not only as spaces of consumption but also as places where rituals are enacted, and communities of taste are constituted. Likewise, it will touch on several levels of analysis, such as the emergence of a new economic rationale based on curation and collecting, the vinyl revival, and the stores' complex relationship with the digital revolution. The methodological approach employed is ethnographic analysis, using direct observation of ten stores and interviews with their owners and customers.[2]

Over the last few decades, the music world has changed profoundly. Large companies such as Amazon have appeared, illegal downloads and software have proliferated, and the reality of streaming platforms has become all-pervasive. The business strategy of record stores has had to change.[3] Record shops started to offer not only objects for purchase but also experiences, with different layers of meaning associated with stories and new cultural practices based on the valorization of the object and craftsmanship.[4]

The chapter is structured as follows. The first section starts with an overview of the subject. The second section explores the emergence of a new aspirational economy based on curation. In the independent record stores, curation strategies have been developed on three levels: spatial, individual, and local. This section analyzes individual and spatial curation. In the third section, independent record stores are examined as spaces of ritual because they combine with the existence of a *totem*, which in this case is vinyl. The relevance of the space is loaded with symbolic density, wherein social actors carry out rituals with subcultural capital in the music scene(s). In the last section, we examine local curation—the legitimacy of belonging to a specific place.

A Biopolitics of Resistance that Dwells in Independent Record Stores

The persistence of independent record stores is closely entwined with the perseverance of vinyl. Discourses on the resurgence of vinyl are well known. However, we can only understand this reality if we observe vinyl as a cultural object that has absorbed a range of symbolic narratives. Vinyl also needs to be seen as an object with growing cultural and symbolic value, both within and outside the musical field.[5]

This leads us to another question. The persistence of vinyl cannot be explained by economic reasons, and it cannot be explained solely through the narrative of nostalgia of postindustrial societies.[6] We must also see this as a strategy of cultural distinction and as an instrument of identity-projection. Taking a Durkheimian approach, we can see that inherent in these processes are rituals and collective experiences that maintain and reinforce a sense of belonging.[7] These rituals require at least a sacred space and a totem. The sacred space is the record store; the totems are vinyl records.[8]

A few factors need to be explained. First, the totem must possess a distinctive meaning. Vinyl is not unique, yet in the face of the intangibility of the digital, it becomes (relatively) exclusive.[9] Second, the totem has to entail some difficulty and requires a certain dexterity. Vinyl does this: it is not enough to press a button; it implies a whole process. Third, the totem has to communicate a sense of belonging, a (historical) narrative of one community.[10] All this makes vinyl an unparalleled subcultural totem. On the other hand, actors with high subcultural capital need to exist in the places of celebration.[11] As we will see, several employees of the stores are actors with subcultural trajectories, who are making a living in these places. It is not enough to have the place and the object; it is also crucial to have someone to "guide" the ritual.

Record Stores as Spaces of Material and Affective Curation

Recently, record stores have had to reinvent themselves. They are facing a revolution in the fields of economics and taste. Andjelic argues that we are in a post-Veblenian aspirational economy.[12] If, for Veblen, status comes from the display of wealth, today status is linked to factors such as ethics, well-being, and morality. It is needed to be *in the know*. A brand that only gives us economic status is no longer valuable; it also needs to give us social, cultural, moral, and even environmental status.

The forms of distinction also change. Bourdieu's work argues that social agents of dominant classes have internalized perception schemes that make them prefer established cultural practices.[13] More recently, Tony Bennett et al. have demonstrated how established cultural styles have shifted in contemporary societies toward omnivorous forms of cultural consumption.[14] However, these studies only focus on the aesthetic dimension, lacking any analysis of the moral dimension. In line with the work of Puetz, we must consider factors such as the moral signaling inherent in certain cultural choices.[15] One of the main moral dimensions is the increasing valuation of the local and the material. This refers to the less tangible value of the purchase in local independent stores.

Andjelic tells us that the most important aspect is to know *what* to buy—to show that we are *in the know*. This is easier said than done, especially in an age of cultural abundance. The answer is curation: a key task in this new economy. It is a filtering process based on an appraisal of what is culturally and morally consecrated.[16] The idea of curation has changed in recent times. It has usually been linked to art, museums, and preservation, but today it means practices such as the selection, evaluation, presentation, and approval of cultural goods.[17] Musical curation, like that in other areas, has gained importance and become a form of distinction, of social and economic competition.

Independent record stores in this economy ultimately become seals of quality. Analyses of curation sometimes neglect the fact that curation is only possible if it is associated with legitimacy based on subcultural capital. This legitimacy may have three levels: spatial, individual, or local. The first level depends on the reputation and history of each record store. There are stores that achieve prominent status within local and national scenes, so their cultural goods immediately bear a seal of quality.

> Where you see alternative record stores you think of a spirit of adventure and love of music. It often means good prices, a wide offering within their music labels, service by connoisseurs. But don't talk about easy money. It was a gamble and a challenge, but it's going well. (Lourenço, 47, university graduate, record store owner, Porto)

> Tubitek[18] helped the musical education of many people, who spent hours in the store with us. We shared conversations about music and bands. It was a constant exchange of ideas. . . . We have a fantastic variety of albums and special editions, back catalogues, with things that disappeared years ago, but we still have it. (. . .) People come into the store and are amazed by the variety. (Pedro, 51, university graduate, record store owner and customer, Porto)

This implies continuous, meticulous work to ensure that the message is perceived as subculturally legitimate. There are barriers that cannot be crossed, otherwise stores and the music scene run the risk of becoming "symbolically polluted."[19] As one record store customer in Lisbon told us:

> Sometimes they are a tourist trap. They have a poor and overpriced selection of vinyl. The most critical thing is sometimes the service—very unfriendly, not geared towards the customer's taste, stuck-up in matters of taste, quite elitist even. (Valério, 58, middle school, record store customer, Lisbon)

Independent record stores are more than just places where you can buy music: they are small communities of taste. Success or failure is based on social ties and bonds of friendship. When the members of a music scene feel a barrier has been crossed the reaction is quick and merciless, as shown earlier.

These stores also engage in curation strategies outside their walls. The best-known means to do this is via digital strategies, such as the creation of a website or social media presence, or classical strategies, involving collaboration with specialized magazines or radio programs. These strategies are used in conjunction with curation in the physical space, which highlights how the idea of a sharp divide between the physical and the digital is too simplistic:

> A few years ago, having a record store meant ordering records from a record company, waiting for them to arrive, cataloguing and displaying them. Not today. Today, there's a much more complex job involved in making a record available on the internet: you have to post all the information, the cover, the line-up, the 30-second track, the synopsis of each album. (Delfim, 50, nonuniversity educated, record store owner, Porto)

> It's not with online commerce that we get by, but with online promotion. (. . .) When we launch the newsletter, we post the newsletter articles on the site . . . the internet is very important, and we put excerpts of the tracks online so people can listen to them. That didn't happen before. (Pedro, 51, university graduate, record store owner and customer, Porto)

The second level of legitimacy is individual. Stores and venues have a history that can be transformed into subcultural legitimacy. In many cases, some stores become synonymous with certain music genres. Legitimacy takes different forms depending on the customers. There is a legitimacy for non-regular clients who know the stores through the media—the top five or ten of the best record stores to visit. Furthermore, there is legitimacy that comes from belonging to the local music scene. Here it is necessary to hold subcultural capital.

Curation gains importance if it takes place in-person. Customers seek the suggestions of store employees due to their subcultural capital recognized by their peers. The stores, and these employees and owners specifically, are understood to be information hubs or music libraries. It is also interesting that in the interviews we conducted with record store owners, they all had a subcultural history and were recognized as important agents in the scene:

> There are sellers who go out of their way to find the record we want. There are even some who are like a human Spotify and suggest something we don't know, but that they somehow guessed we'd like and they're never wrong. Or they just talk and give us kilograms of wisdom about the Lisbon punk scene of the 1980s. (Valério, 58, middle school, record store customer, Lisbon)

Curation is the ability to become a "human Spotify," someone whose subcultural capital means they can convert the immensity of musical possibilities into suggestions appropriate to each customer. It is also the ability to establish connections between the cultural goods they sell and a story, endowing them with broader meanings anchored in tales and mythologies of Portuguese subcultures. Even if they do not find what they are looking for, the customer is always left with "a different experience." These experiences can be extended to the various employees in the stores. Anselmo, owner of Flur,[20] confided that of his four employees, three have a subcultural history, and the fourth, with a degree in human resources, "is working with us more for the music than to put his knowledge into practice"(Anselmo). This is explained by the ability to convert subcultural capital into DIY careers. Many members of subcultures choose jobs that allow them more time for their creative passions than economic comfort. Similarly, we can see here the combination of a love of the work with a reflexive strategy to deal with the precariousness of the labor market.[21]

Finally, individual legitimacy is a double-edged sword for record stores, especially when dealing with customers who are members of the music scene. In many cases, a customer's connection is made with the employee, with whom they end up developing a relationship. When the employee decides to leave their job, this can create problems for the record store. When they decide to quit their job and open a record store themselves, the situation is critical. As one of our interviewees admitted, the departure of two employees meant that a clientele went with them. Consequently, the store had to review its entire musical offering and specialize in other musical genres to not compete with the store where the two former employees now worked.

> When the two Pedros left (...) many customers went with them, obviously, because they had a portfolio of customers who were used to them. They took a lot of them away, but we still managed to hold on . . . So when they left, we concentrated more on alt rock and experimental music. (Martinho, 42, nonuniversity educated, record store employee, Lisbon metropolitan area)

Ritual Practices in Independent Record Stores

Independent record stores can be understood as ritualistic spaces. Three dimensions are required to achieve this: the totem, which in our analysis is the vinyl record; the sacred space, which is the independent record store; and, finally, actors who successfully perform the ritual.

Analysis begins with the totem: the vinyl record. What led to the recovery of this object? It is undeniable that today there is an appreciation of the past. And vinyl responds to that. What consumers get with vinyl is an entry into a specialized mode of musical consumption, a community of tastes that frames vinyl fruition practices as having a higher social status because it implies being able to express musical sensibilities in innovative ways.[22]

There is a need to create alternative modes of consumption that imply a deeper appreciation of music, which cannot be dissociated from the search for materiality in an era of musical digitalization. It is not only the aesthetic beauty of the cover but also the experience itself and the demand for attention. One has to take the record out of its sleeve, listen to the record, hear the little noises, turn it over, put it back. In short, it is a ritual in which even the noises are an integral factor of this experience.

I think vinyl is an important element for purists and the new generation is also picking up on vinyl a little bit. I'm not sure what the reason is, but I think it might be the cult of the object. It's a big, beautiful thing. Normally vinyl editions, nowadays, are beautifully made, they're special. You open them up and they have a lot of information. It's not only about the music itself, but also more about the object. (Martinho, 42, nonuniversity educated, record store employee, Lisbon metropolitan area)

The CD from the moment it entered the large supermarkets and was treated not as a cultural product, but as a product for the masses, certain music lovers hated that. They don't like to mix things; a CD can't be bought in the same shopping cart as potatoes and milk. (Delfim, 50, Store 5, ninth grade, Cedofeita, Porto)

Legitimacy is fundamental to this experience. Delfim tells us about the cultural devaluation of buying music in the same place you buy potatoes; we can also talk about the devaluation of downloading entire albums on Spotify. Vinyl is about taking a step back and seeking a greater engagement in the experience of listening to music.[23]

The sacred space in which this ritual is performed is the independent record store, a space with its own language and practices. Such stores serve as a repository of popular music history. One of the best ways to observe this is through the aesthetics of the stores, the way they seek to create an authentic atmosphere. These stores cannot compete in price with chains such as Amazon, so they have to offer something different. In this case, they are places where experiences are had and rituals performed.

Record shopping is a multisensory experience, and this refers partly to the aesthetic layout of the stores. First, we always have background music, which in many cases signals the musical specialization of the store; second, we have the crates full of records, stacked against the walls, scattered on the floor. These crates are arranged in different ways and are sought after by different types of consumers. Take the case of Carbono.[24] Right at the entrance, the walls are covered in vinyl records, band posters, and stickers; there are record players and a coffee machine. Then we have a store full of crates divided by labels: punk, African, and so on. At Flur, the aesthetic is more organized. It is a large, well-lit space, with vinyl crates on the left, all separated by very specific musical genres, such as Canadian bands and bands from São Tomé, with the rarest and most sought-after records displayed in rows of three on the wall; to the right there are CD racks.

Generally, the most expensive LPs are displayed on the walls or on the highest racks; however, for a crate-digger, the gems are in the crates on the floor or even in the attics. It is here that one sees, as also studied by Hracs and Jansson, collectors spending hours squatting down and looking for bargains. It is not possible to listen to these cheaper LPs before buying them, so some subcultural capital is required to evaluate them. As the crates are rarely organized, it is also possible to see the cohabitation of the physical and the digital. When customers discover an LP that catches their eye, they search on their smartphone, read about the band, listen a little, then consider whether this new acquisition is worth having.

Here we have collectors for whom there is an enormous pleasure to be gained from the search for rare or obscure LPs. Andjelic sees collecting as one of the main activities of aspirational markets.[25] A collection is a narrative, where the symbolic value of the objects is related to background stories and essential to the identity construction of a real fan. In the stores analyzed here, it was mainly in Discolecção[26] in Lisbon and in Matéria Prima[27] in Porto that focused firmly on the collectors' market. There is a concern in these two stores with meeting the specific needs of this group of consumers—be it rare records, special and limited editions, or all the merchandise that exists around music collecting.

> The vast majority of our customers are music collectors and I think that specialized stores work more and more with this type of customer. We have been having an increase in vinyl sales precisely because the collector likes that beautiful object. Sometimes the collector buys the CD and buys the vinyl. The CD is to listen to, and the vinyl is to keep and to enjoy. (Martinho, 42, nonuniversity educated, record store employee, Lisbon metropolitan area)

> I'm a collector, I've never got rid of a single record I've bought. I used to record a lot of music on the radio and didn't buy music and I've never been in the habit of buying singles. I am a vinyl fan, I have never created a relationship with CDs. I also buy whenever I travel; I've been to New York or London for music. It is always a factor to be considered when choosing trips. (Rafael, 42, Musician 48, DJ, producer and programmer, attended university, Lisbon)

Going back to the link between the material and digital, some customers go directly to what they want. We found that this is because they first visit the store's website or Discogs, an important tool for the crate-digger, which allows them to discover the store's stock, prices, and reviews. But many more use the stores as social spaces for a community of tastes. This can be seen in the interactions they establish with the staff and other customers. Note that these interactions are two-way: it is not uncommon for customers to give advice to storekeepers. If Hendricks is right when talking about a ritualistic dimension in entering a store, choosing a record, and going home to listen to it, this is only half the story, as we cannot ignore the close ties that are generated in these spaces.[28]

Like any ritualistic space, record stores must have visual appeal. We can understand this just by looking at the walls of the stores, with photographs of bands, posters, and LPs covering every space. Jansson and Hracs look at record shop walls as an archive of popular culture and knowledge.[29] They are also a way of transmitting a form of subcultural capital. For example, the walls of Porto Calling[30] are decorated with references to the owner's favorite bands; Louie Louie's[31] walls are covered in posters, concert news, and newspaper clippings. If Porto Calling is concerned with creating a more specific environment, Louie Louie is interested in emphasizing the store's discographic diversity. This is a strategy: first as an extension of the records on sale and second as a concern to create a familiar environment, as if a person was in their bedroom, adorned with the posters of their favorite bands.

We can look at the stores as ritualistic spaces of a community of tastes. This community is motivated only in part by nostalgia and the (re)discovery of the appeal of vinyl: the tactile pleasure of the experience of putting the music on the turntable and displaying the collection is associated with discourses of authenticity and cool.[32] In our research, both customers and employees often highlighted that these stores made them feel "at home." Regular customers would be able to find the rarest records on eBay or any other website, but they would lack the sense of authenticity and the sense of belonging those stores offers them.[33]

Record Stores and the (Trans)Local Music Scenes

In this section, we look at local curation. Certain places have an association with specific cultural values or practices. Bartmanski and Woodward found that Berlin has become a key city to visit on the global music circuit.[34] Through multiple references in the specialized or generalist media, the city has become associated with a "seal" of quality and authenticity. It is interesting how this local curation, which depends greatly on the musical history of each city, also depends on what we may call meta-curation—the valorization by the media that establishes a global curation "championship."

In Portugal, although we are analyzing Lisbon and Porto, the capital draws the most attention. Lisbon has quickly become a stopping-off point for any vinyl lover. This has occurred on two levels: first, there is the association of Lisbon with the possibility of discovering "gems" at low prices; second, in a twist of postcolonial irony, Lisbon is associated with African and Brazilian music by the foreign media. It is the place to visit if you want to discover new African music and search for rarities, especially from the former Portuguese colonies.[35] The city of Porto, which is not so associated with postcolonialism, is part of the global DJ circuit. This means that record stores that specialize in electronic music, such as Musak, may have a niche on which to concentrate. They have taken advantage of this by developing strategies to capture this sudden interest of tourists: their websites and social media presence are translated into English, and practically all the employees and store owners are fluent in English.

> Almost all the international DJs that visit Porto come here . . . Incredible as it may seem because it's already on their agenda. The DJ circuit is still more or less operating in Porto and they always end up coming here. But most of the vinyl stores, not only here in Portugal but also in Europe and the USA, don't really cultivate this DJ aspect. We clearly do. (Raul, 49, Store 8, university graduate, Foz, Porto)

However, a store's identity is precarious. There has to be a balance between regular customers and tourists. We might recall the previous quote from Valério accusing a Portuguese record store of being a "tourist trap." In other words, if stores do not strike a balance between these two groups of customers, they run the risk of becoming delegitimized in the local music scene.

Beyond the media, curation is associated with a presence in a local scene. There are areas in Porto and Lisbon with a high concentration of cultural offerings. Areas such as Bairro Alto[36] in Lisbon or Galerias Miguel Bombarda[37] in Porto serve as curation sites for subcultural or cosmopolitan cultural practices. This raises a question: if many stores are concentrated in only one area, will this not provoke competition? We did not find that this was the case but rather resulted in a division of the musical offerings. If we follow an ideal-type approach, we find two main types in Lisbon and Porto. First, stores that invest in having everything. This is the case of Carbono in Lisbon for example. The second is a shop that specializes and focuses on niche markets, whether a certain musical genre or only vinyl. The strategy is to concentrate only on a niche and hence not compete with the large stores such as Fnac or with the other independent record stores. We have the case of Tabatô Records,[38] which specializes in African music.

> The big chains will probably stop selling records and small, specialized stores will appear in various cities (. . .). The small stores are becoming more specialized, and they have to know how to specialize because if they all have the same things it is also complicated. (Martinho, 42, nonuniversity educated, record store employee, Lisbon metropolitan area)

Therefore, we found that competition among Portuguese record stores is all but nonexistent. Anselmo, the owner of Flur, observes that there are friendly relationships among the store owners, who often visit and buy from each other's shops. There is hence no perception of commercial competition, which would be extremely harsh in a small cultural market such as Portugal. Anselmo explains the situation with a gastronomic metaphor: "This is like going to a restaurant: here you can eat fish, there you can eat meat, over there you have vegetarian. Smaller things work like this. Each one has its own specialty."

Conclusion

In a post-Veblenian aspirational economy, record stores have managed to reinvent themselves, having recently experienced a relegitimization. They stopped selling simple cultural objects and started to engage in the curation of cultural objects loaded with stories. These practices are not just a reaction against the digital. Instead, what we can see are practices of a post-digital culture. Like practically everything nowadays, the stores are hybrid spaces between the physical and the digital.

We also found that not everyone can be a curator: curation requires legitimacy. While the spatial and local dimensions of this phenomenon are relevant, the individual dimension is still more important. Recommendations by actors with high subcultural capital are crucial for the success and distinction of record stores. It is interesting, as noted by Jansson and Hracs, how streaming platforms have already started to have this concern and partner with record stores and actors with legitimacy in the scene to periodically create playlists. Although we still have no news of the same trend happening in Portugal, it is an acknowledgment of how necessary it is to individualize curation through recognized actors in the musical field.

Notes

1. The publication of this chapter was supported by FCT—Foundation for Science and Technology, within the scope of UIDB/00727/2020. This chapter results from the research project "Urban Cultures and Youth Lifestyles: Scenarios, Sonorities and Aesthetics in Portuguese Contemporaneity," developed between 2005 and 2010, coordinated by Paula Guerra and hosted at the Institute of Sociology of the University of Porto, which was funded by the Portuguese Science and Technology Foundation. In 2021, this project was resumed in a transnational research project, involving Portugal, the UK, and Australia, called "Lost and Found Sounds: Cultural, Artistic and Creative Scenes in Pandemic Times." It aims to examine and understand the main impacts of the pandemic on the working conditions and daily lives of music-makers and mediators (venues and record stores); it is coordinated by Andy Bennett and Paula Guerra, from Griffith University and the University of Porto, respectively.
2. P. Guerra, "A instável leveza do rock: génese, dinâmica e consolidação do rock alternativo em Portugal," PhD thesis, University of Porto, 2010.
3. N. Zuberi, "The Record Store," in *The Bloomsbury Handbook of Popular Music, Space and Place*, ed. G. Stahl and J. M. Percival (London: Bloomsbury, 2022), 129–38.
4. A. Andjelic, *The Business of Aspiration* (London: Routledge, 2021).
5. D. Hayes, "Take Those Old Records Off the Shelf: Youth and Music Consumption in the Postmodern Age," *Popular Music and Society* 29, no. 1 (2006): 51–68.
6. S. Reynolds, *Retromania* (London: Faber & Faber, 2012).
7. É. Durkheim, *The Elementary Forms of Religious Life* (Oxford: Oxford University Press, 2008 edition).
8. D. Bartmanski and I. Woodward, *Vinyl: The Analogue Record in the Digital Age* (London: Routledge, 2015b).
9. W. Straw, "Sizing Up Record Collections: Gender and Connoisseurship in Rock Music Culture," in *Sexing the Groove: Popular Music and Gender*, ed. S. Whiteley (London: Routledge, 1997), 3–16.
10. Y. Kjus, "Reclaiming the Music: The Power of Local and Physical Music Distribution in the Age of Global Online Services," *New Media & Society* 18, no. 9 (2016): 2116–32.
11. S. Thornton, *Club Cultures* (Oxford: Polity Press, 2017).
12. Andjelic, *The Business of Aspiration*.
13. P. Bourdieu, *Distinction* (London: Routledge, 2010).
14. T. Bennett, M. Savage, E. Silva, A. Warde, M. Gayo-Cal, and D. Wright, eds., *Culture, Class, Distinction* (New York: Routledge, 2009).
15. K. Puetz, "Taste Boundaries and Friendship Preferences: Insights from the Formalist Approach," *Poetics* 86, no. 1 (2021). https://doi.org/10.1016/j.poetic.2021.101551.
16. B. J. Hracs and J. Jansson, "Death by Streaming or Vinyl Revival? Exploring the Spatial Dynamics and Value-creating Strategies of Stockholm's Independent Record Stores," *Journal of Consumer Culture* 20, no. 4 (2020): 478–97.

17 D. Balzer, *Curationism: How Curating Took Over the Art World and Everything Else* (Toronto: Couch House, 2014).
18 An independent record store located in the city of Porto. It opened in the 1980s, and in the following decades it became a crucial place of the musical scene of the city. It closed in 2000 and reopened in 2014.
19 D. Bartmanski and I. Woodward, *Labels. Making Independent Music* (London: Routledge, 2020).
20 Flur is an independent record store located in Lisbon and established in 2001.
21 P. Guerra, "Raw Power: Punk, DIY and Underground Cultures as Spaces of Resistance in Contemporary Portugal," *Cultural Sociology* 12, no. 2 (2018): 241–59.
22 D. Bartmanski and I. Woodward, "The Vinyl: The Analogue Medium in the Age of Digital Reproduction," *Journal of Consumer Culture* 15, no. 1 (2015): 3–27.
23 Ibid.
24 One of the best-known record stores in Portugal. Founded in 1993, and located in Lisbon, it initially specialized in secondhand records.
25 Andjelic, *The Business of Aspiration*.
26 A small record store located in Lisbon, which specializes in selling secondhand vinyls.
27 Located in Porto. It is a record store that sells records, books, magazines, and zines, and curates music events.
28 J. M. Hendricks, "Curating Value in Changing Markets: Independent Record Stores and the Vinyl Record Revival," *Sociological Perspectives* 58 (2015): 1–19.
29 J. Jansson and B. J. Hracs, "Conceptualizing Curation in the Age of Abundance: The Case of Recorded Music," *Environment and Planning A* 50, no. 8 (2018): 1602–25.
30 A record store based in Porto, specializing in vinyl records.
31 Established in 2004, in the city of Porto, this record store has a diverse portfolio, dealing with vinyl and CDs, brand new and secondhand.
32 Kjus, "Reclaiming the Music."
33 E. Harvey, "Siding with Vinyl: Record Store Day and the Branding of Independent Music," *International Journal of Cultural Studies* 20, no. 6 (2017): 585–602.
34 Bartmanski and Woodward, *Vinyl*.
35 V. Belanciano, *Não dá para ficar parado—Música afro-portuguesa* (Porto: Afrontamento, 2020).
36 Bairro Alto could be defined as Lisbon's party district, with its streets packed with bars.
37 A street that has dozens of art galleries and the main cultural neighborhoods of Porto.
38 A record store based in Lisbon and specializing in Portuguese and tropical groove.

10

Music on the Turntables When the Tables Are Turning

A History of Record Stores in Romania from Late Socialism to the Present

Claudiu Oancea

The recent pandemic has made a lot of people aware of the fact that they live more and more in a digital world, where social interaction and cultural activities are performed online. Music is more about online platforms and social media than about physical places, interactions, and experiences. Nevertheless, throughout the past seventy years, record stores have been a focus of attention for music aficionados and record collectors. More than this, they have been a locus of social interaction and local cultural life. Therefore, it is not at all far-fetched to state that the story of record stores in one given place can reveal a much deeper insight into the social, economic, and cultural history of the communities which housed and bred life into them. This story is true not only for developed countries from Western Europe or North America but also for countries whose music and record stories are less known, but equally interesting and revealing.

Record stores in an East European country like Romania have a story (and history) of their own. It differs from that not only of Western record stores but also of music stores from other East European countries. This is a story of official stores, of black markets, of informal connections and networks. A history of ruptures and of continuity, which mirrors the tumultuous history of Romania over the twentieth century. From interwar capitalism to postwar Stalinism, from 1960s and 1970s detente to 1980s consumer restrictions, and from 1990s transition to European Union membership, Romania has experienced the ups and downs of a fluctuating historical trajectory, which is part of my own past, in terms of musical consumption and everyday life habits.

The first time I visited a large record store was in 2011 in Berkeley, California, as a visiting student. I had never seen so many secondhand records crammed together in one place. All kinds of titles, all kinds of genres, and all kinds of people digging through the record stacks. A year later, I would spend a year in Pittsburgh, where I developed a weekly ritual of going to a record store with almost 1,000,000 titles: a real warehouse, where people came every now and then to quench their thirst for music and discovery. At the moment I did not realize that I was part of a global movement, a revival of record music stores, which in itself was part of a larger cultural movement, called retromania.[1]

Upon my return to Europe, and later, to Romania, I realized that, before construing retromania, one needs to look more closely to the multiple meanings of the term "retro." What was and what had been the culture of record stores in a country like Romania? How can one assess the impact of record stores on the emergence and development of various music scenes in Romania? These are larger questions, and any answer would find itself entangled to larger issues of cultural history, as well as to more general problems of historical sources, namely their availability and validity.

Twentieth-century Romanian history represents a series of shifts from one political and economic regime to another. Until 1989, these political regimes often varied from one extreme to another; they were, most often, of an authoritarian nature, with brief moments of liberal democracy. Nevertheless, Romania also underwent periods of rapid economic development, which were most evident in terms of urbanization and industrialization. Before 1939, Romania was a mixture of contradictory tendencies: on the one hand, the industrialization rate was among the lowest in Europe, on the other, the education and health sectors underwent significant improvements.[2] Following the Second World War, Romania became part of the communist bloc, with a centralized, planned economy, based around heavy industries.[3] It developed foreign and economic relations with other East European socialist countries, as well as new countries from the Global South. Closely linked to rapid and intensive industrialization was the urbanization process, which meant the construction of new towns, as well as the development and systematization of existing cities.[4] Ideology lay at the roots of this process, but its actual fruition was a more complex and uneven affair, which was the result of both state enterprise and a private initiative. This latter aspect was most apparent in terms of cultural life.

The life of record stores in socialist Romania is closely linked to these general processes, as well as to the level of everyday life in urban and rural areas, which differed greatly from one region to another. Before looking at record stores, one needs to consider the level of consumption in terms of records and record players.

The starting point of such a discussion is the definition of records, as given by the ideological demands of the communist regime. Unlike capitalist societies, which saw records as a consumer product, state socialism regarded records as modern purveyors of cultural revolution.[5] Which is why records were not supposed to be made accessible primarily through record stores. The socialist state constructed a network of cultural institutions, called houses of culture.[6] At least during the early stages of socialism, these were supposed to be one of the main realms of cultural activity. Likewise, workers' clubs, youth organizations, schools, as well as the national radio network played an important role in popularizing records, as purveyors of music.[7] One needs to take into account the fact that during the late 1940s and early 1950s not only records were scarce and access to them limited, but the technology to listen to them was also highly restricted. Record players were still consumer products that were available mainly in major urban areas, for upper-middle-class people.

Postwar Crisis and Cold War Openness

The early years of state socialism meant the nationalization of record companies and the formation of a sole state-owned record company, which bore the title "Electrecord." This was followed, throughout the 1950s, by an increase in record production, from tens of thousands to one million records for the year 1957.[8] The 1950s also saw the introduction of microgroove phonograph records, eight years after their international introduction in 1948. They varied in size, from 17 to 25, to 30 centimeters, and were joined by the now defunct format of microgroove postcards.[9] The year 1956 signified the beginning of Romanian records exported abroad. Music genres varied from folk music to light music, classical, as well as children's stories, and, last but not least, political records.

This increase in record production also meant the development of a network of music record stores. In Bucharest, the Union of Composers from the People's Republic of Romania decided to build its own music store in 1952, in order to make a profit.[10] Discussions around the store were fierce, as several members of the Union rejected the idea of a store run by their organization and favored an eventual administration by the Arts Committee. Nevertheless, in 1952, the major problem was not who would be running the store but what to sell in it. Domestic supplies were scarce, relations with Western countries almost nonexistent, and the only chains of supply available were those from other Eastern Bloc socialist states. Notwithstanding, the store was opened anyway in downtown Bucharest, on Victory Avenue (Calea Victoriei).[11]

In 1960, the Union discussed the possibility of reconstructing the store. By that time both the record and the music instruments manufacturing industries had undergone significant changes. In 1951, a new music instruments factory had been opened in the Transylvanian town of Reghin, named Hora. Together with the older Doina Factory, in Bucharest, established during the 1920s, Hora would be responsible for a massive increase in music instrument production over the late 1950s and early 1960s.[12] The communist regime supported the development of amateur artistic activities, since it regarded them as an integral part of political and ideological education. Therefore, it also invested heavily in the creation of modern production facilities for instruments, records, as well as in the expansion of an education network that would teach music amateurs how to play instruments, listen to records, and experience a music education more or less according to official requirements.[13]

This production increase, coupled with newly established relations with other foreign countries, meant that the old record stores in socialist Romania were quickly becoming obsolete by the early 1960s. The new record store in Bucharest was to have three stores (a basement, a main floor, and an upper floor).[14] The new Muzica Store was opened in 1963. Over a surface of 1,500 square meters, it featured a modern architecture, mural mosaics, and a special place where customers could sit down and listen to records.[15] The store was to become the template for a network which would expand, in theory, in other major Romanian cities, such as Constanța, Cluj, or Timișoara.[16]

Figure 10.1 Advertising pamphlet for the Muzica Store in Bucharest (early 1970s) (author's personal collection).

Throughout the 1960s and 1970s, the music scene in socialist Romania underwent dynamic changes, influenced both by Romania's developing ties with countries on the Western side of the Iron Curtain and the Global South, and by the international evolution of music genres and the record industry. This is most evident in the record catalogs that Electrecord published during this period, which showcased the company's evolving and diversifying musical output. The 1958 Electrecord Catalog classified its repertoire according to four categories: folk music ("muzică populară" in original, which was a generic term comprising both folklore and an urban type of folk music, which became influenced by Soviet orchestral official folklore), dance music (which included jazz music), classical music (in original "muzică cultă," namely symphonic, chamber, opera, operetta, etc.), and the "various" category (which included the so-called music for the masses, children's stories, etc.).[17] The catalog included incipient influences from African American folklore, slow fox dances, and political anthems, often in chronological order, which indicates that the musical output was a result of negotiations between ideological demands and consumer ones.[18]

The 1961 Catalog classified around 200 long play records, this time according to a slightly different system.[19] The first category included the older "classical music," whose generic name was dropped. A second category was dedicated to choral music. This was due to the increasing importance of the genre for political purposes. Choral music comprised both classical and music for the masses pieces, which were performed on all types of scenes. The third category belonged to the familiar "muzică populară" genre, adding romances as well. The fourth and fifth categories were relegated to light music and to spoken word, theater, and children's stories respectively.[20] The presentation was much more detailed, with credits for all musicians, performers, composers, and songwriters, which also evidenced the increasing variation in foreign influences on Electrecord releases. Equally important was the fact that Electrecord sought to keep its customers informed about its latest releases through a series of catalog supplements. The latter included the very latest releases of Electrecord, which had not been mentioned in the catalog.[21] Furthermore, these catalogs and supplements were joined by informative bulletins, which featured articles and essays about performers, composers, and repertoire.[22] The opening up of socialist Romania in the 1960s and its developing economic ties with capitalist countries had an influence on the record industry as well. Electrecord also started publishing record catalogs in English.[23] Such catalogs were inventoried in international music journals, in order to allow Electrecord's repertoire to reach a wider audience.[24]

The fifth and sixth editions of the Electrecord Catalog maintained the classification system. In terms of music titles, one can notice that the record company focused primarily on classical music. The sections dedicated to folk and light music, as well as that dedicated to spoken word and theater, remained consistent but secondary, compared to the classical music section. Surprisingly, the choral music section was least favored, which shows that during the 1970s, Electrecord made an attempt to focus on consumers' demands, not simply on political ones.[25] This is the same period when Electrecord presented customers in the Muzica Store with pamphlets and questionnaires regarding the quality of its releases. Such was the case in 1975, when Electrecord began releasing cassette tapes and put out questionnaires about Romanian consumers' cassette players, favorite cassette brands, and favorite music genres they wanted recorded on cassette tapes.

Likewise, the Romanian music instruments manufacturing industry also diversified and adapted its output to contemporary requirements, releasing several brands of electric instruments, mainly electric guitars and bass guitars. By the 1970s, the Muzica Store in Bucharest sold records, music scores, amplifiers, music instruments (including electric and electronic ones) and offered repair and maintenance services.

Coupled with the constant new releases of Electrecord, one expects Romanian music stores to have been a focus of attention from music fans. But were they?

Figure 10.2 Front cover of 1975 Electrecord questionnaire regarding the soon-to-be-introduced format of cassette tapes (author's personal collection).

Figure 10.3 Muzica Store in 1979 (Revista Muzica, April 1979).

Alternative Record Stores and the 1980s

From the aforementioned information, one can imply that Romanian record stores during the socialist era were few and that they did not consider the various audiences for the music genres that were developing at the time. One can imply that fans of Romanian folk and light music were generally the most satisfied, as such titles were easy to find and in abundance.[26] Classical music fans enjoyed the historical international repertoire that was performed and recorded by Romanian philharmonic orchestras, as well as Soviet and East European releases, from labels such as Melodiya and Supraphon. In addition, Electrecord paid attention to recording Romanian composers, of either already legendary status for domestic audiences, such as George Enescu, or contemporary musicians, who were members of the Union of Composers.

The situation was more ambiguous when it came to genres such as jazz, rock, or international pop music. It was also acknowledged in official mass media, such as students' magazines. In an article on pop records that were available on the Romanian market, journalist Dora Bunaciu inquired about the inability of Romanian record stores to bring to their consumers the newest records by performers who were actually famous and sought after. Bunaciu wondered about names such as James Royal and Henri Seroka, who were available in Bucharest stores but hardly sold. Others, such as Engelbert Humperdinck, were available but only after a one- or two-year delay.[27] When it came to jazz or rock music, the situation was even more complex. Romanian jazz and rock musicians' memoirs can offer an insight into the music life of socialist Romania and provide an explanation for the paradoxes one often encounters when construing the topic. As jazz and rock musician Béla Kamocsa recalls in his autobiography, rock music audiences were not well regarded in socialist Romania.[28] The reason for this was that jazz and rock music were considered by the socialist state as Western-based music genres. Furthermore, the question was not only ideological but also that of a generation gap. State officials were highly distrustful toward the new music of the youth. However, as jazz and rock evolved and diversified, and new music subgenres emerged, the state sought to accommodate the older subgenres into its official culture. The reason for this was twofold: on the one hand, older genres could counter the influence of new genres, which the state considered to be more dangerous; on the other, as socialist Romania attempted to manifest an image of openness in the late 1960s and early 1970s, it regarded the domestic appropriation of Western-based genres, which had gained international popularity, as such a sign.

This led to a complex situation, in which the state tolerated the development of domestic jazz and rock musicians but never actually encouraged the phenomenon.[29] In terms of record stores, this situation meant that domestic jazz and rock records were available, starting with the late 1960s. International titles were much more difficult to find. In the 1960s, Electrecord developed a policy to record cover versions of international music hits with performers who toured socialist Romania at the time. Furthermore, it used domestic professional musicians to record covers of the Beatles and the Rolling Stones, with lyrics translated into Romanian. Such policies were deemed to cover the gap created by the company's inability to establish ties with international record labels that would have allowed the distribution of original albums and singles under license. The only time when international titles such as Pink Floyd, Led Zeppelin, or Blondie appeared on the Romanian market was through Indian editions, based on a contract between Romania and India.

Censorship issues with Romanian rock records meant that bands could record tracks for Electrecord and then wait for months or years, before finally being approved for the official editing of the album.[30] Given the fact that pop music genres rose or declined in popularity in a matter of months, the delay of domestic rock records created a scene with numerous particularities, most often manifested during the 1980s. Notwithstanding, official releases of popular bands, which took place in Bucharest, at the Muzica Store, were attended by crowds of fans, despite the fact that such events were not always advertised in advance. Word of mouth was efficient enough and Muzica, or other such stores, became occasional places of pilgrimage for those who wanted to buy the record and get an autograph from the band.

At the same time, the lack of international titles led to the creation of alternative networks of manufacture and distribution. Music journalist Florin-Silviu Ursulescu recalls how he created his own network in order to procure records that were impossible to find. He used airline pilots who went abroad and paid them in foreign currency which he exchanged with African students who had come to study in Romania.[31]

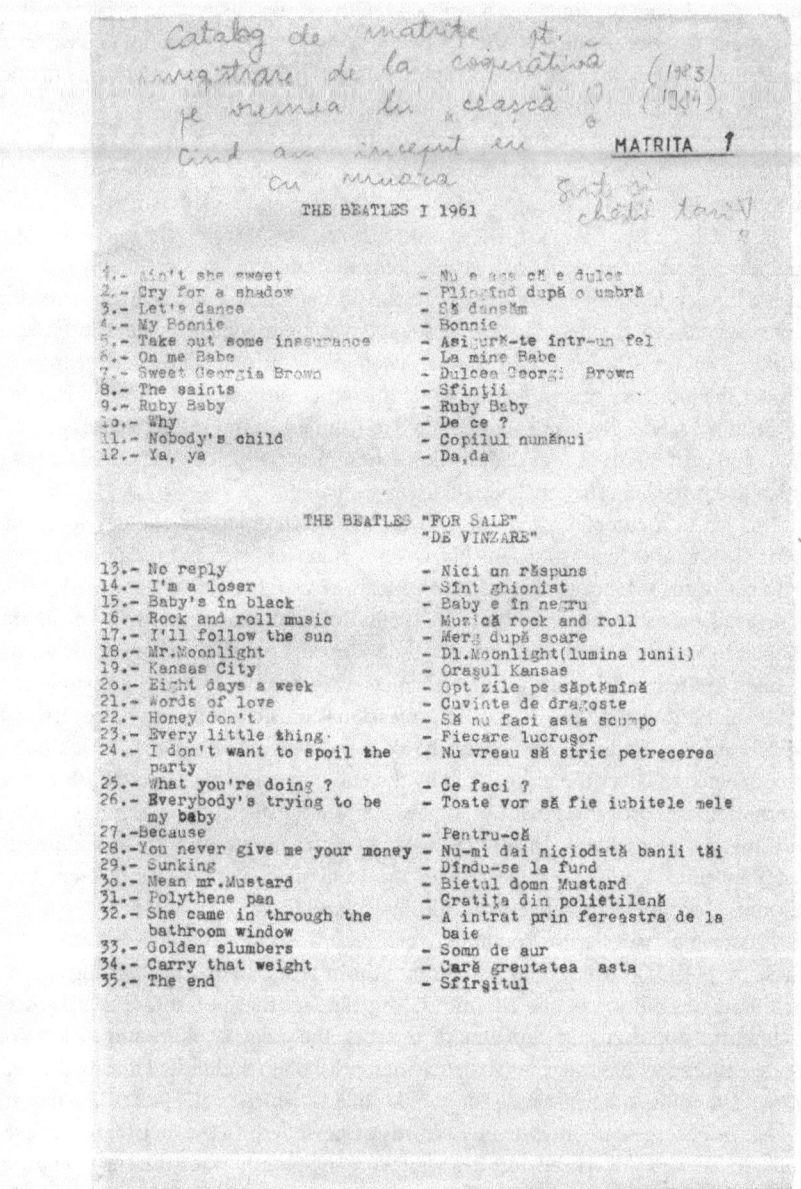

Figure 10.4 Page of "Cooperativa Radio TV" catalog, from the early 1980s (author's personal collection).

Original LP records of famous international bands were extremely expensive on the Romanian market. They could reach as high as 800 lei per copy, at a time when an average monthly salary was around 2,000 lei.[32] This meant that those who afforded the record in the first place either made a copy on magnetic tape and sold the original or kept the original and sold the copy. Ironically, the state contributed to the situation by allowing the editing of music dictionaries that presented foreign bands and artists in minute detail.[33]

In this case, however, there was an official alternative, under the name "Cooperativa Radio TV." These were stores that provided repair services for radio and television sets, but also offered the opportunity to record music on cassette tapes. The network expanded in Romania's major cities. The recording cost was 1 leu per minute, to which one added the cost of the cassette itself, which could be anywhere between 50 and 100 lei, depending on the cassette brand. These stores also came with their own catalogs, which were surprisingly up to date with the latest international releases, not only in terms of famous international artists but also in terms of underground acts. Such a catalog, from the early 1980s, dedicated specifically to "progressive and rock music," contained albums from the Beatles, UB40, Judas Priest, Van Der Graaf Generator, Pink Floyd, and Tangerine Dream, among others.

Besides such stores, which provided the official alternative, there were also smaller stores, in either Bucharest or other Romanian cities, where salespersons could get hold of certain album titles. One customer recalls how, during the late 1980s, he bought an original LP by Romanian progressive rock band Sfinx for the sum of 40 lei (an Electrecord record cost 26 lei). Sfinx had been one of the major Romanian rock bands during the 1970s, under the leadership of guitarist Dan Andrei Aldea. In 1981, Aldea had asked for political asylum while on tour in Belgium and this led to the withdrawal of all 1970s Sfinx albums from the market. Edited albums were still available from Electrecord storages and those who had the necessary connections could get hold of them and sell them on the black market.

The tensions between the socialist state and certain music genres, such as jazz, rock, or international pop music, led to the creation of alternative "record stores." These could be the record stores themselves, when sellers got hold of desired records and sold them at black-market prices. They could also be radio and TV repairing stores, acting as alternative music stores under an official umbrella. But there were also cases of people who got hold of desired records and copied them on magnetic tapes, who became as important as record stores for the smaller communities of music fans scattered around Bucharest's neighborhoods during the 1980s. While this latter situation existed from the early 1970s, it became more common during the 1980s.[34] The explanation lies both in the technological advancements of the era and in Romania's increasing political isolation during the 1980s, which led to a shortage of consumer goods (including records).

1990s and the 2000s: Professionalization, Crises, and the New (Old) Record Store

The year 1989 functioned as a moment of both rupture and continuation for the music and record industry in Romania. The fall of the communist regime brought the opening of the borders and a new position for Romania on the international arena, which affected not only politics but also economic ties and cultural connections. The music industry started to "professionalize," to use the term preferred by music critics.[35] In other words, it had better access to sound and recording technology. But, for the most part, at least during the early 1990s, the scene depended heavily on what had been constructed during the 1970s and 1980s. New record companies relied on Electrecord to record their own artists. LP records went out of fashion in the mid-1990s. At first CDs were available, but CD players were extremely scarce. This meant that, for the first half of the 1990s, the cassette tape ruled supreme as the preferred physical format, regardless of social class or cultural background. Coupled with the absence of copyright legislation until 1996, this led to the development of a prosperous music market, which relied heavily on the black-market lessons of the socialist period but adapted to the new context of the 1990s. In Bucharest, the new Mecca of music aficionados became Piața Romană (the Roman Square), where bootleggers sold tapes of any music genre in an open space market. This was replicated in numerous Romanian cities and towns, in central markets and fairs.

The situation stabilized during the second half of the 1990s. New copyright legislation regularized what could be sold and where. Nevertheless, the cassette tape remained the most used physical format, with CDs steadily stepping in. By the early 2000s, however, both had been sidelined by the avalanche of the internet and music downloading software. MP3s became the new cassette.

When we take into account the economic turmoil caused by the post-communist transition and the financial difficulties encountered by most Romanians from the working and middle classes during the 1990s and early 2000s, it becomes clear that record stores faced a continuous storm during this period.

The most famous store in Bucharest, Muzica remained the property of the newly named Union of Composers and Musicologists from Romania but sold its right to administer the facility.[36] This led to the store losing half of its facility to private enterprises with no music connection whatsoever.

Whereas older stores closed, new ones emerged and formed new communities of music consumers in the process. Throughout the years 2000 and 2010, most music stores in smaller towns closed, in particular, because of illegal music downloading, which made record sales unprofitable. The ones that survived were those in major cities, with a consistent consumer base that preferred physical formats, regardless of the then current fashion.

The 2010s and the resurgence of the LP record as a consumer product led to an increase of music stores in Romania, modeled after a business pattern most entrepreneurs had noticed in West European countries or in North America. However, in order to ensure that their business model enjoyed local success, most adapted and developed their record-selling projects alongside products that supported the main costs of the enterprise (alcohol, T-shirts, etc.). Others also took on the model of offering concerts and performances and turning their record store into a space where socializing was as important as record digging. In a way, they borrowed from the Western model, but also reinstated the former model of the Muzica Store, which allowed for both record listening and music performance. This time, however, the options of music choice were not only much freer than in the second half of the twentieth century but also much more limited from a social point of view.

Personal Views and Conclusion

When gathering sources for this chapter, I also communicated via e-mail (because of pandemic restrictions) with a series of music collectors, of various ages, from thirty-eight to seventy, from various cities of Romania. I was interested in whether my online messengers had any particular recollection of a record store from their childhood, or adolescence, whether they discovered a particular music genre in a record store that changed their life, as well as whether they preferred a certain format for their music. I was also interested in how they got hold of their favorite music. Those from Bucharest recall the Muzica Store, as well as the smaller ones, such as Romanța or Simfonia.[37] One also considered the true Muzica Store to have been in Piața Romană since the 1970s. This recollection is important, as it indicates that the square had already been an unofficial meeting ground for music fans, since the socialist period and that the 1990s were merely a continuation of a trend that had started earlier.[38] Those who lived in smaller cities or towns recall how they only had access to local book stores that also sold records.[39] In most cases, they came across the music that would become their favorite by accident, from bootleggers or from record stores in larger cities.[40] In the case of those who spent their childhood in the post-communist transition, they bypassed the then official record store altogether, as they were already on its way to bankruptcy.[41]

To these recollections, I should add one of my own: I started buying and collecting cassette tapes at my local record store in a small province town in southern Romania in the late 1990s. The store closed down eventually but, as I went to bigger cities, I found various record stores which I started to view in the same way as I had my first record store: the place to go to, the place that provided a unique cultural identity to the place where I lived. Based on the recollections that I gathered and on the sources that I found on the history of record stores in Romania, I consider that regardless of social background, historical context (socialist, post-socialist), or musical genre, this view of the "record store" as a marker of cultural identity has known similar construction patterns, which cross iron borders and years of rupture and continuation.

This chapter is published as part of the project "Rocking under the Hammer and the Sickle: Popular Music in Socialist Romania between Ideology and Entertainment (1948-1989)", supported by a grant of the Romanian Ministry of Research, Innovation and Digitization, CNCS – UEFISCDI, project number PN-III-P1-1.1-PD-2021-0244 within PNCDI III.

Notes

1. See Simon Reynolds, *Retromania. Pop Culture's Addiction to Its Own Past* (London: Faber&Faber, 2012).
2. Bogdan Murgescu, *România și Europa. Acumularea decalajelor economice (1500-2010)* (Iași: Polirom, 2010), 308–10.
3. Ibid., 342–6.

4   Ibid., 349.
5   Edgar Elian, *Discul—Istoric, sfaturi pentru discofili* (Bucharest: Editura Muzicală a Uniunii Compozitorilor din R.P.R., 1964), 53.
6   Claudiu Oancea, "Claiming Art for Themselves. State Artists versus Amateur Artists in Art Exhibitions before and during the 'Song of Romania' Festival (1976-1989)," in *The State Artist in Romania and Eastern Europe. The Role of the Creative Unions*, ed. Caterina Preda (Bucharest: Bucharest University Press, 2017), 267–8.
7   Elian, *Discul*, 52.
8   Ibid., 57.
9   Ibid.
10 Octavian Lazăr Cosma, *Universul muzicii românești. Uniunea Compozitorilor și Muzicologilor din România (1920-1995)* (Bucharest: Editura Muzicală a Uniunii Compozitorilor și Muzicologilor din România, 1995), 238.
11 Ibid., 238–9.
12 Robert Matei, "Peste 4 milioane de instrumente produse la Reghin," *Reghinul Nostru.ro*, February 4, 2014. http://online.reghinulnostru.ro/articole/reportaj/peste-4-milioane-de-instrumente-produse-la-reghin.html (accessed May 10, 2022).
13 Petre Brîncuși and Nicolae Călinoiu, *Muzica în România după 23 august 1944* (Bucharest: Editura Muzicală a Uniunii Compozitorilor din R.P.R., 1964), 235.
14 Octavian Lazăr Cosma, *Universul muzicii românești*, 343.
15 Ibid., 362.
16 Ibid., 343.
17 *Catalogul general Electrecord 1958*, Bucharest, 1958.
18 Ibid., 66.
19 *Catalog general de discuri microsillon Electrecord*, Bucharest, 1961, 5.
20 Ibid., 7.
21 *Supliment de catalog Electrecord—octombrie, noembrie, decembrie*, Bucharest, 1961. Also, *Supliment de catalog Electrecord—iulie, august, septembrie*, Bucharest, 1965.
22 *Buletin informativ Electrecord 21*, Bucharest, 1966.
23 See, for instance, *Romanian Records Electrecord Bucharest-Romania*, Bucharest, 1970.
24 Laura E. Newsom, "Current Catalogues from the Music World," *Notes* 27, no. 1 (1970): 29–32.
25 See *Catalog general Electrecord 1973*, Fifth Edition, Bucharest, 1973 and *Catalog de discuri Electrecord*, Sixth Edition, Bucharest, 1977.
26 The term "light music" was preferred to that of "pop," for reasons that had to do more with ideological than musical reasons. See Andrei Tudor, "accesibilitate și popularitate: o privire panoramică asupra muzicii ușoare românești," in *Noi istorii ale muzicilor românești II. Ideologii, instituții și direcții componistice în muzica românească din secolele XX-XXI*, ed. Valentina Sandu-Dediu and Nicolae Gheorghiță (Bucharest: Editura Muzicală, 2020), 329. See also Edmond Deda, *Parada muzicii ușoare românești* (Bucharest: Editura Muzicală a Uniunii Compozitorilor din Republica Socialistă România, 1968), 1114.
27 Dora Bunaciu, "De ce nu se vinde Humperdinck?" *Viața studențească* 40, no. 554 (1974): 10.
28 Béla Kamocsa, *Blues de Timișoara, o autobiografie* (Timișoara: Editura Brumar, 2010), 64–9.

29 For the intricacies of starting out in a rock band in the early 1970s in socialist Romania, see Sandu Albiter, *Rebel și liber* (Bucharest: Akakia, 2021), 18–27. See also Béla Kamocsa, *Blues de Timișoara, o autobiografie*, 33–78.
30 Ilie Stepan, *Șapte zile plus una/Mircea Mihăieș în dialog cu Ilie Stepan* (Iași: Polirom, 2018), 242–3. See also Nelu Stratone, *Rock sub seceră și ciocan* (Timișoara: Ariergarda, 2016), 157–70. And Virgil Mihaiu, *Jazz Connections in Romania* (Bucharest: Institutul Cultural Român, 2007), 120–3.
31 Florin-Silviu Ursulescu, *FSU/Florin-Silviu Ursulescu în dialog cu Doru Ionescu* (Bucharest: Casa de Pariuri Literare, 2015), 41–3. For a similar situation regarding sound equipment, see also Sandu Grosu, *Lumea mea, lumea sunetului/Sandu Grosu în dialog cu Nelu Stratone* (Bucharest: Casa de pariuri literare, 2018), 25.
32 Claudiu Oancea, "Rocking Out within Oneself: Rock and Jazz Music between Private and Public in Late Socialist Romania," in *Outside the Comfort Zone: Late Socialist Eastern and East-Central Europe between the Private and the Public*, ed. Tatiana Klepikova and Lukas Edeler (Berlin, Boston: Walter de Gruyter GmbH, 2020), 164.
33 Daniela Caraman Fotea and Florian Lungu, *Disco Ghid Rock, Ediție Revizuită și Adăugită (*Bucharest: Editura Muzicală, 1979). A first edition was published in 1977.
34 Again, this was also influenced by officially published guide books that taught amateurs how to repair and maintain audio systems. See, for instance, D. D. Grigorescu, *Redarea discurilor de gramofon* (Bucharest: Editura Tehnică, 1959), or Ervin Șervan, *Picupuri și discuri mono și stereofonice* (Bucharest: Editura Tehnică, 1968). Also, George Dan Oprescu, *Instrumente muzicale electronice* (Bucharest: Editura Tineretului, 1965).
35 Nelu Stratone, *Rock în Timpuri Noi. Povestea fascinantă a anilor '90 (*Timișoara: Hyperliteratura, 2018), 39.
36 *Mugur Mihăescu a cumpărat cu 1,7 milioane de euro dreptul de a administra magazinul Muzica*, Mediafax.ro, May 19, 2010. https://www.mediafax.ro/life-inedit/mugur-mihaescu-a-cumparat-cu-1-7-milioane-de-euro-dreptul-de-a-administra-magazinul-muzica-6135616 (accessed May 21, 2022).
37 A. M. personal e-mail communication, November 22, 2021.
38 D. I., personal e-mail communication, May 7, 2022.
39 G. T., personal e-mail communication, November 19, 2021.
40 G. E., personal e-mail communication, November 21, 2021.
41 H. B., personal e-mail communication, November 19, 2021.

11

Jazzhole

How a Record Store Became the Lone Priest of Nigerian Oldies' Pop Culture

Eromo Egbejule

In the heart of Ikoyi, arguably the most upscale district in Lagos, the commercial and entertainment capital of Nigeria, stands a record store called Jazzhole. Sandwiched between a petrol station, banks, furniture stores, and boutiques, it is the best known of the pocket of record stores nationwide and the lone signpost to bygone eras. Ordinary-looking on the outside, its insides are a repository of cultural memorabilia. Little wonder, then, that the Vinyl Factory added it to a list of the world's best record shops, back in 2017.[1] It describes the store as "a welcome escape from the mile-a-minute hustle and bustle of downtown Lagos." Jazzhole was also listed as one of the world's best record stores in *Around the World in 80 Record Stores*, a book by British journalist and DJ, Marcus Barnes.

The shelves running across the length of the left and right of the store are stocked to the brim with classics from older and contemporary artists, from Ebo Taylor and Cesária Évora to Asa, Bob Marley to Fatoumata Diawara, from Frank Sinatra to ABBA. It is situated less than 2 kilometers from the initial location when it opened in 1991: a small kiosk-style operation just a few yards away from Dodan Barracks, the seat of power in Nigeria.[2] "How can you come to Lagos, the largest Black capital of the world, and not have a decent record store to buy jazz and music from all of Africa?," Kunle Tejuosho, one half of the couple who run the business, asked this writer rhetorically during an April 2019 interview. That was his vision for the store. And over the years, Jazzhole has fulfilled that vision.

The store is now an essential piece of pop culture and sociocultural history in a rapidly gentrifying metropolis, Africa's largest city by population. Across Lagos and parts of the south, there are still small collectors and retailers of vintage records who are a faint reminder of a time when vinyl reigned supreme in the world's most populous Black country. But Jazzhole remains firmly etched as a lone custodian of the culture of yesteryears, in the upper echelons of that fading scene. In that capacity, it has effectively become the high priest of throwback pop culture in Africa's de facto entertainment capital, preserving dear music legacies from colonial times to early post-independence Nigeria and up until, say, the mid-2000s.

Its catalog also includes music from West, Central, and parts of Southern Africa, as well as African American music and Black music forms from the Carribbean and Latin America. Unsurprisingly, it is now a historic reference point for music of Black origins. The records it holds and curates, including several LPs and certain genres of Nigerian music, are not readily seen in our digital libraries and digital streaming platforms (DSPs) like Apple Music, Boomplay, Deezer, Pandora, and Spotify, or in some cases, even the much more diverse YouTube. Effectively, it is an extremely reliable repository and the last bus stop indeed for many a quest for so many classics of a bygone era. "It has possibly the best collection of African artefacts—records, books, artworks—on the continent," the Vinyl Factory said in its listing. "If you're looking for palm wine music from Sierra Leone, soukous from Congo, Afro-diasporan music like Afro-Cuban jazz, salsa, we have it," Tejuosho told this writer. It remains the only place this writer has seen copies of *The New Yorker* magazine stocked, in all of Anglophone West Africa. It also has room for comics and fashion magazines.

As the name suggests, Jazzhole's primary concern is curating jazz music and other related music genres like highlife, Fuji, juju, and the Fela-kind of Afrobeat. The store doesn't concern itself with Afrobeats, the trendy genre that descended from Fela Kuti's Afrobeat or hip-hop and its pop records. Its one weak spot is that it barely stocks Nigerian music not native to or not popular in Lagos or southwest Nigeria. Maybe that is because Tejuosho is a native of the region and also because Lagos has long enjoyed a monopoly of attention and baseline music infrastructure, a situation that is still the case today. For decades and even today, musicians have had to move to Lagos to properly stand a chance of being at the top of their game and being noticed by viewers and entertainment industry stakeholders across Africa, due to the city's strong media presence. Even most of the other record stores and indigenous music labels in Nigeria have either shut up shop or moved to Lagos to continue the business or explore other opportunities.

—

To understand how record stores came to be almost extinct across Nigeria, one must divide the timeline for the rise and disappearance of vinyl records into two: before the civil war and after the civil war. By 1903, McCaslin (2018) writes, the British controlled the territory that comprises modern-day Nigeria but as three separate administrative blocks of Northern Protectorate, Southern Protectorate, and the Lagos colony. The latter two were eventually merged into one.[3]

Eleven years later, the colonial administration joined the two protectorates together. The amalgamation, arguably the singularly most important act in the country's recent history, was done by Frederick Lugard, who became the first governor-general of the new country. As British colonial officers and other expatriates moved to Nigeria, they brought in their worldly possessions and in some cases, their families too. This led to aspirational behaviors for the Nigerians in the administration, as the colonial officers were seen as the goal for social mobility.

In those times, the colonial government was easily the largest employer of labor across Nigeria, according to Ed Keazor, a Lagos-based historian. Consequently, as more people were drafted into the public service nationwide and earned consistent salaries, their standard of living improved. So also did their taste for the comfort that they saw the foreigners acquire, including smoking pipes and buying record players.

In 1925, an Anglican priest called Josiah Ransome Kuti became the first Nigerian to release a recorded album—a Yoruba language hymn project on the gramophone with Zonophone.[4] And that opened the door for other recording projects from the country.

In the late 1930s and early 1940s, the first indigenous record labels were pioneered by men like C. T. Onyekwelu and Badejo Okusanya, who would later head the Nigerian Recording Association. In 1948, the Nigerian Association of Gramophone Record Dealers (ANGARD) was chaired by one O. A. L. Araba, a heavyweight in those days. Both organizations played key roles in organizing the entry of indigenous record dealers into the business.[5] But it was not until the 1970s that Nigeria became an emerging market on the international scene, and it was due in part to the oil boom that happened at that time.

That boom was triggered by oil shocks in the Middle East. During the Yom Kippur War, Western countries like the United States and the United Kingdom supported Israel against Egypt. Arab oil producers responded in solidarity by placing an embargo on the lifting of crude oil and that instantly caused prices to soar from $3 per barrel to $12 by 1974.[6] The skyrocketing oil prices were good for members of the Organization of the Petroleum Exporting Countries (OPEC), which Nigeria had joined in 1971 and remains a part of to date. So, the situation catapulted Nigeria into the league of wealthy nations as it became the world's seventh-largest oil producer. This led to a huge rise in public expenditure as the country undertook postwar economic recovery.

Nigeria's newfound wealth also enabled it to do something which it had been struggling with for years: civil service reforms. Eleven commissions were launched for the purpose of reviewing administrative issues for the colonial civil service and its postcolonial iteration from 1934 till 1960, the year of independence.[7] In 1972, the most popular one of them all, the Udoji Commission, was established and tasked with a similar purpose: improving the efficiency of the civil service. Officially known as the Udoji Public Service Review Commission, it was itself the recommendation of an earlier commission. It was headed by Jerome Udoji, a bureaucrat who had made history as the first Nigerian administrative officer to serve in the British colonial government.[8]

The Udoji Commission advocated for a unified structure and smaller government, standardization of conditions of service, and an increase in public sector wages, among other things. The government selectively implemented the recommendations in the report, focusing on pay rise. Inevitably, this led to an expansion of the middle class because an increased purchasing power for civil servants meant that more families were able to afford more television sets and record players. The progression from gramophones to tapes to radiograms and video players and cassette tapes would continue onwards into the 1990s as what was termed the golden age of Nigerian cinema petered out.

In 1979, Nigeria transitioned to democracy, ushering in a new constitution as Shehu Shagari was elected president. And then oil prices began to fall worldwide in the early 1980s. Oil revenues fell to $10 million, a fraction of the $26 million it had been generating from its oil wells in the previous years. Public expenditure rose, along with the country's foreign debt portfolio. Meanwhile, the economic strangulation of the middle class was in motion. *The New York Times* painted a picture of a nation in crisis. "The naira was maintained at greatly overvalued rates of exchange, making imports unrealistically cheap and domestic products uncompetitive," its correspondent wrote.[9]

To reverse the economic downturn and root out corruption which had become a mainstay in government affairs, Shagari introduced austerity measures in the 1982 budget. This spectacularly backfired in a country facing high inflation and unemployment rates. The president also expelled millions of foreigners who had come to Nigeria in search of a better life, an incident that came to be known in pop culture as "Ghana Must Go." The state of the economy and the rampant corruption were excuses that were cited by the military after it overthrew the Shagari government in a New Year's Eve Coup in 1983. The coup leaders installed Muhammadu Buhari, a stern-faced general who thought that making Nigerians queue and closing the borders would break the fall.

It didn't. Instead, Nigeria's per capita income fell from $800 in 1985 to $380 in 1988.[10] This paved the way for a palace coup d'ètat eighteen months later, led by Ibrahim Babangida, a serial coup planner who took office in his stead. Under pressure to revitalize the economy, the new head of state implemented the structural adjustment program reform to be able to get loans from the International Monetary Fund (IMF). But even that didn't work. The naira rapidly devalued as the gap between the official exchange rate and that of the black market widened. Soon enough, foreign companies began divesting from their operations in Nigeria. Big record labels exited too, so production and distribution of vinyl slowed down drastically.

The economic situation in the 1990s became so grim that foreign investors downed their tools and left the country, unable to cope. Consequently, the ownership change led to name change; CBS Records evolved into Sony Music, Polygram Records was renamed Premier Music, and EMI Records became Ivory Music. Record stores and pressing plants began to disappear too. The stores and dealers did not disappear but the products changed, says Keazor. Many diversified into films, computer software sales, and CD processing equipment importation. In 1999, democracy returned, and the deregulation of many Nigerian sectors, including telecoms and broadcast media industries under the Olusegun Obasanjo administration, invariably helped hatch the evolution of popular Nigerian music—and new structures.

The number of higher revenue-generating business opportunities—and government contracts—increasing as democracy and its accompanying capitalism widened the sphere of wealth redistribution for more Nigerians than had been the case under the military, even if most of the public still weren't beneficiaries. With that and the increasing digitization of the music business, many record stores went out of business and their owners either turned to other ventures or found new avenues for revenue within the music industry—including owning enterprises for rental, sales, and distribution of CDs, VCDs, and later DVDs. Alongside the evolution of popular Nigerian music was the evolution of the indigenous taste for music machines. This period coincided with the return of cinema culture to Nigeria. Cassettes and videotapes began to disappear as there was a surge in the use of computers and the MP3 format.

As teledensity and internet connectivity speeds increased over the years, children of the upper middle class and the elite got introduced to Limewire, the free peer-to-peer file-sharing software. Then there were music blogs hosted by members of the diaspora at first and then Nigerians at home, allowing free downloads of singles. The infrastructure that had come in handy during the days of the record industry's boom became repurposed. Record stores became video rental stores and audio equipment stores, with many of them relegating vinyl to smaller and smaller sections of their shops.

—

A few record stores are still open for business in parts of southern Nigeria including Onitsha, which remains at the heart of the physical distribution networks for music. But there is probably no other record store in Nigeria primarily focused on selling records or preserving that heritage, on the same scale as Jazzhole. Jazzhole was Tejuosho's dreams made a reality in 1991—imbibing lessons learned from working in his mom's bookshop chain Glendora as a young Lagosian. And even though it remains as a lamp post to yesteryears, the store has also had a metamorphosis of its own.

Jazzhole now has a record label and artist management arm that has helped it revive the careers of a few musicians and helped paved the way for some others to rise. In 2005, Keziah Jones, the French-based Nigerian guitarist and singer-songwriter, collaborated with Jazzhole record label on a number of collaboration tracks and in March of that year, performed at a Musical Society of Nigeria (MUSON) event for the store. But perhaps the most striking example of its role in music reconnaissance so far is the case of the Faaji Agba group, a palm wine and juju music ensemble. Jazzhole acted as their record label and intermediary in facilitating their collaboration with other jazz artists, culminating in, most notably, a 2011 performance at a Brooklyn concert alongside Seun Kuti, Fela Kuti's son and bandleader of his father's Egypt 80 band for more than two decades now.

The Faaji Agba group was a loose collaborative of some veteran musicians who first began performing around the 1940s. Their work inspired other musicians from around that time, through the 1960s to the 1980s—some of whom became even more popular than them. For example, there was the case of Ebenezer Obey, one of the most celebrated Nigerian acts ever, who was a dutiful student of Prince Olayiwola Fatai Olagunju aka Fatai Rolling Dollar, a member of the Faaji Agba group. Tejuosho met the latter in 2009 and soon enough convinced seven other musicians to work as part of a group. They were SF Olowookere, Chief Seni Tejuosho, Niyi Ajileye, Pa Oguns, Prince Eji Oyewole, Alaba Pedro, and Sina Ayinde Bakare. In 2015, with the help of British Nigerian filmmaker Remi Vaughan-Richards, Kunle Tejuosho and Jazzhole record label produced a documentary chronicling their reemergence from years of inactivity. It was also titled "Faaji Agba." The documentary went on to win the African Magic Viewers' Choice Award in 2016.

Tejuosho's purpose of documenting their story was to not just preserve their memories and legacy but highlight a major historic point in the history of Nigerian music specifically, and also African music in general. And that purpose has led what started as a record store to evolve into having sections for a bookshop and coffee shop as well as hosting small concerts as part of its offerings a few times a year.[11] There are also intermittent monthly events like artsy soirees and film viewings. Its patrons are mostly middle-aged or older collectors from the middle class and the elite as younger citizens busy themselves with streaming companies and alternative genres on the internet. But still, it is a perfect location for a date for lovers, old or young, with coffee and cake to boot. In its bookshop are books of all genres in African literature and popular novels from elsewhere across the globe. One of those books is *Americanah*, the critically acclaimed novel by Nigerian author, Chimamanda Adichie—which has a scene set in the store. In the novel, the story served as a meeting point for the characters Ifemelu and Obinze when they both returned to Nigeria from the United States. According to popular Nigerian daily *Business Day*, Jazzhole is a personal go-to store for books for the Nigerian author.

In 2018, Nigerian author Teju Cole had a reading session with fans and other members of the public for his book, *Blind Spot*. Apart from reading sessions, Jazzhole is also known for its live stage performances from local and international artists over the years. Some of the Nigerian artists who have performed on its iconic stage include Jones and alternative soul singer Brymo. In 2018, European jazz couple Celine Rudolph and Lionel Loueke also performed at the store as a part of their Obsession Tours.

Quite a number of notable figures in the literary and entertainment sector across West Africa and even beyond have also visited the store. These include the legendary playwright J. P. Clark, novelist Sefi Atta, and superstar highlife guitarist, Ebo Taylor.

In 2011, legendary Malian singer-songwriter Salif Keita walked into the store during a visit to Lagos. Seeing in flesh a man whose works he'd sold for ages elicited excited screams from Tejuosho. It was the coming-of-age story for a man who'd had the heart to go against the grain and take the road less traveled, to ignore more trusted profitmaking ventures and stick to curating "jazz in a hole" in Lagos. And African culture is all the better for it.

Notes

1 Amar Ediriwira, "The World's Best Record Shops #055: The Jazz Hole, Lagos," The Vinyl Factory, 2017. https://thevinylfactory.com/features/the-worlds-best-record-shops-055-the-jazz-hole-lagos/.
2 Coincidentally, Jazzhole launched just a few months before the official relocation of the federal capital to Aso Rock in Abuja, still the federal capital today.
3 J. M. McCaslin, "Lord Lugard Created Nigeria 104 Years Ago," *Council on Foreign Relations*, January 2, 2018. https://www.cfr.org/blog/lord-lugard-created-nigeria-104-years-ago.

4   I. Sansom, "Great Dynasties: The Ransome-Kutis," *The Guardian*, February 14, 2017. https://www.theguardian.com/lifeandstyle/2010/dec/11/fela-kuti-great-dynasties-ian-sansom.
5   C. Waterman, *Juju: A Social History and Ethnography of an African Popular Music* (Chicago: University Chicago Press, 1990), 91.
6   T. Macalister, "Background: What Caused the 1970s Oil Price Shock?" *The Guardian*, February 14, 2018. https://www.theguardian.com/environment/2011/mar/03/1970s-oil-price-shock.
7   Olaopa, a professor of public policy and former federal permanent secretary, penned this op-ed. T. Olaopa, "1974: How Reformed Is the Civil Service?" *The Nation*, October 14, 2018. https://thenationonlineng.net/1974-how-reformed-is-the-civil-service/.
8   Odinkalu is a former chairman of Nigeria's National Human Rights Commission.
9   C. D. May, "Nigeria's Search for Recovery," *The New York Times*, January 30, 1984. https://www.nytimes.com/1984/01/30/business/nigeria-s-search-for-recovery.html.
10 Under Buhari and Babangida, the naira plummeted heavily against the dollar.
11 E. Egbejule, "Visiting the 'Hole' That's the Jazziest Record Shop in Lagos," *OZY*, May 15, 2019. https://www.ozy.com/around-the-world/visiting-the-hole-thats-the-jazziest-record-shop-in-lagos/93762/.

12

The Influence of Imported Records and Their Stores on the History of Popular Music in Japan

Ken Kato

The impact that records and their stores have had in the history of Japanese popular music is immeasurable. They have influenced Japanese music culture in various ways, not only as places to purchase music but also as portals into different cultures and spaces for socializing with like-minded people. Japan's music software market has long been one of the largest in the world,[1] with over 100 years of history in the record industry. Japan is also a country with an uninterrupted love affair with physical media, having experienced a slow shift to streaming compared to the rest of the world. The following discussion examines the influence of imported records (called *yunyū-ban* in Japanese) and their stores within this history. Japanese record stores that deal mainly with imported records not only have always been small in number but also have been spaces with a special role to play. Specifically, this chapter considers the birth of "Western records pressed in Japan (*kokunai-ban*)"; music fans' longing for imported records; friction between domestic and imported records; the increase of import-focused record stores after the Nixon Shock in 1971; megastores built by domestic and foreign capital since the 1980s; and "Record Village," which formed in Tokyo's Shibuya district in the 1990s. Furthermore, the impact of imported records on the domestic music scene will be analyzed from both aesthetic and commercial perspectives.

As a supplementary note, the Recording Industry Association of Japan (RIAJ) defines "*Hō-ban*" as records pressed by Japanese companies that are produced using domestic masters and "*Yō-ban*" as records produced using foreign ones.[2] The term "*kokunai-ban*" used in this chapter is synonymous with the latter. The names of Japanese people appearing in this chapter are given in the order of family name—first name, following the regulations of the Ministry of Foreign Affairs of Japan.

The Birth of the Record Industry in Japan

Japan's music industry history dates back to 1879 when an Edison phonograph was assembled by James Alfred Ewing, a Scottish physicist who taught at Tokyo Imperial University. In 1888, the wax cylinder graphophone was also introduced to Japan.

These were collectively called *chikuonki* (sound-storing machines) and attracted curiosity. In 1899, Matsumoto Takeichirō opened Sankōdō in Tokyo, Japan's first store specializing in records and phonographs, and in 1903, sales of disc-shaped, Berliner-type gramophone records manufactured by Columbia began.[3]

As Japan's music market grew, UK and US gramophone companies like Victor and Columbia expanded their investments in Japan, but Western music was the preserve of a small elite nationwide and unfamiliar to the general public. Therefore, these companies sent recording engineers to Japan to record a repertoire of popular music and popular entertainment, which was pressed into records in their home countries and exported to Japan. During this period, when the technology to press records domestically was not available, imported music included music from one's own country.[4]

American F. W. Horn became involved in the management of Sankōdō through the import and sale of phonographs, and in 1907, in cooperation with Matsumoto, launched Japan's first record company, Nichibei Chikuonki Kabushikigaisha (The Japan-American Phonograph. Mfg Co., Ltd.). This company became Nippon Chikuonki Shokai (Nipponophone Co., Ltd.) in 1910 and began producing and selling records and phonographs under the Nipponophone trademark. The birth of this company marked the first time that Japanese music records were pressed and released in Japan. A wide variety of records were released all of which were recorded by Japanese performers. Local productions made records more accessible to the general public and the Japanese record industry grew steadily, with hit songs such as "Kachūsha no Uta (Katyusha's song)" (1914). However, recordings by foreign performers still relied on imports.[5]

Kokunai-ban and the Yearning for *Yunyū-ban*

Japan's recording industry changed with the Great Kanto Earthquake of September 1923. The magnitude eight earthquake followed by a massive fire devastated Tokyo. After this unprecedented disaster, the government imposed a 100 percent import tax on luxury goods such as records and phonographs to stimulate domestic economic recovery. Foreign record companies, which had previously imported records and phonographs to Japan for sale, now set up branches in Japan and devised ways to reduce their taxes by pressing Western music in Japan. Subsequently, in 1927, Nipponophone sold 47 percent of all its shares to British and American Columbia to become the Japanese subsidiary of this corporate group, and JVC was established with 100 percent ownership by Victor USA. Furthermore, Nippon Polydor was founded through a technical cooperation and master record contract with Deutsche Grammophon with the Japanese record importer Anan Shôkai. Records of Weingartner, Mengelberg, and Stravinsky by Nipponophone, Stokowski, Toscanini, Heifetz, Thibaud and Casals by JVC and the Berlin Philharmonic by Nippon Polydor were pressed and sold on shellac discs nationwide.[6] At this time, records pressed in Japan using foreign masters, that is, *kokunai-ban*, were born.[7]

Although more record companies were quickly established in Tokyo as well as in Osaka and Nagoya during this period, the repertoire of music on domestically pressed records remained limited. Therefore, imported records were expensive and rare, favored by wealthy students and enthusiasts. The shops that carried them, such as Tokyo's Yamano Gakki and Jūjiya, became pilgrimage sites for these audiences. To reach those without record players, "record concerts" were organized where audiences gathered to listen to records. Ongaku kissa (cafés that played the latest imported records) also emerged. In 1927, Japan's first record magazines, such as *Gramophile* and *Meikyoku*, were launched. Aoki Kaneyuki, born in 1901 and editor-in-chief of *Record* magazine, which was renamed from *Gramophile*, recalls that in the early 1920s Victor red seal records cost around 5 yen each.[8] Since *kokunai-ban* shellac discs cost 1.50 yen,[9] imported records were more than three times as expensive. Although the volume of imported records was far less than that of domestic ones, they were a means of accessing up-to-date, authentic performances and continued to have a comparative advantage as a medium in this respect until the 1940s, when the war interrupted the supply of masters and records from overseas.[10]

Frictions between Domestic and Imported Records in Postwar Japan

After the country's defeat in the Second World War, Japan was placed under US-led Allied occupation. For some time after 1945, trade and prices were strictly controlled and the import of new discs and masters was impossible, as all use of foreign currency required government authorization. The only ways to listen to new Western music were to obtain the few records which were imported, purchase them at PX shops on American army bases, or listen to the US forces' FEN (Far East Network) military radio. Price controls on records were lifted in 1948, and in November that year Nipponophone, which had signed a license agreement with US Columbia, changed its name to Nippon Columbia. The industry gradually began to recover.[11]

With 1949's Dodge Line policy, Japan moved to a fixed exchange rate system of 360 yen to the dollar. In 1951, the San Francisco Peace Treaty was signed, marking Japan's return to the international community. The first *kokunai-ban* LP released that year, *Beethoven's Symphony No. 9*, conducted by Bruno Walter, was 2,300 yen.[12] Since the starting monthly salary for high school graduates in 1952 was 7,165 yen,[13] records remained a luxury item. Additionally, the quality of Japanese-pressed LPs at the time was still poor and the catalog was not well stocked. Thus, expensive imported LPs were still popular. Watanabe Shigeru, who worked in the Western music department of Nippon Columbia at the time, noted that Münchinger's *Four Seasons* (1951), released by Decca in the UK, sold for 3,000 to 3,500 yen and was well-received for its sound quality. He usually learned about overseas records from *High Fidelity* and *Billboard* magazines and remembers that record shops such as Nippon Gakki and Harmonia in Tokyo's Ginza were among the first to import such records.[14]

Alongside the domestic record industry's revival, friction between imported and locally produced records began to emerge. In post–World War II Japan, the price of domestic records rose very slowly. This was largely due to the 1953 amendment to the Anti-Monopoly Law, which made books and records subject to a resale price maintenance system. This law made it mandatory for them to be sold at a fixed price, and retailers were no longer free to sell them at a discount. Therefore, as Japan's postwar economic recovery progressed, records became increasingly accessible. In these early days, though, domestic LPs were still expensive and importers continued buying records in large quantities to sell in Japan. Moreover, the price reduction of American records in 1955 made imported LPs cheaper than Japanese ones, leading to calls from domestic record companies to ban their sale. In response to this, Tanabe Hideo, then an employee of Nippon Columbia, recognized the usefulness of imported records: their quality was still superior, the catalog was extensive, and the understanding of foreign culture could also be enhanced. It would be impossible to listen to works published by labels that had no contract with Japan if only domestic records were available. Tanabe proposed a compromise: importing only a certain percentage of works per month—those not already released in Japan or by labels that did not have master record contracts with Japanese companies.[15]

Such decisions made imported LPs expensive again, but no one could stifle the passion of true music lovers. By 1966, even Yamaha, which had become internationally successful as a musical instrument manufacturer, opened a shop in Shibuya, Tokyo, where it sold imported rock and folk records—a store attracting young consumers.[16]

The Unique Development of *Kokunai-ban* and the Formation of *Yunyū-ban* Culture

In August 1971, the world moved to a floating exchange rate system when US president Richard Nixon suspended the convertibility of the US dollar to gold, ending the Bretton Woods system (known as "Nixon Shock"). Afterwards, the yen rapidly appreciated against the dollar, reaching 271 yen to the dollar in 1973 and 210 yen in 1978. In 1973, partly due to the first oil crisis, Japan experienced its first negative growth in the postwar period. Soon thereafter, Japan's rapid economic growth ended. The sharp depreciation of the dollar, however, made it easier to travel abroad: the number of Japanese traveling abroad increased to about four million by 1979. Furthermore, in 1972, the Import Trade Control Order was amended to liberalize imports up to 308,000 yen (approximately $1,000 at that time), which also accelerated commercial transactions.[17] Subsequently, many young Japanese traveled abroad, encountering myriad new cultures and subcultures doing so. In Osaka's downtown area, these travelers began selling vintage clothing, records, and surfing goods bought in California in low-rent warehouse districts, which eventually became a shopping area known as "Amerika-mura." This area later became the center of new wave and noise music in Japan.[18] These economic trends triggered an increase in the number of *yunyū-ban* record shops nationwide.

The revision of the Import Trade Control Order, which liberalized the procedures for personal importation, spurred an increase in *yunyū-ban* imports. The appreciation of the yen made imported records cheaper than their domestic counterparts by the late 1970s. Even by the late 1960s, though, Japan had been shifting away from protectionist trade, and several joint ventures with foreign companies such as CBS Sony, Toshiba EMI, and Warner-Pioneer were created. Intensifying competition among Japanese record companies led to a rise in royalties on contracts with foreign labels, and this was a factor in keeping the price of *kokunai-ban* high: in 1974, a Japanese-pressed LP cost around 2,300 yen, while an imported one was 30 to 50 percent cheaper.[19]

Yet price was not the only reason for the imports' popularity. As Takahashi points out, *kokunai-ban* had been subjected to a variety of adaptations. These included original Japanese titles and artwork; "*obi*" (paper strip) wrappers; liner notes and bilingual lyrics written by Japanese music writers; and changes to the recording and song order—all designed to make the music easier for Japanese people to understand. However, the emergence of concept albums by the Beatles and other artists helped spread the idea of respecting the original versions among music fans, while diversification of musical tastes led to an increased demand for genres still not readily available in Japan (R&B, soul, disco, Latin music, minor singer-songwriter works, etc.). This increased demand for imported records not only as a specific kind of product but as a fetishized object.[20]

December 1978's issue of *Takarajima*, a magazine about Tokyo subcultures, featured imported record shops. In an article titled "Imported Records Are 'Authentic' No Matter What You Think," author Komatsu Motoi describes feeling elated when seeing a record jacket without Japanese advertising or *obi* and the touch of an imported record jacket. He states:

> Why are imports "the real thing" and domestic releases cheap reproductions? [. . .] He's singing in a studio in Nashville, he's not going to think that his songs will be played from a Sony in a four-and-a-half-tatami-mat room in Setagaya, Tokyo. Jamaica's message songs to their fellow countrymen were never sung for us. This realization suddenly becomes apparent when I pick up an imported record with no commentary and sing in a language I don't understand.[21]

These words show that customers who flocked to import record shops at the time felt the "authenticity" of the country of origin in imported albums. Music fans were buying imported records driven by a longing for the West, which had come closer economically but was still geographically and culturally far away.

Imported Record Shops in the 1970s and 1980s: The Example of Pied Piper House

Unlike *kokunai-ban*, which were supported by a systematic distribution system and could be purchased anywhere in Japan, *yunyū-ban* were distributed by small and medium-sized businesses and private importers. For this reason, import record shops were concentrated in cities. A succession of record shops opened in Tokyo during the 1980s, including Cisco, Disc Road, Merurido, Sumiya, and Disc Union. Pied Piper House, an import record shop in Tokyo, stood out from the crowd for its tasteful selection of goods and its friendly staff.[22] Iwanaga Masatoshi, who was born in 1948 and came from the student movement and folk scene, opened the shop in 1975. In late 1977, Nagato Yoshiro, manager of the band Sugar Babe, joined the staff, giving the store a more sophisticated feel. Although the store principally traded in imports, it also sold records by young Japanese musicians, from both major and indie labels. In other words, it placed Japanese and foreign musicians in the same category. Before, Japanese record shops had made clear distinctions between the two groups, separating foreign albums from those by Japanese artists. Many musicians and critics gathered in the shop's small café and discussed music until all hours. The artists who met there included many musicians who would later be collectively known under the moniker "city pop" and respected as legends of Japanese rock music. People with a sensibility based on the culture of Tokyo's private university students also gathered there and introduced Japan to 1960s folk-rock, the California sound, and sunshine pop, all of which were then little known there. Additionally, this group sometimes interacted directly with these musicians and organized concerts.[23]

Peterson and Bennett have identified three types of "music scenes" that emerge from a solidarity between members with similar styles, norms, and musical orientations: "local" scenes, which form during and in a particular time and place; "trans-local" scenes, where connections are established beyond national borders; and "virtual" scenes, where networks are formed on the internet.[24] Based on this theory, the interaction of people in Pied Piper House suggests a local scene, but the spatial and consumerist characteristics of this imports-oriented record shop made it a trans-local one.

Since *kokunai-ban* recordings were to be sold at a fixed price under the law, the Fair-Trade Commission established a time-limited resale system in 1992, which allowed retailers to set prices for records and CDs after two years from their sale (in 1998, the time period was shortened to discounted sales after six months).[25] Thus, in this age of vinyl, treasure hunting for cheaply priced "cutouts" was an experience specific to *yunyū-ban* stores. Pied Piper House adeptly created hit albums from cutouts, most notably *Roger Nichols & The Small Circle of Friends* (1968), a solo album by Roger Nichols, who wrote hit songs for the Carpenters. It became so popular that it was rereleased on CD in 1987 in Japan only.[26] This reissue was made possible by Pied Piper House's proposal to the record company for the CD release of the A&M catalog. In the liner notes for this *kokunai-ban*, Pizzicato Five front man Konishi Yasuharu said of the album: "I want to take it to a desert island, it's the best record I've encountered in my lifetime."[27] Konishi's band was later known as the symbol of the Shibuya-kei movement for their excellent sense of sampling. He, too, accumulated his musical knowledge within this media ecology. Pied Piper House closed in 1989 due to urban redevelopment, but thanks to those who remembered it fondly, it was revived in 2016 as an in-store shop at Shibuya's Tower Records and remains an important symbol of Japan's record store history.[28]

Saison Culture and the Shibuya Record Village

By the mid-1990s, Shibuya, Tokyo, was home to the world's largest concentration of music media stores, with around fifty crammed into a 0.4 square kilometer area and known as "Record Village." The 1985 Plaza Accord, which led to a sharp appreciation of the yen, and excessive monetary easing as a countermeasure, gave rise to the "bubble economy." This was also a period of transition in media from records to CDs. While the number of imported records fell for a time, they then began to rapidly increase. Even after the bubble had burst by the early 1990s, the music industry continued booming, and the number of import record shops continued to increase.[29]

Pioneering the Record Village was Roppongi WAVE, a super-sized media shop with one basement floor and seven floors above ground, built by the Saison Group in 1983. Saison had already developed a sales strategy that went beyond department store management and catered to culturally sophisticated urbanites. Roppongi WAVE was the flagship shop of this corporate group, with a record shop comprising its first four floors and recording studios on the sixth and seventh floors. Its catalog was the largest of its kind in Japan and[30] the store's imported vinyl covered not only British and American pop music but also ethnic music, film soundtracks, and ambient and experimental music. Moreover, sales staff used their musical knowledge to stock the imports section. For Japanese youth with diverse musical taste, this place was a utopia.[31]

Consequently, the opening of foreign megastores such as Tower Records, HMV, and Virgin Megastores from the 1980s onwards greatly contributed to the concentration of record shops in Shibuya.[32] Tower Records, which opened its first shop there in 1981, was one of the largest shops selling imported records and attracted music fans with its low prices, wide selection, and "American" atmosphere. Shibuya's HMV, opened in 1990, came to form a one-of-a-kind brand image, thanks to its European luxury feel and the respected musical taste of its staff. The continuous import of records, the expansion of a secondhand market, and reissues of back catalogs on CD created a music inventory in Shibuya on an unprecedented scale.[33]

Indeed, Record Village's ecosystem was central to Shibuya's music culture and helped shape it. For example, Shibuya-kei was an exceptionally transnational and transcultural genre of Japanese popular music, remixing retro French pop, bossa nova, jazz, easy listening, film soundtracks, Motown, disco, house, and 1980s UK indie-pop (e.g., Aztec Camera, the Pastels.). Pied Piper House, WAVE, and HMV served as a cultural backdrop for these styles, while fans frequented record shops because of their admiration for the Shibuya-kei musicians. Importantly, the term "Shibuya-kei" originally referred to the work of Japanese musicians who were presented exceptionally well in HMV Shibuya, an import record shop. These domestic artists' records were sold with the same authenticity as *yunyū-ban*. Shibuya-kei was seen to display cultural hybridity and evoke global resonances between subcultural scenes.[34] However, to consider this movement only from an external perspective overlooks more than a century of history in the pre-internet world, where people who were relatively geographically and culturally isolated imported Western culture via records and consumed it independently within their own countries, and with exoticist longing. The reason for this cannot be explained without mentioning the historical role played by *yunyū-ban* stores in Japan.[35]

The emergence of Record Village also encompasses the rise of Japanese hip-hop culture. In the 1990s, hip-hop was not yet popular in Japan and new hip-hop records were not often pressed in the country. Therefore, Japanese youths with a yearning for this music and its culture frequented dance music import shops, such as DMR and Manhattan Records, and tried to learn more about hip-hop through purchasing these records. This shares similarities with a structural pattern of cultural reception that has been repeated in Japan since the early 1900s. However, hip-hop trackmakers, such as DJ MURO, continued to find so many records that their collections were overflowing with them, eventually developing a unique sound. One star to emerge was Hashimoto Toru, who established Japan's "rare groove" music scene. Moreover, Nujabes, who ran Shibuya's Guinness Records and was strongly influenced by Hashimoto, remains a legend of lo-fi hip-hop, even after his death at age thirty-six in 2010. Whether linked to the genres of Shibuya-kei or hip-hop, these musicians connected with the global music scene through the singular space of the import-heavy Record Village, and, as a result, they have both played influential roles within Japanese youth culture. In turn, it is difficult to imagine how such music cultures would have established themselves in Japan without the pivotal, cultural role of these record stores.[36]

Conclusion: The Legacy of Imported Music in Japan's Record Stores Today

Japan's music market peaked in 1998 and began to shrink given the record industry's confusion over the crackdown on illegal downloading, a prolonged recession, and tax increases due to the Great East Japan Earthquake in 2011—which had sapped people's purchasing power. The music industry, which had become too rigid to abandon its success story, had missed the shift to the internet age. Indeed, the continuing success of CDs in Japan meant that record shops and rental shops remained, and Tower Records' Japanese subsidiary survived after its U.S. headquarters went bankrupt in 2006. Nonetheless, they declined with time. Roppongi WAVE disappeared in 1999 as a result of urban redevelopment and HMV Shibuya closed in 2010.

In the 2010s, however, two interesting phenomena occurred. One was that vinyl was booming again and the other was that old Japanese records were being traded abroad at high prices. Manufacturers such as Toyo Kasei, which produces vinyl and operates Record Store Day Japan, are behind the boom. Unlike in Europe and the United States, records have yet to overtake CD sales, but data for 2020 showed a steady revival, with an annual CD production of 103,932,000 copies, compared to 1,095,000 in vinyl.[37] Sony Music Entertainment revived its vinyl production in 2018 after twenty-nine years, and HMV is back in Shibuya and Shinjuku as a dedicated record and cassette shop. Meanwhile, record shops such as Disk Union continue to expand.[38]

That Japanese records are now being bought and sold abroad is also completely unexpected. In the past, Japanese music records were seldom exported, excepting those by Shibuya-kei artists, which were distributed overseas and released on foreign indie labels. The unexpected export boom is due to the development of global marketplaces and platforms such as eBay and Discogs, which have made it easier to purchase records across borders. The Japanese language barrier is lower today given digital translation, and, due to Japan's economic stagnation, Japanese records are now relatively cheap. City pop records are particularly popular among collectors, and there is a phenomenon of overseas fans who have discovered the genre's songs through YouTube.[39]

This chapter has discussed the history of popular music in Japan, focusing on imported records and their shops as crucial sites of development. *Yunyū-ban* is a product that has allowed Japanese people to mentally connect with the places where the music was created and recorded. The record shops dealing in them *were* and *are* rare spaces that have allowed access to the wider world. It is impossible to predict what the future holds for Japanese record shops, but the rich cultural footprint they have created over the past 100 years will not disappear.

Acknowledgments

This work was supported by JSPS KAKENHI Grant Number JP 20J11249. English editing was supported by Christine Feldman-Barrett, Fukuda Teresa Megumi, Moritz Sommet, and Nagatomi Mari.

Notes

1. IFPI, *Global Music Report* (London: IFPI, 2021), 11.
2. RIAJ, "Audio CD no Hyōjijikō oyobi Hyōjihōhō" [Labeling Matters and Labeling Methods for Audio CDs]," The Recording Industry Association of Japan. https://www.riaj.or.jp/f/pdf/issue/ris/ris204.pdf (accessed March 15, 2022).
3. Kurata Yoshihiro, *Nihon Rekōdo Bunkashi* [A Cultural History of Japanese Records] (Tokyo: Iwanami Shoten, 2006), 6–8, 41–9; Azami Toshio, *20 seiki Nihon Rekōdo Sangyōshi: Gurōbarukigyō no Shinkō to Shijō no Hatten* [A History of the Japanese Record Industry in the 20th Century: The Invasion of Global Companies and the Development of the Market] (Tokyo: Keisō Shobō keisōshobō, 2016), 44–53.
4. Kurata, *Nihon Rekōdo Bunkashi*, 53–64; Azami, *20seiki Nihon Rekōdo Sangyōshi*, 54–76.
5. Azami, *20seiki Nihon Rekōdo Sangyōshi*, 77–83; Nippon Columbia, *Columbia 50nen Shi* [50 Years of Nippon Columbia] (Tokyo: Nippon Columbia, 1961), 66–73.
6. Utasaki Takahiko, *Shōgen Nihon Yōgaku Kurashikku Rekōdoshi Senzenhen* [Testimony—A History of Japanese Classical Western Music Records: Prewar Edition] (Tokyo: Ongaku no Tomo Sha, 1998), 50–84; Azami, *20seiki Nihon Rekōdo Sangyōshi*, 100–12.

7 Azami, *20seiki Nihon Rekōdo Sangyōshi*, 98–112; Yasuda Masahiro, "Japan," in *The International Recording Industries*, ed. Lee Marshall (Oxfordshire: Routledge, 2013), 155–6.
8 Utasaki, *Shōgen Nihon Yōgaku Kurashikku Rekōdoshi Senzenhen*, 87.
9 Morinaga Takuro, *Bukka no Bunkashi Jiten: Meiji・Taishō・Shōwa・Heisei* [Dictionary of the Cultural History of Prices Meiji, Taisho, Showa, Heisei] (Tokyo: Tembōsha, 2008), 331.
10 Hosokawa Syuhei and Matsuoka Hideaki, "Vinyl Record Collecting as Material Practice: The Japanese Case," in *Fanning the Flames: Fans and Consumer Culture in Contemporary Japan*, ed. William W. Kelly (New York: State University of New York Press, 2004), 152–5; Azami, *20seiki Nihon Rekōdo Sangyōshi*, 122–6; Mori Masato, *SP Rekōdo Nyūmon Kisochishiki kara Shiryōkatsuyō made* [Guide to SP Records: From the Basics to the Use of Historical Documents] (Tokyo: Stylenote, 2022),64–79.
11 Utasaki Takahiko, *Shōgen Nihon Yōgaku Kurashikku Rekōdoshi Sengohen1* [Testimony—A History of Japanese Classical Western Music Records: Postwar Edition1] (Tokyo: Ongaku no Tomo Sha, 2000), 21–5; Mitsui Toru, *Popular Music in Japan Transformation Inspired by the West* (New York: Bloomsbury Publishing, 2020), 80–1.
12 Utasaki, *Shōgen Nihon Yōgaku Kurashikku Rekōdoshi Sengohen1*, 81–2.
13 Morinaga, *Bukka no Bunkashi Jiten*, 444.
14 Utasaki, *Shōgen Nihon Yōgaku Kurashikku Rekōdoshi Sengohen1*, 82–3; Takagi Shinji, *Conquering the Fear of Freedom: Japanese Exchange Rate Policy since 1945* (Oxford: Oxford University Press, 2015), 36–7.
15 Tanabe Hideo, "Yunyū LP to Nihon-puresu LP [Imported LPs and Japanese Press LPs]," in *Rekōdo Dokuhon* [Introduction to Record Discs], ed. Oki Masaoki (Tokyo: Kawade Shobō, 1955), 42–6.
16 Makimura Kenichi, Fujii Takeshi, and Shiba Tomonori, *Shibuya Ongakuzukan* [Dictionary of Shibuya Music] (Tokyo: Ōta Shuppan, 2017), 76–9.
17 Japan National Tourism Organization, "Visitor Arrivals, Japanese Overseas Travelers by Years (1964–2020)," under "Statistics & Data," https://www.jnto.go.jp/jpn/statistics/marketingdata_outbound.pdf (accessed March 15, 2022); Japan, *Import Trade Control Order*, Cabinet Order No. 414 of December 29, 1949, Supplementary Provisions [Cabinet Order No. 406 of November 24, 1972].
18 Masubuchi Toshiyuki, *Yokubō no Ongaku Shumi no Sangyōka Purosesu* [Music of Desire: Industrialization Process of Hobbies] (Tokyo: Hosei University Press, 2010), 155–6; Takagi, *Conquering*, 127–46.
19 Mitsui, *Popular Music in Japan*, 131–5; Yasuda, "Japan," 159; Azami, *20seiki Nihon Rekōdo Sangyōshi*, 175–7.
20 Takahashi Sota, "Nihon-ban Rainānōtsu no Bunkashi [Cultural History of Japanese Disc Liner Notes]," in *Afutā Myūjikkingu: Jissen suru Ongaku* [After Musicking: Music in Practice], ed. Mori Yoshitaka (Tokyo: Tōkyō Daigaku Shuppankai, 2017), 96–111.
21 Komatsu Motoi, "Bokuha Yunyū-ban Furīkusu Yunyū-ban ha Dōkangaetatte 'Hommono' Nandayo [I'm an Import Record Freak: Import Record Is 'Authentic' No Matter What You Think]," *Takarajima*, December 2004, 18–21.
22 Oshima Tetsu, "Bunka Kūkan to Shite no Rekōdoten—Yunyūurekōdoten 'Pied Piper House' no Jirei wo Chūshinni'" [Record Shops as a Cultural Space: A Case Study of an Imported Record Shop 'Pied Piper House'], *Ongaku Kenkyū* 21 (2009): 111–25.

23 Iwanaga Masatoshi, *Yunyūrekōdo Shōbai Ōrai* [Textbook for the Import Record Business] (Tokyo: Shōbunsha, 1982), 75–6; Nagato Yoshiro, *PIED PIPER DAYS: Shiteki Ongaku Kaisōroku* [PIED PIPER DAYS: Personal Musical Memoirs] (Tokyo: Rittō Myūjikku. 2016), 90–3, 132.
24 Richard A. Peterson and Andy Bennett, "Introducing Music Scenes," in *Music Scenes Local, Translocal, and Virtual*, ed. Richard A. Peterson and Andy Bennett (Nashville: Vanderbilt University Press, 2004), 6–12.
25 RIAJ, "Ongakuyō CD tō no Saihanseido no Danryokūn-yō no Jōkyō ni Tsuite [Status of Flexible Operation of the Resale System for Music CDs]," The Recording Industry Association of Japan. https://www.riaj.or.jp/f/leg/saihan/ (accessed March 15, 2022).
26 Nagato, *PIED PIPER DAYS*, 133–4.
27 Konishi Yasuharu, *Koreha Koi Dehanai* [This Can't be Love] (Tokyo: Gentōsha, 1996), 211–12.
28 Amano Ryutaro, "Pied Piper House Nagato Yoshiro to Furikaeru Pied to Tawareko 4nen no Ayumi [A look back with Nagato Yoshiro of Pied Piper House on four years of Pied and Tower Records]," Mikki, December 15, 2020. https://mikiki.tokyo.jp/articles/-/27148 (accessed March 15, 2022).
29 Oshita Yoshiyuki, "Ongaku Kontentsu Sangyō no Jiremma 'Ushinawareta 10nen' Kara no Saisei no Kanōsei [The Dilemma of the Music Content Industry: Possibilities for Revival from the Lost Decade]," *UFJ Institute REPORT* 9. no. 4 (September 2004): 52–5; Gakuyō Shobō Henshū and Hon no Shuppansha, eds., *Rekōdomappu '96* [Record Map '96] (Tokyo: Gakuyō Shobō, 1995), 62–80.
30 Hon no Shuppansha, *Rekōdomappu '89* [Record Map '89] (Tokyo: Gakuyō Shobō, 1989), 49–50.
31 Washio Go and Dobashi Kazuo, *Yumemachi POP DAYS: Ongaku to Shoppu no Katachi* [Dream Town POP DAYS: The Form of Music and Shops] (Tokyo: Rutles, 2016), 190–230.
32 Yasuda, "Japan," 160–1.
33 Ibid.; Kato Ken, "Shibuya ni Shōkan Sareru Shibuya-kei: Popyurā Ongaku ni okeru Rōkaritē no Kōchiku to Hen-yō [Relocalizing Shibuya-kei Music in Shibuya: The Construction and Transformation of Locality in Popular Music]," *Popyurā Ongaku Kenkyū/Popular Music Studies* 24 (2020): 21–2.
34 Simon Reynolds, *Retromania: Pop Culture's Addiction to Its Own Past* (London: Faber & Faber, 2011), 169–71; Martin Roberts, "'A New Stereophonic Sound Spectacular': Shibuya-kei as Transnational Soundscape," *Popular Music* 32 no. 1 (January 2013): 117–20.
35 Kato, Shibuya ni Shōkan Sareru Shibuya-kei, 19–20.
36 Yasuda Masahiro, "Tokyo no Street Generation [Tokyo's Street Generation]," in *Watashitachi ha Yōgaku to Dō Mukiatte Kitaka: Nihon Popyurā Ongaku no Yōgakujuyōshi* [How We Have Faced Western Music: A History of the Reception of Western Popular Music in Japan] ed. Minamida Katsuya (Tokyo: Kadensha, 2019), 184–205.
37 RIAJ, *The Recording industry in Japan 2021* (Tokyo: RIAJ, 2021), 7–8.
38 Uesaka Yoshifumi, "SONY ya Audio Technica: Rekōdo Fukken Hibiku Shin Anarogu no Shirabe [Sony and Audio-Technica: The Return of Vinyl Records, a Resounding New Analog Sound]," *Nikkei Business*, July 16, 2021. https://business.nikkei.com/atcl/NBD/19/00117/00161/ (accessed March 15, 2022); Rekoma Writers club and Hariya

Junko, *Rekōdo+CD mappu 21–22* [Record+CD Map 21–22] (Tokyo: Henshūkōbō Tama, 2020), 44.

39 Max Brzezinski, *Vinyl Age: A Guide to Record Collecting Now* (New York: Black Dog & Leventhal Publishers, 2020), chap. 1, Kindle; Moritz Sommet and Kato Ken, "Japanese City Pop Abroad: Findings from an Online Music Community Survey," (March 2021), under "Fribourg Open Library and Archive," https://folia.unifr.ch/unifr/documents/309133 (accessed March 15, 2022).

13

Recording the Irish Experience
The Record Shop and Fair as Archive

Paul Tarpey

In 2022, as independent Irish record shops and fairs facilitate renewed interest in vinyl, an awareness of the historic connotations associated with used records has developed. This text proposes that the national history surrounding these products encourages the reception of records as cultural objects and retail sites as archives. It is possible that shops and fairs will now generate new perspectives on twentieth-century Irish history.

Recent projects that have positioned records as signifiers of time and place have sought to excavate meaning from these pre-owned musical artifacts. Through this perspective, records can document the reality of Irish culture that has come from reaching out internationally. Visualizing the tension caused by Catholic interests managing imported culture is a task common to Irish historians and artists interested in interrogating the cultural boundaries brought by modernism—and it is key to this discussion. Examining the contribution of record shops and fairs in this context will highlight a culture that has been shaped for decades by censorship. In the activity associated with the browsing that happens in record shops, this text emphasizes that the act of digging for records is a means of visualizing understated but significant junctures in Irish history.

A Brief History of Irish Record Stores

When giving context to the Irish experience of record shops in postwar pop culture, there are similarities to neighboring Britain, but they are minimal. Irish researchers do not have the advantage of the British when the task is to insert a music-led narrative into their national universal history.[1] One can communicate the impact of important British record shops in the manner that commentators use to describe acclaimed music venues, which may result in easier access to the testimonies of those who worked in shops at peak times of change. In this regard, likely only two Irish shops could approximate the significance of Rough Trade in London in the late 1970s or the influence of Eastern Bloc on Manchester in the 1990s. These are Pat Egan's Sound

Cellar in Dublin (opened in 1969) and Terri Hooley's Good Vibrations in Belfast (1976–2015). The following illustrates how this environment and market came to be.

From the outset, a unique culture resulted when Irish record shops became conduits for imported culture. This interest was driven by the Irish listening to foreign broadcasts and their curiosity about the secular freedoms signified by "modern" sounds. In the 1920s, there was a perception of jazz and other modern musical strands by Catholic authorities as signifying of the liberated lifestyles that were increasingly projected by imported Hollywood films. Official censorship with a moral focus to uphold Catholic values persisted as postwar musicals appeared later soundtracking even more secular activity. Records were a less visible threat but by this time returning immigrants, or on occasion the children of Irish Americans, would have brought modern sounds and gramophones with them. Altogether, this was a modernity to be feared. In 1926, Catholic authorities set up a national radio station, in part to address moral concerns encircling popular culture. Traditional and classical records were deliberately programmed to counter secular soundtracks and establish a cultural template, where respectable and native music exposed what was essentially a church state. This apprehension remained in place long after the Republic actively sought international recognition by 1960.[2]

Tension grew in the 1960s with the rise in pirate radio and independent broadcasts from Radio Luxembourg, especially those celebrating the idea that certain types of popular music could now identify a generation. Imported records heard on the radio gave Irish youth the opportunity to think of themselves as something other than predestined young adults in an insular Catholic society. Although pop trends took hold nationwide, the reception of music and style in the capital and rural areas differed. While music trends were serviced by dedicated record outlets in Dublin, the sale of music in smaller towns took place in shops that also sold electrical goods and groceries. Dublin's McCullough Pigott (established in 1823) is worth mentioning as it is one of the longest standing music stores in the world to still be trading instruments and sheet music. Understandably for both retail and cultural histories, much of the focus on the history of record trading is on the Republic's capital. It was always the first destination of imported culture.[3]

Moving forward in time, and regarding the development of rock and punk in the 1980s, the pioneering efforts of Hooley's Good Vibrations (albeit in Belfast) and Egan's Sound Cellar in Dublin were followed by other outlets in the capital, such as Advance, Base X, Record and Tape Exchange, and the still trading Free Bird Records. These featured in the nationwide list of almost 100 shops selling record and music-related material in the *Hot Press* music magazine's yearbook of 1979. Forty shops were listed for Dublin. By 1994, the capital hosted approximately fifteen specialist and secondhand shops trading alongside chain or branded shops, such as HMV. In the 1990s, the radical nature of Irish DJ-based dance culture was serviced by Abbey Discs (closed in 2008) in Dublin and Comet Records in Cork (an offshoot of Comet in Dublin, which closed in 2011).[4]

A standout group of shops is an Irish-owned chain that has served the mainstream since 1962. Golden Discs peaked with eleven Dublin outlets in 1979. While other Irish chains from the 1960s that closed, such as Zhivago or Dolphin Discs, who sell music by Irish artists solely online Golden Discs is notable for keeping a presence in Irish cities with its current focus on selling reissued vinyl. Other stores still in business include the now independently run Tower Records and Claddagh Records (established in 1959). Both are Dublin based, with the latter managing a record label, online trading, and a physical shop in the city center. In Kilkenny, Rollercoaster Records also manages a record label. Other notable mentions nationally include Cool Discs in Derry, South East Records in Wexford, Wingnut Records in Galway (located in a bookshop), and Plugd as well as the Thirty-Three Record Shop, both in Cork. Additionally, and at the time of writing, Bigmoon, which offers "pre-loved" and new vinyl, is open in Drogheda. Moreover, Steamboat in Limerick is possibly the largest independent stocker of vinyl in the Republic. Also in Limerick the legacy of Black Spot Records remains strong while the Dingle Record Shop in Dingle Kerry promotes itself as being the smallest record shop in Ireland. While all shops currently trading have an online presence including social media, the legacy details on those that had a significant presence and were instrumental in nurturing folk, punk, or dance scenes are minimal.[5]

With pre-owned vinyl being an important part of the identity of most shops currently trading, this product has a closer relationship with other record-selling environments. For example, charity shops and record fairs are usually regarded as the opposite of conventional record shops, but now, by dealing in used vinyl, they have come to mirror mainstream sites.

Ostensibly, record fairs in particular invite more of a participatory response from visitors due to their communal setup and the priority given to used stock. Additionally, in contrast to the long-standing tradition of most record shops being located in Dublin, the majority of Irish fairs now take place outside the capital.

Black Spot Records: Finding Meaning in Surprising Encounters

My interest in examining records as artifacts for perspectives on Irishness came after a decade of sorting through vinyl as a customer of Limerick's Black Spot Records. Opened in 1994, this shop was a hub for communities that formed around diverse styles of music imported by the owner, a middle-aged man known as Doc. The rented building in which the store was housed had seen better days, but it was a functional space with a storeroom at the back holding various collections and car boot hauls. Doc had the measure of every warehouse and charity shop in the county and would regularly travel even further afield for stock. Open from Thursdays to Saturdays in the late 1990s, he shared counter duties with a younger enthusiast who liaised with independent record companies for music sought after by the DJ community.[6]

The shop was defined by a unique visual display that combined new and used records, and it was this mix that provided its distinct identity. There was no retail rationale to use this type of layout or to give each genre equal status. In fact, the more striking the record sleeve, the more it was featured in the display, with the top contenders taking pride of place on the wall behind the counter. The other walls showcased a vintage set of posters of Limerick gigs documenting activity organized by bands and DJs over the years, and—as a salute to Limerick's long-standing mod scene—a moped once stood proudly in the shop's window.

This idiosyncratic curation of stock prompted curiosity and encouraged customers to decipher the meaning behind the clash of music genres in the racks. Perusing the shop's records, browsers might ponder a list of friends' names written on the back of a 1970s folk LP or question why a techno record and a 1970s country western LP were placed together. For someone like me, who was interested in recontextualizing the provenance of records, the act of finding a relatively obscure New York hip-hop record with traces of a stamp showing it had been played on local community radio in Tipperary during the early 1980s encouraged me to seek out the history of local radio.

After Black Spot closed in 2004, Doc introduced me to record fairs, an environment where the socio-historical possibilities presented by the displays in Black Spot resonated in the present. Fairs are a relatively recent phenomenon in Ireland. They were introduced by an Englishman named Steve Taylor who, according to Doc, constructed a network of traders in the early 2000s. Upcoming dates for events are usually announced by the traders through social media and fairs are often held in the open air under canvas. With each visit, I saw how former shop owners coordinated their stock, giving the impression that the event existed to present one whole collection. The experience of digging through records with a focus on the subject of Irishness at one particular fair will be discussed in a subsequent section of this chapter. For now, however, and to put that activity into context, it is worth reviewing some exhibits that foreground records as artifacts and signifiers of time and place.

Outside the Shop: The Displayed Record as Artifact

In the last decade, historians, designers, and socially engaged artists with an interest in subcultural history have explored the significance of records when focusing on specific junctures in social history. This is seen in innovative video projects by UK artists Mark Leckey's *Fiorucci Made Me Hardcore* (1999) and Jeremy Deller's exhibit *Everybody in the Place: An Incomplete History of Britain 1984-1992* (2019), which followed from his 2018 documentary film of the same name. Most prominently, it is apparent in the work of US artist Theaster Gates, a self-described artistic archivist, whose display, *A Song for Frankie* (2017–21), is constructed from 5,000 records, a DJ booth, and a record player.[7]

Their works foreground the concept of music being linked to a time and place and as a device for putting certain eras into context. In 2010, Theaster Gates reactivated an abandoned collection of vinyl from a record shop in his hometown of Chicago called Dr. Wax. Visitors to a foundation he manages in the city called the Rebuild Foundation are now encouraged to engage with the rehoused records and consider the relevance of all that is involved with this interaction. In this case, the reactivated vinyl represents its original location, as well as contests a familiar capitalist narrative where a closed shop is symptomatic of either personal failure or failure to adapt. In 2019, Gates underlined this activist take on record stores by having elements of the Dr. Wax shop be available to display elsewhere.[8] In 2019, curated records were featured in a small but carefully organized display about the war years in Northern Ireland in Manchester's British War Museum. The exhibit's designers used punk and protest LPs to offset rubber bullets and other articles of war that once visualized the conflict in British media.[9] Like all these artists' use of albums to reflect upon time and place, the following Irish examples are also socially engaged displays serving to commemorate an unfixed sense of Irishness—one that is caught between tradition and modernity.

As institutions and commissioning bodies popularize exhibits which have the physicality of curated records direct discourse on understated junctures, artists and designers are responding in kind. There is also the growing trend of artworks in this vein transferring online after being physically displayed and validated intuitionally. Both Leckey's and Deller's projects are shown in galleries before being made available online. This allows comments to be posted furthering public discourse on how memory is provoked. In a work such as *A Song for Frankie*, Theaster Gates has the public physically engage with legacy vinyl becoming temporary curators in the process. This installation directly references a record shop to commemorate the memory of a famous Chicago DJ Frankie Knuckles (1955–2014), whose vinyl archive is the basis of the work. Collectively, these are fluid forms of commemoration that use the essence of certain records to unravel memories. They are monuments that stand in opposition to the tradition of something solid deployed to direct memory, a very Irish condition. Any Irish music-based researchers interested in applying fluid forms for commemoration will automatically be contesting a long-standing Irish trait to have monuments and pageants simplify complex history.[10]

In 2015, the understated history of the two record pressing plants that existed on the east coast of Ireland in Dublin and Wexford was commemorated with a display of excavated vinyl. Musician and artist Stephen Rennicks, who was on residency in the McGee Repository Museum at Sligo Folk Park, used discarded Irish-pressed pop records he found in the museum's storage for the exhibit. The intention was to pay tribute to an industry many have forgotten once existed in Ireland: manufacturing vinyl. Rennicks took advantage of the fact that such a display was unexpected in an official setting. In doing so, it restored the dignity of the records and gave them heritage status. The museum validated the symbolic value of the excavated items, and the setting acknowledged that records could exceed their artistic worth. In addition, Rennicks's decision referenced core displays of vernacular commemoration of records elsewhere. For example, Irish records pressed in Ireland with Irish themes are a stable of vernacular collections often tied to tourist sites.[11]

In 2017, a project curated by Dr. Ciarán Swan and Niall McCormack in the National Print Museum used an installation design to present an overview of LP covers designed and printed in Ireland between 1955 and 2017. Documentation of the exhibition, *Green Sleeves: The Irish Printed Record Cover*, lives online as a tribute to the range of visual imagery produced by under-recognized Irish designers and printers of the

1960s. This project portrays how Irishness was rendered for local music genres as they became established and includes a section on LPs that dealt with political, religious, and cultural identity. Most notably on the website, there is a volume of Irish acts who produced country-style, middle-of-the-road LPs in the 1960s and 1970s. Another standout is a reference to a 1980s LP by the punk-style Irish band the Golden Horde. It was designed by a band member who held a day job printing work for the Irish police force. Although this story of minor rebelliousness is slight, it resonates for Irish punk history, as police in Dublin took an instant dislike to this subculture.[12]

Evidence of the growing interest to unpack strands of ongoing conflicted Irishness and probe the material in the present prompted a 2022 exhibition of music and discussion in the Ormston House art gallery in Limerick. Titled *Engine of Hell*, it included documentaries on punk in Northern Ireland and a history of pirate radio that examined music's complex relationship with Irish political history. The title of the exhibition refers to a phrase used in 1934 by Catholic priest Peter Conefrey as part of the anti-jazz campaign of the 1930s. At the time, a group of clergymen and politicians endeavored to control congregating at open-air dances where roads met. When gramophones and records were available, jazz often entertained unsupervised couples. An exhibit specifically focusing on that phenomenon was titled *Crossroads: A Proposal for an Artwork that Never Existed*.[13] Here, musician and artist Steve Maher asks audiences to imagine an Ireland where a legitimate fusion of jazz and traditional Irish music could have existed. The work points to the legacy of the anti-jazz campaign that established licensing laws still used by the Dance Hall Act.[14]

Arguably, the most significant project regarding Irish identity and popular music was undertaken in 2019 by a Dublin record shop called All City. Opened in 2001, the shop began by pressing records by Irish acts to support the DJs of Dublin's small but passionate hip-hop community. In 2019, owner Olan O'Brien began to highlight underrepresented Irish independent music from the 1960s to the 1980s and created a label called Allchival. The first release commemorated unconventional Irish rock groups from the 1970s and 1980s whom he thought had been sidelined. The title of this project, *Quare Groove*, used an Irish slang term for "unusual" to play on the phrase "rare groove," shorthand for record collectors to describe unjustly underrepresented sounds.[15]

Conscious of the first impact of a record cover and its role in a shop display, O'Brien made sure that the sleeve of his record signified the archival nature of the project while alluding to the conditions that had initially sidelined this music. O'Brien used four images of an Irish nun playing table tennis and framed the sequence in papal yellow. The images are seen in an American magazine article about Ireland in the 1970s and with a 1970s styled typeface in cooper black italic, the cover more so resembles a religious artifact from that time than a 1980s LP. It is safe to say that the bands resurrected in this compilation would never have chosen religious imagery for a record cover in the 1980s, let alone a quirky one. At the time, even mainstream Irish artists and musicians kept their distance from religion after experiencing the clergy through the educational system. It can be speculated that an image of a nun at play would have possibly alienated some in the industry and invited unwanted commentary from the remaining church state. Thus, the *Quare Groove* cover brought underappreciated music to the fore via a design that evoked the depth of tension regarding the hegemony which had it relegated to the sidelines of culture.[16]

Each in their own way, these projects have records affect an environment—whether a museum, art gallery, or record shop—to encourage us to focus on understanding the unfixed Irishness that came from modernism being filtered through a church state.

Rennicks's tribute to the ambition of Irish record plants includes his telling of when one Dublin plant refused to press a punk song deemed blasphemous by the staff as late as 1978, while the digital archive of the *Green Sleeves* exhibit has a section dedicated to Irish religious records. Outside the museum, O'Brien's and Maher's use of the record shop and art gallery space in a commemorative sense was a means of drawing attention to the tension generated by church supervision on generations of modern music.

The Record Fair as Archive: Benefits of a Psychogeographic Drift

With the decline of record stores and the rise of digital streaming, record fairs have become a new site for reflecting upon social history through vinyl. Moreover, the community found at Irish record fairs suggests that the act of "drifting" between sellers is a critical experience within these spaces. A methodology that recognizes this experience is known as psychogeography. First introduced in France by the Situationists during the late 1950s and described by member Guy Debord as "The Study of the specific effects on the geographical environment, consciously organised or not on the emotions and behaviour of individuals," it has been used as a strategy for artists in the form of an emotional compass to record unplanned drifting through spaces and places. It is useful for communicating the absent history of places, the significance to be found in liminal space, and is regarded as an open-ended methodology embracing the element of chance and subjectivity to challenge orthodoxy in mapping projects.[17] The following is a documentation of an exercise to explore an Irish record fair with a psychogeographic focus on records. The intention was to regard each display as an element of a whole archive and have a drift excavate material to further understand records as historic artifacts. The result would confirm that these fairs—like traditional record stores—are recourses for interrogating conflicted Irishness.

At the June 2021 Limerick Record Fair, I was greeted by Doc, former owner of Black Spot Records, in his role as an organizer. I began to "drift" and explore the stock with a psychogeography focus. Strong memories of discovering my own Irish identity in rural Ireland in 1980 were evoked by looking at a secondhand copy of the Clash LP, *Sandinista!*, from that year. For my generation the universal appeal of UK punk and post-punk made up for the absence of a strong Irish movement of bands who questioned personal and national identity. Irish Clash fans appreciated that the band had insisted that this LP, a triple set of records, would be priced low for them. Decades later the original sticker on the sleeve still said, "Pay No More than 9.99," but now the dealer had the LP on sale for 100 euros, explaining that it was the current price for an Irish copy in good condition.

Another serendipitous moment occurred after finding an LP by Simon and Garfunkel near a record by the traditional Irish group, the Chieftains, a group synonymous with raising the profile of native Irish music internationally. The album *Chieftains 3* came with the first record player I bought in rural County Mayo, and *Bridge Over Troubled Water* was the second. Moreover, looking at the once risqué cover of Simon and Garfunkel's soundtrack for the 1968 film *The Graduate*, I thought of the legacy it carries with Irish censorship. As a film that questioned changing social attitudes to morality, *The Graduate* was scrutinized by a Catholic censor who was tasked to make sure all mainstream films served a moral and didactic purpose. In the archives of the Irish Film Institute, we see it was recommended that the vernacular use of the words "Jesus" or "Christ" to be deleted.[18] Before moving to another stall, I marked what provoked this encounter by arranging a display where the venerated Irishness embodied in the Chieftains' LP was placed next to *The Graduate* soundtrack, which symbolized an awkward modernism.

A trigger for a wider conversation about Ireland's unfixed modernity came with finding another soundtrack. This time, it was the soundtrack for the 1959 film *Mise Eire*, seven pieces of orchestral music on a seven-inch record. Unexpectedly lodged in a box of ska singles, it was for sale by a trader named Wally Cassidy, who remains the sole documentarian of the 1980s Irish punk scene. As a photographer, Wally has published essential histories of outsider communities, including those connected with the Irish antiwar protests, that tested the bond between Ireland and the United States for decades. Looking at everything represented by this artifact, we agreed that in terms of all the curious records to visualize unfixed Irishness through a clash of the new and old, *Mise Eire* was possibly the definitive artifact.

Indeed, the *Mise Eire* (*I am Ireland*) record is a unique marker of Irishness.

When the record was released, it embodied the reactivated nationalism of 1966, the fiftieth anniversary of the rise against Britain. The 1959 film itself relied heavily on newsreel footage to tell the story of Irish independence and was scored by Sean O'Riada. His arrangements were praised for their modern approach to traditional music's own terms. This resetting of traditional music for modern sensibilities in a film to be shown internationally was in line with the new Irish government policy to end economic protectionist policy. Though not made explicit officially, *Mise Eire* seemed to visualize the 1958 publication of *The First Programme of Economic Expansion* regarded as the foundation of modern Ireland.[19]

The record's sleeve graphics and text written in both Irish and English celebrate this ambition styled with international almost pop graphics. On the front, a rendering of a site of the 1916 rebellion is complemented with one of the mythical Irish warrior Cú Chulainn and finished with complementary advertising speak on the back. This would not be the last time allegory and facts would be deployed to identify the nation.

I wondered why I had never come across this record before. Wally suggested that I most certainly had and just never registered it. According to him, many copies remain in charity shops throughout Ireland that were possibly donated by families clearing possessions after an older family member had passed, someone who had originally held the record as a talisman. This experience of a psychogeographic drift through a record fair as a heightened auto-ethnographic dig confirmed that an appreciation of record fairs as vernacular archives can ground personal narratives in a wider context. Importantly, these environments keep the legacy of Irish records shops, such as Black Spot, alive.

Conclusion: Archives for the Future

While it remains to be seen how a resurgent interest in vinyl and a new appreciation of Irish record shops will last, it is obvious that the situation has brought mainstream and outlying traders closer than ever expected. In retail terms, the independent shops and fairs that have survived while mainstream shops folded in the 1990s now prompt conversations about the record trade led by a renewed interest in vinyl. Though reanimated mainstream shops salute the paradigm shift, this is not a continuation of pre-1990s business. The majority of (expensive) reissued vinyl stocked by mainstream Irish record stores (mainly Golden Discs) consists of reissued pop and rock classics with a predictable emphasis on famous national acts. Here, when a record's value is determined, it is by the retail history it has accrued rather than its cultural significance.

New pressings of old records are also validated by how well they sit within the mainstream media platforms that rely on the legacy of music produced during the 1960s, 1970s, and 1980s. History has never before been featured in trading records in the way it is now. Depending on the focus, one could process reflective nostalgia with an LP in a revitalized main street shop or be challenged by the same artifact elsewhere. It may be speculative, but it is still worth stating that as the display and presentation of records in independent spaces become more familiar, the historical significance of vinyl artifacts grows. With this, the critical aspect of digging through records will become more apparent—especially if stores permit artists and researchers to include them in projects that visualize the reality of Irishness in the twentieth century. Opportunities also remain for a combination of physical and digital forms of music to present a timeline depicting the tension caused when the church state sought to manage imported culture. It is a timeline that begins with Irish-themed records sent home from America by immigrants in the early twentieth century and ends with forgotten Irish rock music being released on vinyl by a Dublin record shop in the twenty-first century.

Notes

1 See Dave Halsam, *Life after Dark: A History of British Nightclubs and Music Venues* (London: Simon & Schuster, 2016), and Paul du Noyer, *In the City: A Celebration of London Music* (London: Virgin Books, 2010).

2   In 1938, the station became Radio Eireann and was seen to promote the cultural nationalism and the nurturing of a hegemonic Gaelic Catholic state. See Gerard Dooley, "Entertainment in Independent Ireland: Evolution of Irish Parochial versus Commercial Dance Hall Culture," *History Studies* 13, no. 61 (2012): 47–60. See also Ruth Barton, ed., *Screening Irish-America: Representing Irish America in Film and Television* (Dublin: Irish Academic Press, 2009), 29; Kevin Rockett, *Irish Film Censorship* (Dublin: Four Courts Press, 2004), 115.
3   Ibid.
4   For some online traces of these stores, see Dublin Music Trade. https://www.dublinmusictrade.ie/node/371 (accessed April 18, 2002); Freebird Records, http://www.freebird.ie./history (accessed April 20, 2022); Fiona Garland, "Swansong for Abbey Discs as Digital Music Delivers Vinyl Nail in the Coffin," *Irish Times*, November 25, 2008. https://www.irishtimes.com/news/swansong-for-abbey-discs-as-digital-music-delivers-vinyl-nail-in-coffin-1.914398; Jim Mahoney, "Comet Records Cork," Blackpool Sentinel, April 20, 2015.
5   https://theblackpoolsentinel.com/2015/04/20/comet-records-cork/; "Lost in Music: Dublin's Record Shops 40 Years On," 909 Originals. https://909originals.com/2019/03/16/lost-in-music-dublins-record-shops-40-years-on/ (accessed April 20, 2022); "Golden Discs: About Us," Golden Discs. https://goldendiscs.ie/pages/about-us (accessed April 20, 2022).
6   See mention of Black Spot in Jim Carroll, "Rock around the Shops; Independent Record Stores Are Keeping Their Doors Open by Beating the Drum for Those Hard-to-Find Albums," *Irish Times*, October 25, 2002.
7   For Leckey, see Laura Cumming, "Playing to the Gallery," *Guardian*, April 19, 2009. https://www.theguardian.com/music/2009/apr/19/art-and-music-andy-warhol; Jeremy Deller, "Everybody in the Place: An Incomplete History of Britain 1984-1992," The Modern Institute. https://www.themoderninstitute.com/exhibitions/everybody-in-the-place-an-incomplete-history-of-britain-1984-1992-2019-03-16/6701 (accessed April 25, 2022); The Frankie Knuckles Foundation, "Who We Are—The Frankie Knuckles Foundation," YouTube video, 2:31, April 14, 2017,. https://www.youtube.com/watch?v=3VptxldZRmY&t=115s.
8   For the foundation and Dr. Wax project, see Hesse McGraw, "Theaster Gates: Radical Reform with Everyday Tools," *Afterall* 30, no. 1 (2012): 86–9.
9   "Frankie Knuckles Foundation."
10 Ibid.; "Decade of Centenaries, 2013-2023," Decade of Centenaries. https://www.decadeofcentenaries.com/about/ (accessed April 26, 2022).
11 Stephen Rennicks, "Ireland's Vinyl Record Manufacturing Industry (1950s-1992)," Abstract Analogue, July 16, 2015. https://abstractanalogue.tumblr.com/post/124261912599/irelands-vinyl-record-manufacturing-industry. The website for National Treasures functions as a peoples' archive that encourages the Irish to contribute any type of material that can help represent 100 years of cultural heritage. "A Peoples' Archive," National Treasures. https://nationaltreasures.ie (accessed January 28, 2022).
12 "Green Sleeves: Sleeve Notes," National Print Museum. https://www.nationalprintmuseum.ie/green-sleeves/ (accessed March 2, 2022).
13 Stíofáin Ó Meachair, *Crossroads: Context for Socially-Engaged Artistic Strategies in Dealing with the History of Censorship in Post-Civil War Ireland* (Limerick: Ormston House, 2022). https://ormstonhouse.com/wp-content/uploads/2022/04/crossroads-by-steve-maher.pdf (accessed April 14, 2022).

14 John Porter, "The Public Dance Halls Act, 1935: A Re-examination," *Irish Historical Studies* 42, no. 162 (November 2018): 317–33.
15 "Irish Artists and Labels," *All City Records*. https://www.allcityrecords.com/product-category/irish-artists-and-labels/ (accessed April 3, 2022); "Various Artists—Quare Groove Vol. 1," The Thin Hair, accessed April 6, 2022.
16 Ibid.
17 Guy Debord, "Introduction to a Critique of Urban Geography," in *Situationist International Anthology*, ed. Ken Knabb (Berkeley: Bureau of Published Secrets, 1981), 5.
18 In the 1960s, the Republic referenced British standards of classification. See "Irish Film Censors' Records," Trinity College Dublin, https://www.tcd.ie/irishfilm/censor/show.php?fid=3 (accessed January 15, 2022).
19 The film was described as "an anti-imperialist polemic whose release in 1960 was a national event." See "Reel Hero of Irish History," *Irish Times*, January 4, 2008. https://www.irishtimes.com/culture/reel-hero-of-irish-history-1.925660; Martin Mansergh, "The Political Legacy of Seán Lemass," *Etudies Irlandaises* 25, no. 1 (2000): 142.

14

The Revolution Will Not Be Televised, It Will Be Taped

Western Music Acquisition in Pre- and Post-Revolution Iran

Lily Moayeri

Introduction

This chapter explores the role of popular music in both pre- and post-revolutionary Tehran. In it, the author reminisces about her consumption of popular music via cassettes as a child in Tehran and discusses the way cassettes were used to circulate sermons but were banned and burned when distributed for secular purposes, as well as speculating on the ways that modern technologies like the internet have, forty years after the Revolution, allowed music lovers to collect and consume popular music that has unalterably interfered with the process of censorship—in the best possible way.

I.

When I was nine years old, I watched the cassette store across the street from our home in Iran, decimated in the riots that accompanied the demonstrations of the 1979 Islamic Revolution. As the windows were smashed and all the merchandise was destroyed, I felt my insides shatter. I pictured the cassettes I would stare at in the glass display cases, obliterated, and I felt frantic. I wanted to rush out and rescue those little plastic cases and everything they held. My parents told me to get off the balcony where I was watching this music massacre and to stay away from the windows in case of stray bullets—and because the trauma I was experiencing was obvious to them.

Today, I am a full-time music writer, and I remember that cassette store with fervor and sadness, but also with awe. So much has changed in Iran since before the Revolution that it is now unlocatable on Google Maps: even the names of the street which I lived on has changed, as have the businesses that existed nearby. I have tried in vain to find it, but it's as if it never existed.

And yet, that cassette store is where my childhood obsession with music was nurtured and fed, stocking as it did all the popular Western artists of the time—the Bee Gees, ABBA, Donna Summer, the Osmonds, Jackson 5—and as such, epitomized everything the demonstrators were protesting *against*. They were guided by the words of the founder of the Islamic Republic of Iran, Ayatollah Khomeini, who prohibited Western entertainment and directed the arrest of dissidents, who were then imprisoned or shot.[1] In July of 1979, Ayatollah Ruhollah Khomeini banned all music from Iranian radio and television because, he said, it is "no different from opium."[2] According to *The New York Times*, the 79-year-old revolutionary leader said, this music "stupefies persons listening to it and makes their brain inactive and frivolous."[3] Since that time, no musical instruments have been seen on screen in Iran. To this day, they might be heard but never seen being performed.

It's ironic that cassette stores like the one I called "mine" were a target of these riots, since cassettes were integral not only to the consumption of Islamic doctrine but specifically to Khomeini's rise to power.[4] In the mid-1970s, when the Ayatollah was exiled in France, his speeches were distributed via that medium and that medium alone. As Simon Adler, speaking recently on "On the Media," explained it, because the shah had blockaded TV and radio and only allowed sanctioned speakers, "(the leftists, and the Ayatollah) needed an underground, analog way in . . . and cassettes filled this need."[5] Once a week, while in exile, Khomeini would preach a sermon, recorded on cassette, and his handlers—a network of exiled leftists who were unhappy with the shah's regime—would then play that cassette over a transatlantic line. Back in Iran, those who'd conference-called in would record it on cassette and then distribute it hand to hand, playing it in cafés, cabs, and restaurants, spreading it from house to house.

According to Adler, at its height, 90,000 mosques were duplicating and distributing his message, via answering machine. But these cassettes were spreading the Ayatollah's message of rebellion and religion, not music, and the leftists who propagated their circulation did not realize the power of the Ayatollah's primarily religious message. They thought of him as a tool to remove a corrupt regime of the reigning shah; what they got was a religious fanatic who's hold on Iran can still be felt today. In an interview with journalist Oriana Fallaci in September 1979, speaking specifically about Western music, Khomeini said Iran "got many bad things from the West" and that he didn't want Iranian youth to study in the West and become corrupted "by the music that blocks out thought."[6]

Eliminating Western music must have seemed like a no-brainer to him, but eliminating *music*, and the love of it, is impossible. In my case, it was too late: my parents came of age during the shah's reign and that of his father, Reza Khan, when Iran was moving toward modernization, with the latter using "Western nations as his role model."[7] The shah's White Revolution, which gave women the right to vote, nationalized forests, reformed land use, updated worker and employee regulations, and ended illiteracy, was met with majority approval. But it also created a backlash from the Islamic fundamentalists who claimed the White Revolution would destroy traditional values.[8] These reforms caused an even greater economic divide and increased the influence of the West in Iran, particularly on the middle and upper classes, including my family, particularly my parents' generation, the so-called silent generation, and even more so on the boomers.

My parents, like many Iranians of their class at that time, had lived outside of Iran for a while before the Revolution, which is how they became the catalyst for my music obsession. When I was four years old, we were living abroad, and my father gave me a side-loading turntable and a stack of his 45s from the 1950s and 1960s. After the family moved to Iran when I was seven, I was presented a portable record player in a red case with a moveable tone arm. A couple of years after that, my parents gifted me a small tape player for my birthday. It was the mid-1970s, and my mother tasked my cousin (who is my senior by fifteen years and was a frequenter of Tehran's discos) with collecting mixtapes from the DJs. These mixtapes soundtracked everything we did as a family, particularly every car trip. The songs ran the gamut from Earth, Wind & Fire's "Fantasy" to Bob Marley & the Wailers' "Is This Love," 10cc's "I'm Not in Love," and Roberta Flack's and Donny Hathaway's "The Closer I Get to You." At the time I didn't know the artists' names, these were mixtapes after all, but I remembered all the songs.

II.

Some Westerners may think the Ayatollah was being unnecessarily hard on popular music. But in fact, he wasn't underestimating its importance to Iranians. Before the Revolution, music was a big part of the three television channels in Iran, two in the native language, Farsi, and one in English. The former two had performances by popular Iranian artists: Googoosh, Leila Forouhar, Sattar, Aghasi, among many others, as interstitials. The third channel is where *Donny & Marie* appeared every Thursday night, the variety show filled with skits and music. Through the brother-sister duo, I was introduced to the other Osmonds. Once I was allowed to cross the wide avenue our home was situated on, I made a beeline for the cassette store where I found numerous choices from the Osmonds and many others.

There was so much music in that small store. Cassettes lined the walls to the ceiling, most of them Iranian artists but many Western artists as well. I wanted to take all the Western pop music home. But my mother allowed only one cassette each for me and my sister. She made it abundantly clear that these cassettes belonged to her, not us, but we could listen to them as much as we liked.

Each cassette my mother bought—the choices of which I negotiated with my sister—I picked carefully from the store's displays. I would make mental notes of them, but there were so many, I panicked that they would be gone by the time my mother permitted another purchase. She bought two velvet-covered portable cassette carriers that held twelve tapes each for our growing collection. Both would go with us from car to home and back again, no matter how short our journey.

There were other ways we found out about music in addition to television and that store, though. For example, at parties with my young schoolmates, I was exposed to more music, since music and dancing at parties is an Iranian cultural tradition which even the Ayatollah couldn't prevent. My aunt, Haideh, told me that in 1965, when she was in high school, it was the height of the Beatles' and the Rolling Stones' popularity. Some of her close-knit group of classmates who went back and forth to Europe (England in particular) would bring home their music on cassette. She said, "We had parties every weekend and we had a new set of songs or bands."[9]

Along with the Beatles and Rolling Stones their favorites included James Brown, whose dance moves they attempted to imitate, the Supremes, James Taylor, Italian crooner Peppino Gagliardi, and French chanteuse Françoise Hardy. "None of us listened to Persian music at the time," she said, but they did frequent Beethoven Music Center, a record store established in the 1950s, which is still open today.[10] My cousin Farid, seven years my senior, remembers very small record stores in the early 1970s, in the Tajrish neighborhood. These establishments carried vinyl, rather than cassettes. Here, he purchased albums from bands like Black Sabbath and the Eagles. "I remember having little record players in cars that would play 45s," he says. "The records would get ruined."[11]

Another small record store is where Farid purchased low-quality cassettes of both Western music and Iranian musicians such as Dariush and Viguen, along with perennially popular singers, and sisters, the late Hayedeh and Mahasti. At this point, no Iranian music was available on vinyl, only cassettes. Farid said of the Western music in these stores:

> The cassette covers were never the actual album covers, They were generic covers with the band's name. I remember buying the band America's debut album on cassette, one of my favorite albums of all time, but the quality of the recording was terrible. It had been recorded over and over again, poorly. The small stores selling vinyl were more legit, it was real vinyl, in shrinkwrap. They had all the mainstream music of the time: Carole King, Kiss, The Rolling Stones. It was all there.[12]

One of my other aunts, Sussan Safari, lived in the port city of Ahvaz, in the Persian Gulf during the 1960s, where she says, "We lived more out of the country than in. Anything that was coming from abroad would come to Ahvaz first, then get distributed to the rest of the cities. Europeans, Americans, people were coming in from everywhere. We listened to music from all parts of the world."[13] The offspring of an employee of the National Iranian Oil Company, at the NIOC's headquarters, Sussan was able to watch foreign films in their original language and be exposed to Western music in the company store, long before these items made their way to Tehran. Additionally, record stores in Ahvaz and neighboring Abadan brought Western music, from which Sussan and her friends would purchase their music.

"We didn't gather in record stores like you do in the U.S.," she says. "We had country clubs with a lot of facilities that we would take advantage of. After school or on weekends and holidays, we would go to the country club, hang out and listen to music, then maybe go to one of our houses in the evening. We had a lot of freedom from when we were 13 or 14."[14]

All of this stopped after the Islamic Revolution "when strict interpretations of Islam were instituted in all facets of public life"[15] and, as Salisbury and Kersten say, "Islamic states such as Afghanistan, Iran, and Saudi Arabia maintained legitimacy through social regimentation and oppression."[16] Truthfully, for many Iranians, particularly those of the middle and upper class, the reality of not having access to the Western entertainment they had gotten used to for the last half-century wasn't the first cause for concern post-revolution. But, after the initial shock of the change the Islamic Revolution brought politically and societally, after the executions of friends and colleagues, and the fleeing of others from Iran, for those remaining, like our family, the absence of Western music, movies, television programs regularly being funneled to us started to make itself felt.

For a few years after the Islamic Republic of Iran was instituted in 1979, there was a strict travel ban. Its gradual lifting throughout the 1980s didn't mean hassle-free passage, but rather that Iranians weren't prisoners in their own country, and Iranians abroad were not banished from returning home. But it also meant that bootleg audio and videotapes began circulating through an underground network, via the pilots of the Iran Air flights who would smuggle back albums on cassettes they had bought in Europe and videos of programs they had taped off European television. Most notable of these was the UK's long-running music chart program, *Top of the Pops*. These items were then duplicated onto blank cassettes, packed into unmarked briefcases, and delivered by an affable fellow who would only come to your home on referral from a trusted client. My aunt Sussan—who was introduced by one of her close, music lover friends—connected our family to this underground network, which allowed us to mainline Western entertainment once again. Thus, the cassette store of my childhood was replaced by door-to-door service. *Top of the Pops* episodes informed my choices for the albums I would purchase from the selections in the briefcase: Culture Club, Duran Duran, Wham!, Howard Jones, Tears for Fears, Spandau Ballet, Thompson Twins, Simple Minds, A-ha, the Style Council, the Human League, Madonna, Cyndi Lauper, Bananarama. The list was long.

I petitioned my parents for a boombox with a double cassette player so I could make mixtapes from the albums. They agreed, in exchange for a good report card, and with the stipulation that I would have to charge my friends for the mixtapes I made them, to offset the wear-and-tear on the boombox and the cost of the blank cassettes.

Now, instead of doing homework, I would painstakingly make record/play/pause mixtapes. My mixtapes soundtracked the parties my high school friends would throw. They were turned down just low enough so as not to be heard outside, inadvertently alerting the Revolutionary Guard (aka the Pasdars), who under the Islamic law had the right to raid the party, confiscate our "contraband," any other "inappropriate" items, and take us all to jail.

Perhaps one reason I was willing to risk this low-level cassette trafficking business was that sharing music with friends, talking about the songs and dancing, is not just natural human instinct but, as mentioned earlier, a huge part of Iranian culture. We practiced the dances in heavily choreographed videos like Michael Jackson's "Thriller" and music-driven films like *Flashdance* and *Footloose*. We performed for each other at parties and were intoxicated with the camaraderie and connection it created.

Of course, Iranian music, both traditional and modern, was always popular and more easily passed around, which made finding people who were interested in the *Top of the Pops*-inspired music I was into both harder and more important to me. This was in the mid-1980s, when the Islamic Republic was still getting a foothold in Iran, establishing its structure and punitively enforcing its extreme laws. To quote Khomeini in his Speech on the Uprising of Khurdad 15 (1979): "But as for those who want to divert our movement from its course, who have in mind treachery against Islam and the nation, who consider Islam incapable of running the affairs of our country despite its record of 1400 years—they have nothing at all to do with our people, and this must be made clear."[17]

III.

Today, things are much different, thanks to new technologies that can be curbed but not altogether stopped. So, while Islam is still the guiding force for Iran, technology and a very slight, inconsistent loosening of restrictions allow for Western music to circulate via various mediums. In 2017, an Iranian musician, Kasra Vaseghi, professionally known as Kasra V, recalled growing up "on a musical diet of VH1 and MTV while living in Tehran, Iran's capital city, thanks to satellite TV." But what was shown on state-controlled television was surprising, since the staff of many of these TV stations was captivated by Europe's avant-garde electronic music scene. The article names Jean-Michel Jarre, Kraftwerk, Vangelis, and Tomasso Albinoni as some of the artists played on IRIB, Iran's national state broadcaster. And with the advent of the internet, however, there was no limiting Vaseghi's access to Western music. He told *Wired*, "There weren't any record stores, but there were computer shops that would download music on your behalf. You turned up with a list and either they had it or would go and download it for you." Vaseghi also downloaded music from internet radio shows such as BBC Radio 1's *Essential Mix*, "The internet at the time was so slow. I used to get up before school, start downloading the radio shows, and by the time I came home they were done," says Vaseghi in *Wired*.[18]

As Vaseghi's story indicates, improvements in high-speed internet access and more and more technology platforms have made finding Western music easier for people in Iran. For members of Generation Z, such as Salma and Parsa, whom I spoke to for a 2020 *Los Angeles Times* article, virtual private networks, the dark web, Spotify, Torrent, and the messaging app Telegram are some of the methods they use to get around the Islamic sanctions against all forms of Western culture, particularly entertainment.

This is not to imply that the Islamic Republic's attitude toward popular music has gotten lax in any way. In order to perform, release, or sell music legally, your work must be officially sanctioned by the government via *Vizarat-i-Farhang va Irshad-i Islami* or the Ministry of Islamic Culture and Guidance. The Ministry allows only traditional Iranian music and poetry—even though some of these poems are quite subversive, their subtle nature allows the messages of the poems to remain hidden to the untrained ear, as well as instrumental music and Western classical music to be produced and distributed to physical shops. Some of these are music stores, like Beethoven Music Center, but the primary distribution is through sundries stores like the chain *Shahr-e-Ketab*, or "city of books," alongside reading materials and stationery supplies, not unlike Barnes & Noble.

According to many Iranians today, if you are a decent musician with new ideas or if they don't like what you're saying lyrically, the Ministry will not issue a permit for your product. This has been Mohammad Reza Hariri's experience as a member of the Iranian experimental pop band, SheCan. There is no way SheCan's über-modern, electronic-driven sounds, topped by despair-laced Farsi lyrics that speak truthfully to the feelings of its creators about their lives, will ever be greenlighted by the Ministry. SheCan is fronted by a woman, Shadi Tabibzeh, which is another roadblock in Iran as if you are a woman, even if they approve your content, you are never allowed to perform with your voice unaccompanied by other voice.[19] SheCan's music is, however, distributed on all the digital streamers internationally, and Hariri himself states Spotify as his platform of choice for music discovery—a stance which, at the very least, casts an interesting light on the criticism that platform garners from Western musicians and fans. Most Western music fans acknowledge that Spotify is taking advantage of the market, but for Parsa, a 22-year-old multidisciplinary new media artist who grew up in Tehran and currently lives and studies in Vancouver, British Columbia, Spotify, in particular, has served as a catalyst for Iranian music fans.

> Spotify is best. They have playlists, and it's easier to expand your taste. You've got the whole selection there, otherwise, you have to search for specific songs online to download. That's why people would rather pay for Spotify or even Apple Music. If you can't find someone outside of Iran to pay for it, you can buy gift cards from the sites and pay for your account that way. (Moayeri, 2020)[20]

Similarly, Salma, a 20-year-old college student living in Tehran, who also takes advantage of Spotify's capabilities, but differently than Parsa, uses a VPN to access her premium account, which a relative living abroad set up for her. She names The 1975, Panic! at the Disco, Mac DeMarco, My Chemical Romance, Fall Out Boy, and Girl in Red as some of her favorite artists. She curates her own playlists but is particular about who she shares them with, wanting to make sure they are worthy of the share. As I wrote in 2020, "Her 'legal' Spotify Premium is a recent acquisition. Before that, she'd hack into Spotify users' accounts abroad and try to listen to music as anonymously. Free Spotify only works for about a week in Iran then flashes the message that it is not available in that country. Most people in Iran don't have access to Spotify Premium. Instead, they can join any number of well-known music channels created by users on Telegram and follow the content. There is also a Telegram bot that can search for a song or an artist, scanning the web and finding where it can be downloaded for free, then send you the link."[21]

Parsa, Salma, and Tabibzeh's experiences show the many ways that new technologies, particularly streaming technology like Spotify, have been quite beneficial for Iranian musicians and music fans alike. Being able to see what they can't have on the internet is, in large part, responsible for those living in Iran to want access to it. At the same time, it has provided an accessible way to circumvent government-imposed restrictions. Interestingly, the restrictions imposed by the government have created gender equity among Iranian musicians, who are finally all facing the same barriers to entry. If anything, the restrictions have provided a robust support system among all musicians, who not only defend their female contingent but now take them very seriously.

In short, musical artists in Iran have created their own subcommunities, which live beneath the surface of the Islamic Republic. And now, in addition to allowing kids to dub cassettes, technology helps young Iranians see new music, as well as hear it. Once again, there is an underground network—much like we had in the 1980s—made all that much easier to tap into because of technology. Now when artists in Iran perform—which can be at someone's home—they spread the word to their fans online.[22] Says Parsa, "You get in touch with the artist, find out where they're performing, go to the space with flasks of bootleg booze, buy tickets at the door and enjoy the concert." It may well be a concert that's more or less illegal. But according to Salma, "we learn from a very young age how to get around the rules and how to break the rules."[23]

As all fans of rock music know, rules were made to be broken. Perhaps even the leaders in the Islamic Republic have grasped this simple truth—and why their grip on the dissemination of music has loosened. During my post-Revolution years in Iran, which ended permanently in 1986, the thought of a musical performance of any kind beyond the rhythmic reading of the *Quran* was unheard of. Now, there are licensed cafés where live music is performed, theaters and art galleries with plays, musicals, and operas, electronic music concerts with techno, deep house, and ambient music. Apparently, as long as it's not hip-hop, there is a chance it will be greenlighted by the authorities. "That is one genre they're afraid of," says Parsa. "Rapping is about arguing and sharing your beliefs."[24]

Hip-hop notwithstanding, it has been heartening for me to hear all these stories of Iranian young people gathering, collecting, and disseminating popular music, creating the same kind of underground networks that I participated in myself, many years ago. On the night I watched my cassette store burn, I imagined that the music itself was being taken away forever. But music doesn't burn; it's indestructible. Moreover, the impulse to collect it and consume it and create it is clearly something no revolution or religion can prevent; even the act of trying to do so is counterproductive. I may have been heartbroken on the day the store burned, but taking the long view, the role of cassettes and popular music in Iran is a case study in just how powerful—and how utterly enduring—popular music is. Moving forward into a more technological age makes it seem as if all the artifacts that bear its mark—the records, the cassettes, and CDs—are transient, as are the stores that distribute it. But the truth is, they have already served their purpose—and it was revolutionary.

Notes

1. "Ayatollah Khomeini." In *World History: The Modern Era*, ABC-CLIO, 2022. https://worldhistory.abc-clio.com/Search/Display/314864 (accessed January 17, 2022).
2. John Kifner, "Khomeini Bans Broadcast Music, Saying It Corrupts Iranian Youth," *New York Times*, July 24, 1979.
3. Ibid.
4. Hirschfield?
5. Simon Adler, "The Tapes that Sparked the Iranian Revolution," *On the Media*, November 26, 2021. https://www.wnycstudios.org/podcasts/otm/segments/tapes-sparked-iranian-revolution-on-the-media (accessed February 15, 2022).
6. Oriana Fallaci, "An Interview with Khomeini," *New York Times*, October 7, 1979.
7. Neil Hamilton, "Reza Khan." In *World History: The Modern Era*, ABC-CLIO, 2022. https://worldhistory.abc-clio.com/Search/Display/316363 (accessed March 27, 2022).
8. "White Revolution." In *World History: The Modern Era*, ABC-CLIO, 2022. https://worldhistory.abc-clio.com/Search/Display/310041 (Accessed March 27, 2022).
9. Haideh Adab, personal communication, January 1, 2022.
10. Ibid.
11. Farid Jalinous, personal communication, January 9, 2022.
12. Ibid.
13. Sussan Safari, personal communication, January 16, 2022.
14. Ibid.
15. Nancy Stockdale, "Postwar Tension: The U.S. and the Middle East." In *American History*, ABC-CLIO, 2022. https://americanhistory.abc-clio.com/Search/Display/376703 (Accessed January 17, 2022).
16. Joyce E. Salisbury and Andrew E. Kersten, "Islam in the Modern World: Overview," In *Daily Life through History*, ABC-CLIO, 2022. https://dailylife.abc-clio.com/Topics/Display/1427144?cid=41 (accessed January 17, 2022).
17. "Ayatollah Khomeini: Speech on the Uprising of Khurdad 15 (1979)." In *World History: The Modern Era*, ABC-CLIO, 2022. https://worldhistory.abc-clio.com/Search/Display/354614 (accessed January 17, 2022).
18. Wired, "I'm with the Banned: Kasra V," *Partner Content. Wired Insider*, 2017. https://www.wired.com/brandlab/2017/07/im-banned-kasra-v/.
19. Yara Elmjouie, "Alone Again, Naturally: Women Singing in Iran," *The Guardian*, 2014. https://www.theguardian.com/world/iran-blog/2014/aug/29/women-singing-islamic-republic-iran.
20. Ibid.
21. Ibid.
22. Ben Gittleson, "Inside Iran: Music Proliferates Underground," *ABC News*, November 5, 2013. https://abcnews.go.com/International/inside-iran-music-proliferates-underground/story?id=20794160.
23. Lily Moayeri, "Tech Is Music to Their Ears," *Los Angeles Times*, March 8, 2020.
24. Ibid.

Part III

Sites for Fandom and Performance of Subcultural Capital

15

Making Indie Noises in the Corporate Outlet

How Hanging Around and Working in Small Record Shops in Aotearoa New Zealand Changed My Life

Roy Montgomery

Introduction

In the present day with instant 24-hour electronic connectivity, it often requires some imagination or recollection to understand how just decades ago the local record shop acted as a principal, if not critical, clearing house or information bureau for connecting producers of independently created and manufactured music with consumers. In the case of Aotearoa New Zealand (ANZ) the situation was more extreme. Until the mid-1980s import restrictions and tariffs meant that it was difficult and expensive to bring records into the country. Major labels such as EMI, CBS, Polygram, and WEA either had their own vinyl pressing plants, or they subcontracted pressings from supplies in Australia. But it was still relatively expensive to do special orders from Australia and marketing managers and sales representatives here were rarely interested in the promotion of small quantities of punk or post-punk or other specialist music such as reggae unless they had been included in local releases—even if they were offered to record shops with little or nothing in the way of promotion.

Independent radio stations were limited to a half-dozen broadcasters around the country. While they might play the odd obscure import, they still depended on what the big record companies chose to release. Music magazines such as *New Musical Express* and *Sounds*, which were then fully involved in indie music writing, were effectively the only "listening post." Importing of magazines and vinyl was slow, whether privately or via a small record shop, often taking up to three months to arrive by surface mail. Small-run, independent records released in the United Kingdom, the United States, and other countries were often simply unobtainable here, and it was frustrating reading about exciting things you could not hear.

I was an indie fan, musician, record shop assistant, and manager in both commercial and independent stores during the late 1970s and sporadically across the 1980s and 1990s. This chapter is an auto-ethnographic account of trying to champion alternative music in an often conservative commercial setting. At times I felt like I was trying to beat capitalism at its own game by slipping indie music through the corporate back door of an otherwise thoroughly commercial and corporate retail outlet. This may have been something peculiarly "kiwi" or "antipodean" as both New Zealanders and Australians have a tendency to favor the underdog in any competitions.

Otautahi Christchurch in Context

Otautahi Christchurch is a city of some 400,000 people located near the eastern coast in the middle of the Te Waipounamu or the South Island in Aotearoa New Zealand. The Canterbury settlement was established in 1849 by the Canterbury Association, based in London, and Otautahi Christchurch was its capital town. It owes its existence to the millions of acres of farmland bought or seized from the indigenous Māori, depending upon one's view of how colonialism and European colonization of the Pacific have tended to operate by European settlers to establish an agricultural economy and society. The general rhetoric around colonization in Australia, New Zealand, and North America was that it would be a great social leveler, one's social status being determined by merit and hard work rather than inheritance or tradition. However, the most well-off settlers to this country tended to see themselves as a kind of southern gentry transplanted from the English upper or merchant classes and nowhere was this more pronounced than in Canterbury[1] (Eldred-Grigg, 1980). That said, there has always been a sense of do-it-yourself or making-do with limited resources as a kind of levelling of pre-existing class differences. This independent or "can do" spirit has permeated all sectors of society and culture has not been confined to mainstream economic activities.

Over time the city has developed, along with Otepoti Dunedin, a reputation in the indie music world since the 1980s as the birthplace of the Flying Nun label, and it has enjoyed a high yield of major artists, for its size, most recently producing international songwriters and performers such as Lorde and Aldous Harding. But in the 1960s and 1970s Otautahi Christchurch was still a small and rather conservative main center. It had a population of only 295,000 in the late 1970s.[2] It was often referred to as, and it remains for many, the "garden city" of Aotearoa New Zealand reflecting a sense of pride in the domestic single-family home and garden. It was later branded as "The City that Shines" by the local city council as if the main virtues of the city were to be found in the blue skies and blue waters outdoors rather than within arts and other cultural sources, including music. In essence it was a small, relatively young urban center that served the rural hinterland.

Up until the 1950s the social veneer of the city was highly conformist and very white European (or *pakeha* in the language of indigenous Māori) compared to other towns and cities in the North Island. Cultural life, for men at least, revolved around rugby, horseracing, and beer. Urban centers tended to service the rural economy and were ingrained with conservative attitudes and tastes. Right through into the 1960s news traveled slowly through a limited number of government-controlled or owned media outlets. Television broadcasting was limited to an average of nine hours per day, and there was effectively only one national television channel and very limited regional radio broadcasting funded by national government.

This is not to say that progressives or liberals did not have their counterparts here, and ironically, perhaps because Otautahi Christchurch was so conservative, leftist political and antiwar organizations often set up shop in the city before doing so elsewhere out of frustration or defiance (Locke, 1992). From the mid-1950s, this city was also the staging and supply base for the US Navy's Operation Deep Freeze Antarctic research and exploration program. Their base at Christchurch International Airport was not subject to New Zealand immigration and customs controls, which meant that goods and materials of various kinds, such as records and controlled drugs, could be brought in without border inspection. Also, some of the best writers and artists of the twentieth century in this country, such as painters Rita Angus and Colin McCahon, novelist Ngaio Marsh, poets James K. Baxter and Denis Glover, and composer Douglas Lilburn, have come from Otautahi Christchurch so the city has always punched above its weight and has a reputation as a cultural hub but often only for those in the know[3] (Simpson, 2016). Yet for all intents and purposes, it looked like a conformist place, a reputation that survives to this day.

The Record Bar and Beyond

The independent record store as a purely owner-operated concern did not exist in Aotearoa New Zealand until the 1970s. There were stand-alone record shops, but they were the retail branches of international companies such as HMV with their World Record Club outlets or small record shops that sold mainly 45s. There were musical instrument and/or Hi-fi stores which had a number of record bins, but mainly sold a range of instruments or stereo equipment. For youngsters in Otautahi Christchurch such as myself, the go-to places for records were the "record bars" housed within department stores, the EMI Record Shop and its companion store World Record International, or a small shop called the Record Factory, which was independently owned but mainly had general release records and chart toppers. A common ritual was to go into town after school or on a Friday night and hang out in a record bar and listen to singles in a listening booth. There were a number of secondhand record shops and hippie emporiums that dealt in secondhand goods including records, but I did not dare go into them as they seemed too weird for my tastes. At this stage I was not aware of alternative or independent music. I was exposed to popular culture through mainstream radio and television and until the age of around eleven did not question it.

I found my way to fully alternative music through a combination of exposures. At twelve I bought a mail-order subscription to the Philips Record Club in 1971. This "club" had been initiated here by a Philips senior manager in 1962 in order to cut out retailers and it meant that you were sent titles which were chosen for you and which you could return free of charge within a certain period if they were not to your taste. As a "teen hits" category subscriber for me this meant exposure to bands like Black Sabbath, T-Rex, and the gender-bending explorations of David Bowie and Lou Reed. Second, my listening to the weekly relayed overseas radio broadcasts of *Top of the Pops* by Brian Moore of the BBC (John Peel's shows were not available) introduced me to art-rock bands like Roxy Music. This also piqued my interest in things that were not mainstream, and I could correlate these listening experiences with written material such as reviews and interviews published in British music magazines such as the aforementioned *New Musical Express* and *Sounds*. These arrived only by sea mail in the 1970s meaning that they were always three months out of date, but they were critical lifelines to a world of different music. But after a couple of years feverishly devouring such magazines my interest flagged in the mid-1970s as rock music descended into the self-congratulatory, bloated, stadium rock dominated by the Eagles, Pink Floyd and Emerson, Lake and Palmer, and music magazines fawned over these self-indulgent bands and artists. I stopped buying the magazines, though I was still a long way from knowing what "indie" was at this point.

I did not fully discover the music of obscure or alternative cultures until I went to university in the late 1970s. I began a bachelor of commerce degree in 1977, which I promptly set about failing. Bored with accounting and economics courses I drifted toward the record bar section of the University of Canterbury Bookshop, part-owned by the university students' association, where I received free lectures on my poor, incomplete musical tastes. To my eyes the bookshop appeared to be staffed entirely by an assortment of weird, eccentrically dressed "older" people in their twenties and thirties.

This weirdness reached its apogee with the two men who, albeit in a desultory fashion, seemed to "run" the record counter. Tony Peake and Warren Pringle looked like close relatives of Lou Reed and Johnny Thunders respectively. Their punk-style clothes and hair appeared to change every week in what seemed like acts of intentional provocation. It was, however, clear that they had been hippies in their youth. There was often incense on the counter and they smoked Indian cigarettes while talking to customers. Because it was 1977, Tony and Warren played at being punks and, again in a desultory fashion, played in punk bands, but they had eclectic tastes which they felt obligated, as is the case with almost every person who has ever worked behind an indie store counter, including myself, to pass on to all and sundry.

On the face of it having two men wise-cracking behind a record counter fits only too well the Nick Hornby/*High Fidelity* model of male-dominated indie record shop culture. Yet Tony and Warren were both very camp, if not openly gay, which gave the place a much more gender-neutral, non-trainspotter vibe, and they championed a wide range of alternative music and culture that included dub and African music. They also had the unusual advantage as record shop assistants of using the bookshop's generic import license to bring in overseas pressings from a variety of sources. They could also get airmailed copies of the *New Musical Express* and *Sounds*, the editorial approach of both music weeklies having been transfixed and transformed by the emergence of punk rock.

As an eighteen-year-old without older siblings this was like finding a magical news desk from an alien planet. I took to heart their recommendations far more seriously than I did the prescribed academic course work. I did not realize then that the tables would be turned within a couple of years and I would find myself working behind record counters at least three times over the next twenty years, where I would be an advocate for music I considered alternative or indie.

The Industry Environment in ANZ in the Late 1970s and into the 1980s

My first entry into working in a record shop rather than just hanging around one dates only from September 1979 and ends in January 1982, but it is worth providing some context for the recording industry, record manufacturing, and the record and tape import and export situation in Aotearoa New Zealand at this time. Import restrictions, although exclusive to ANZ, were regarded in the 1960s and 1970s as overprotective. Balance of payments crises in 1974–5 and 1979 and a central government move to establish light manufacturing industries drove or entrenched some of these protectionist measures. Furthermore, many products were subject to import licensing and this extended to luxury goods such as records. The way that the major commercial record companies overcame this was to set up their own manufacturing plants, which EMI and Polydor did. Alternatively, labels could license that manufacturing to EMI or Polydor which WEA, RCA, and the Australasian conglomerate of A&M and other labels, Festival Records, did. Or, they could use the import licenses already granted for musical and recording technology on which to piggyback record products. Imports of records and tapes in small quantities by private individuals were permitted, but they were still subject to a considerable customs tax. Some institutions, such as universities, all of which had bookshops attached to them for the sale of imported textbooks, were able to use a liberal interpretation of what constituted "educational materials" to bring in records. Whichever way one looked at it through, bringing in imported recordings was expensive, and titles ended up priced at well over twice what it would have cost in Europe or North America. Small, independent labels, therefore, had little hope of selling directly into the market here.

Not only were strict import controls a fact of life for would-be importers, tariffs applied to commercial importers of records and tapes and customs duties were applied to individuals. And while tariffs, duties, and sales taxes were not unique to this country, they tended to be dauntingly high—the latter at a rate of 20 percent in the early 1970s. However, tightening of import restrictions and an increase from 20 percent to 40 percent sales tax on records and music from May 22, 1975, under a Labour government were nothing short of punitive. The government wanted to discourage the exportation of overseas funds, and music products were mistakenly assumed to be imports when in fact most records and tape cassettes were manufactured in Wellington at EMI and Polygram pressing plants.[4]

There was a great deal of pushback from artists, supporters of the arts, and the record industry but an incoming conservative national government late in 1975 did nothing to correct the error and, in fact, doubled down on the matter with Prime Minister Robert Muldoon famously saying that pop and rock music were not "culture" and therefore should not be given any leeway. So the tax remained until 1984 when, ironically, an incoming highly neoliberal Labour government was reinstated and was keen to remove all tariffs and barriers to free trade. The 40 percent sales tax was abolished and a Goods and Services Tax on retail goods and services of all kinds was implemented, initially at 12.5 percent and later raised to 15 percent.

The Rise of Independent Stores

The removal of the punitive sales tax and import tariffs opened the door for new players entering the market as record shop proprietors, and, importantly, it allowed existing secondhand record shops to augment their stocks by bringing in their own supply or ordering from independent distributors. The secondhand shops had done well under the high tax regime because many people could not afford to buy new records and had to wait until they recirculated. Now secondhand shop owners were able to sell local new releases and import new records themselves, which many of them did. Some of them transitioned from retailers specializing in hippie tie-dye T-shirts, incense, candles, and clothes with a few record bins into fully fledged indie record stores. Riverside Traders and the Record Joynt, both started in 1975, morphed into Galaxy Records and Echo Records, respectively, in the mid-1980s. Indeed, the mid-1980s saw a proliferation of new and second retailers entering the market up and down the country and Otautahi Christchurch was no exception[5] (Gilbert, 2017).

We were still dependent on corporate retailers such as EMI and the one genuinely independent record store, the Record Factory, which opened in 1973 for what could be bought, but these shops sold only new, mainstream product. But slowly, even these shops cottoned to the fact that something was happening in the alternative music world and there was money to be made. What they needed were young people in funny, skinny clothes and not just ex-hippies to select and peddle new alternative wares to customers.

The Crown Crystal Glass Factory and Small Wonder Records

My faltering university studies could not continue forever and at the end of 1978, I decided to quit pretending to be a student and look for a job. To force the issue, with my last student allowance and what was in my savings account I bought a genuine 1970s Fender Telecaster Custom Deluxe guitar, an expensive 100-watt amplifier, and a distortion pedal with the idea of starting a punk band that sounded like Wire, the Buzzcocks, and the Saints. At the start of 1979 through a friend from high school days, Tony Green, whom I didn't know that well, I heard there was an evening shift job at a glass manufacturer and packing plant. Although this may seem tangential to an indie store history it was through working the shifts and killing otherwise monotonous time we talked almost exclusively about music. While feeding glass products along conveyor belts into baking or annealing furnaces, only to pack them up again, we soon recognized our shared, deep interest in alternative music.

Over the deafening sound of furnaces and forklifts we talked about the latest reviews in the *New Musical Express* and *Sounds*, UK rock magazines that had taken on a new lease on life in the late 1970s, thanks to punk rock. We looked enviously at the classified ads at the back of these magazines wherein indie stores and indie labels ran lengthy lists of 45s and LPs which were unavailable in New Zealand. Eventually, the frustration became so great reading about the "Singles of the Week" in both magazines that we targeted a couple of English mail-order places with enquiries about shipping 45s to us. Once we got over the pain of unfavorable exchange rates, the need to buy foreign currency in closely controlled and limited amounts, the steep shipping costs, and exorbitant import duties that would be imposed upon arrival, we set about doing largely complementary orders with one particular shop and mail-order business, Small Wonder Records in Walthamstow, London, which also ran a label.[6] Tony and I routinely bombarded the mysterious "Mari," who seemed to handle all mail orders and who was unfailingly helpful and always wrote chatty notes to go with shipments, with bulk orders of 45s.

Although the releases on Small Wonder's label were, for the most part, too old-school punk for our tastes, we were both struck by what it stocked and the professionalism and enthusiasm of those running its mail-order business. It demonstrated that there was a fully functioning independent industry which was often diametrically opposed to the big labels and big distributors. Our orders with Small Wonder delivered 45s to us from bands like the Buzzcocks, the Desperate Bicycles, Scritti Politti, and the Gang of Four. Some of their cheaply made sleeves provided full details of the production costs for the record. Such packaging functioned as a call to arms to self-release material. It was a deliberate goad to the major record labels and at times the art work went as far as critiquing capitalism directly. Some of the sleeves carried slogans or images promoting the downfall of capitalist economic structures. Others railed against neo-fascist groups and right-wing governments. Having mail-order indie stores on the other side of the world like Small Wonder and Rough Trade, which effectively curated the titles they stocked according to philosophical and political principles, resonated with us. And while these records were expensive in New Zealand, whether locally pressed or imported, such DIY rhetoric and information found its mark locally. The idea of a self-released record in this country now seemed feasible. It occurred to me at the same time as it did my friend from school days, Roger Shepherd, who went on to found Aotearea New Zealand's most well-known indie label Flying Nun Records while he was working as an assistant at a relatively mainstream retail outlet called the Record Factory in central Otautahi Christchurch.[7]

Apart from sharing our mail orders and copies of the NME, starting a band which lasted into 1980 under the names Murder Strikes Pink and Compulsory Fun, we both ended up working in record shops or shops with record sections. I left first in late

1979 for a job at EMI and in 1980 Tony started on what was to be his long record shop career in a general electronics department store chain called Noel Leemings, which had a record and stereo equipment section. In 1987 Tony worked in, and later bought, a local secondhand record store called Galaxy Records. He moved to California in 1993 and, after brief stints at CD Research in Davis and the Sacramento Virgin Megastore, he joined Amoeba Records in Berkeley in 1995 and then moved to its new branch in San Francisco, where he managed more than 150 employees. Currently, he is both manager and co-owner of Amoeba Records, one of the biggest, longest-running indie stores in the United States.

Working for EMI

While Tony and I were content to exchange banter and our latest import purchases working at the glass factory, we both knew this was a transitional phase for two college dropouts. As mentioned earlier I left first after seeing an advertisement for a record shop assistant position in an EMI Record Shop in the central city in September 1979. At the time, EMI had a chain of some dozen stores throughout the country and had few rival corporate retail competitors. Although EMI's image had been tarnished for punk rockers by the whirlwind signing and dropping of the Sex Pistols, I saw this as an opportunity to earn money for doing what I loved: playing records and talking about them with people, whether staff or customers.

The shop where the vacancy existed was no more than a hole in the wall, but it was in the very center of the city on the highest foot traffic street. The staff were relatively young. The manager and buyer Steve was in his late twenties and his two female assistants, Donna and Julie, were around the same age. I was twenty years old and although my sympathies lay with alternative music like punk and post-punk, I had a good general knowledge of pop and rock music. I was hired for reasons I never fully understood and, despite my dissimilar musical tastes, we got on well. The balance of two male and two part-time female staff with varying tastes was in some respects as good as one could hope for in a mainstream record store. These were the days of having glasses of wine on the counter toward the close of business on a Friday evening at the end of each week in most record shops, indie or mainstream.

Within nine months of my appointment, Steve was moved to manage another branch in the central city. Neither Donna nor Julie were interested in taking on the role of manager, and I replaced Steve as manager and buyer. While the music industry was aimed primarily at the youth market, I became one of the youngest managers in EMI's nationwide chain. Although my musical tastes lay well outside of the mainstream, I did not immediately set about turning the shop into an alternative music outpost. That would have been impossible as it was one of the most profitable stores in the chain with a stock turnover and revenue not far behind that of the leading one in the capital city, Wellington. My store was on the city's main street, Colombo Street, less than 100 meters from the city's main square or gathering place, Cathedral Square, and the high daily foot traffic was every retailer's dream. To the south, on the other side of Cathedral Square was the Record Factory, which, as mentioned earlier, was where Roger Shepherd worked. The shops were clear rivals for the disposable incomes of any potential customers whether they bought commercial or top forty releases or obscure indie titles, but it was reassuring to know that such business places could be helmed by people with noncommercial personal values.

In finding my way through the management structure of EMI and in doing bulk sales order deals in combination with some of the other branch managers, I quickly found some kindred spirits. Chris Caddick, who was marketing manager at EMI's head office in Lower Hutt at the time, just north of metropolitan Wellington, Darryl Parker, manager in the Lower Hutt retail store, and Steve Clansey, the manager of the Cuba Mall shop in central Wellington (the shop with the highest turnover in the chain), all proved sympathetic when it came to trying to raise the profile and sales of alternative bands. As shop buyers Darryl, Steve, and myself would occasionally pool our orders with other record companies for a particular indie title and receive large discounts that were passed on to customers. Our most notable success was with the local release of Joy Division's album *Unknown Pleasures* and the 12" single of "Love Will Tear Us Apart," which were released simultaneously in June 1981. With a combined order numbering in the thousands, which allowed a generous discount to buyers, we virtually guaranteed the number 1 spot for both titles in the national sales charts on June 21, 1981, with or without returns of sales being sent in from any other shops in the country.

At the day-to-day level of working behind the counter at EMI with my colleagues (whom by 1980 were close friends and associates), we set about trying to shape or improve the tastes of customers stopping well short of the kind of open sarcasm or overt nerdiness displayed by the characters in Nick Hornby's *High Fidelity*. My main partner in this mission was a very close friend, Chris Owens, who joined the staff in 1980. Chris was the cheerful music enthusiast with eclectic tastes and love of dub and reggae music while I was the arrogant and slightly off-putting guy who always wore black. In short order, the store had developed a reputation as a post-punk hangout. One morning in 1981, I arrived at work to open up and found the large glass door spray-painted in fluorescent green with the words "Roy Division." This was more a jibe about my derivative vocal sound in my band the Pin Group than it was about the shop, but it illustrates how that EMI shop had lost its status as just another chain store and had now come to resemble the record counter at the University of Canterbury bookshop where, ironically, I would return to as manager while studying for an arts degree several years later.[8]

Another key advantage to working for EMI, apart from staff discounts on records, was access to the company's pressing plant in Lower Hutt, one of only two pressing plants in the country.[9] After learning about the DIY ethos and process, I was convinced that paying for your own pressings and sleeves and selling records independently was crucial for an indie band. I arranged via phone calls to the EMI plant for the Pin Group to have its first two seven-inch 45s pressed there in 1981 and shipped directly to us to assemble and sell either through mail order or through Flying Nun Records that agreed that our first single would carry their first catalog number.[10] I also arranged for the Pin Group to record their next release in the lavish EMI studios, graced by the likes of Kate Bush and Cliff Richard, in Lower Hutt at a generous discount rate using the most experienced recording engineer in the country at the time, Frank Douglas.

These aspects of my time at EMI were all serendipitous but one of the main motivations for working there was to earn enough money to travel to the UK, a place I considered a musical Mecca. The EMI job provided the direct financial backing for that trip; it also provided me with some contacts via colleagues in other shops and I left for England in February of 1982. By this time I had already been ordering vast amounts of records from the Record and Tape Exchange in Notting Hill in London. Their chief buyer, Steve Cross, was a New Zealander who had gone to the same high school as me (as had Tony Green and Roger Shepherd) and was the partner of my girlfriend's cousin.[11] Steve and Linda put me up on my arrival in London and helped me find my feet.

Just as importantly in my web of connections to the indie store world was Steve Clansey, from the EMI shop in Wellington. He'd told me time and time again that if I ever went to Liverpool, where he was from, that I had to look up his great friend Geoff Davies who ran the most iconic indie store in the Northwest, Probe Records in Button Street, Liverpool 1, just around the corner from the site of the Cavern Club made famous by the Beatles. I did and I not only ended up helping out at the shop occasionally and traveling with Geoff as he took records on the Probe label around shops in the Northwest, I also lived with Geoff and his wife, Annie, and son, Jessie, for months at a time throughout 1982. I felt like part of the great indie family in the most literal sense, and it has left a deep impression on me to this day. The fact that I came from an out-of-the-way place like Otautahi Christchurch seemed to make no difference: if you passed through the doors of an indie record shop anywhere you effectively belonged to the same tribe of circumspect consumers wary and weary of the excesses of capitalism.

Conclusion

This autobiographical narrative illustrates the paradoxical nature of "indie" and the creation of cultural objects (records) for public if not mass consumption. In some parts of the world the line between the independent record shop or label and corporate capitalism may indeed be more clearly delineated and oppositional. In a small country like Aotearoa New Zealand, where at some level, everyone did (and does) seem to know everyone else, it was impossible and impractical to be purest about cultural production. Perhaps the out-of-the-way location here afforded more opportunity for blurring the lines as the real corporate head offices of record stores and labels were in London or New York or Los Angeles and the people in charge in the colonial outposts could turn a blind eye to a little "indie-fying" of the corporation.

At the heart of the concept of indie lies another paradox and in the case of the indie record shop it is vividly demonstrated in the relationship between staff and customer. In all my transactions in this environment, whether as buyer or seller, what was at stake was an independence or freedom in thought but that could only play out interdependently through, then, face-to-face discourse. I realize now that my indie store days were just another form of intense, university-like study, where I was trying to learn to think and act more independently but through over-the-counter dialogue. And just as attending university is to some extent merely a rite of passage, albeit a privileged one, this rite of passage has left a much greater mark on me. Since my first forays into alternative bands and indie record shop employment, I have retained a presence in indie record stores until the present day through the release of some thirty solo or band recordings on indie labels such as Flying Nun, Majora, Drag City, Kranky, Drunken Fish, VHF, Badabing, and Grapefruit Records. So although I no longer stock the record bins of indie stores as a manager or buyer, my own creative outputs are available in these outposts and that is where I would prefer them to be found. I still have no interest in "making it big."

Notes

1. S. Eldred-Grigg, *A Southern Gentry: New Zealanders Who Inherited the Earth*, X (Wellington: A.H and A.W. Reed, 1980).
2. R. Shepherd, *In Love with These Times: My Life with Flying Nun Records*, X (Auckland: Harper Collins, 2016).
3. P. Simpson, *Bloomsbury South: The Arts in Christchurch 1935-1955*, X (Auckland: Auckland University Press, 2016).
4. https://www.audioculture.co.nz/scenes/sales-tax-on-lps-in-1970s-nz (accessed October 21, 2021).
5. P. Gilbert, *The Sad Demise of the Christchurch Record Shop*, 2017. https://nostalgiablackhole.webs.com/christchurchrecordshops.htm.
6. http://www.musiclikedirt.com/2013/07/22/pete-stennett-of-small-wonder-records-interview-part-1/. Interview with one of the label's founders (accessed December 25, 2021).
7. Roger started working at the Record Factory in late 1976. For a description of his time there, see Shepherd, *In Love with These Times*, 25–31.
8. I worked there from 1987 to 1988 while studying for a bachelor's degree in Russian language and literature. After completing the degree, it was suggested that I study for a master's degree in natural resource management (which I did) and has led to a largely unintentional career in academic as a lecturer in environmental management.
9. Roger Shepherd used his favorable connections while working at the Record Factory to establish favorable terms for getting records pressed when he launched the independent label Flying Nun Records in 1981.
10. The first Pin Group single "Ambivalence" carries the catalog number FN001, and the second one "Coat" carries the number FN003. Original copies sell for huge sums.
11. In another turn of the indie store wheel when I was on another overseas sojourn in late 1994, I found myself in London looking for work. I walked into the Notting Hill Gate headquarters of the Music and Video Exchange and applied for a job. You had to sit through a lengthy written test of your musical knowledge. My scores were low on current music but very high on earlier decades, especially obscure indie bands whose original releases changed hands at very high prices. I was hired and on a number of occasions for the six months I worked there, I was on the counter of the rare records section where Steve Cross had worked more than ten years earlier.

16

Rip Off Records (Hamburg) and the Microhistory of Capitalism

Karl Siebengartner

There were interrelations within "Rip Off". It was a strange bunch of initiatives, undertakings under one roof. It once was the production of badges, the distribution of badges, individual mail delivery of records, the shop, gigs. Afterwards, we also did wholesale; everything was somehow completely entangled. From the outside, it seemed like a big enterprise. When people arrived in Hamburg, they were flabbergasted because this all happened in a rattrap, and I booked the gigs in my bed.[1]

This is how the journalist and label owner Alfred Hilsberg described what he and Klaus Maeck, the owner of Rip Off Records, were faced with in 1982. Some saw the record store as a bad corporation. Yet, Hilsberg and Maeck hardly made a living from this set of DIY actions. Maeck recalls that he did not save up money to pay taxes, which finally led him to declare bankruptcy in 1983.[2] Despite the tragic ending, Rip Off Records in Hamburg was one of the first record stores in West Germany that solely focused on punk and new wave. Punk in West Germany arrived roughly at the same time as the left-wing terrorism of the Red Army Faction (RAF) culminated in the German Autumn of 1977. Some punks sympathized with this group. However, these relations were complex.[3] In this climate, punks in Hamburg often clashed with the police.[4] They did not know how to deal with the new and strange-looking youth culture. There existed one other record store in Hamburg with a similar idea that was called *Unterm Durchschnitt*.[5] Stores like these partly profited from a vacuum after the disco boom in the late 1970s. The West German record industry could not come up with a new trend, and various new and more affordable media like VCR and home computers posed a threat to general music consumption in the country. They lost a fourth of their market share from the late 1970s until the mid-1980s.[6] That is probably partly why the owner Klaus Maeck sold records which he imported from England. Furthermore, he stocked new singles and LPs from the formative punk scenes in the *Bundesrepublik* and a vast array of fanzines from all over the country. The shop soon became a hangout for punks and well known in West Germany and beyond.

As already sketched, Maeck and Hilsberg were highly controversial figures among punks in West Germany because they, on the one hand, provided the actors with much sought-after goods. On the other hand, punks perceived them as being part of capitalism and mass culture. In that way the ironic store name Rip Off turned out to be true for some punks, although Maeck tried to do the opposite by building up an alternative infrastructure. This punky joke was not funny anymore. That is why this chapter will argue how the idealized and physical space of the record store itself was highly important for punks in the *Bundesrepublik* but at the same time the entrepreneurial spirit of the people involved with it posed a threat to self-conceptions of the punks. For them, the name Rip Off became synonymous with everything they hated about capitalism. In that way, I link the subjects and their perceptions with a history of mass culture through the lens of the record store Rip Off.

Implicitly, Hilsberg also stated that punks saw him and Maeck as capitalists who highly profited from the goods they sold. It is interesting to follow that idea and see how views of capitalism were negotiated between punks and the likes of Maeck and Hilsberg. In order to tackle such a perspective, we must conceptualize capitalism not only as economic relations within societies but also as a cultural phenomenon itself. Capitalism is a cultural formation that ties producer and consumer together. The historian Thomas Welskopp rightly sees consumption linked to forms of labor because choosing and buying items consume time, and effort must be put in hunting down goods. Consumption is far from being a passive pastime.[7] That is what German punks needed to go through to obtain records. By focusing on these everyday practices of tracing down records, buying them in Rip Off Records or ordering them via mail, we can observe capitalism on a micro level. By zooming in on this one shop and the different perceptions and practices of it, I will draw on the ideas of the method of microhistory.[8] A "microscopic analysis"[9] will help to uncover the cultural nature of capitalism. Reducing the scale of observation to Rip Off Records also grants us a glimpse into the beginnings of punk in West Germany and its contested meanings. The combination of a broad understanding of capitalism and a focus on Rip Off Records will show how controversial profit-oriented business practices were in the early days of West German punk. Hence, the case of Rip Off Records reveals how ambiguous perceptions of capitalism get once we focus on a small scale and try to enclose many voices and sources.

Beginnings

Prior to beginning his shop, Klaus Maeck already sold books. He also imported badges from London. On one of his trips, he got a travel ban for the possession of pot. He speculated that the real reason for the ban was that he was somehow seen as being connected to the left-wing RAF terrorism in the late 1970s. In an interview, he assumes that his magazine *Cooly Lully Review* was on a "blacklist" and that the reasons for the ban were in fact political and not because of drug possession.[10] In 1978, he placed a small ad in the Hamburg-based German music magazine *Sounds*, in which Hilsberg started to write about punk from that year onwards. Maeck advertised "a big variety of punk-badges, printing of own motives +++ PUNK-maga-fanzines—also for distributors—sells RIP OFF Feldtstr. 48, 2 Hamburg 6. Request a list by sending sending DM 1,— return postage."[11] The magazine was an important medium of disseminating punk in West Germany.[12] Therefore, it is no surprise that Maeck advertised in this magazine and not in the glossier *Musikexpress*. Maeck never hid his business interest in selling punk accessories and already made that clear in this ad. In fact, punks at first even embraced the opportunity to buy badges. The Bremen-based fanzine *Der Schunt*, for example, lists Rip Off as a cheap and nice alternative to obtain them instead of buying punk gear at shops which only wanted to cash in on punk. Funnily enough, they also list *Cooly Lully* as the go-to alternative for fanzines, which was also run by Maeck.[13] Clearly, at this point, the punks saw Rip Off as a genuine enterprise which helped to spread the word of punk. Some, like Hollow Skai in Hanover, also wrote about the choice of German fanzines one could buy at Rip Off.[14] *Preiserhöhung* advertised that Rip Off was the sole distributor of the fanzine.[15] Rip Off and therefore Klaus Maeck were already associated with punk before he even started the record store. In this early stage, he provided punks with fanzines and badges and those valued his service.

The Opening of the Record Store

The rise in popularity of the goods among punks eventually led Maeck to open a store in 1979. The news spread and Hollow Skai in Hanover was very excited: "ON 1 April (no joke) MAX RIP-OFF [as Maeck was sometimes called, K.S.] opens his shop full of tin (-badges) & other shit (books, songbooks, magazines, fanzines, poster). But also, cloth (T-shirts and so on) and a small selection of plastic discs (singles + LPs)."[16] Nevertheless, Skai also thought that opening up the shop was a bit of a joke. He ironically wished Maeck "all the money that you deserve—which is plenty."[17] In that statement, the future accusations of capitalism were already hinted at. Later that year, ads of the record store in different fanzines cropped up. There the possible buyers could read that Rip Off was selling "NEW MUSIC PUNK + NEW WAVE FROM GERMANY ENGLAND + USA."[18] The store opened from Monday until Friday from 2:00 until 6:30 p.m. and from 11:00 a.m. until 2:00 p.m. on Saturdays. In an ad for the store appearing in the Hamburg fanzine *Pretty Vacant*, Maeck mentions names of the first German punk bands who already had a single at that point.[19] By placing the ads for his store in 1979, he intensified the effort to gain a share in the punk market. His advertisement was very specifically catered to punks in West Germany. Ads in fanzines directly spoke to those who eagerly searched for the new music.

That was only one strategy of how Klaus Maeck sought to promote records and his store. He also placed ads in the music magazine *Sounds*. There he praised new German records with superlatives: "Brand-hot! Brand-new! Brand-urgent!"[20] Furthermore, he had vending tables with his products at gigs and festivals, most famously at events that were put on by Alfred Hilsberg in the *Markthalle* in Hamburg. For example, one punk from Augsburg, in southern Germany, remarked that Maeck had a stand at the

Geräusche für die 80er festival on December 29, 1979.²¹ This was also a way to sell vinyl records, badges, and other punk-related commodities. The entrepreneurial spirit of Maeck led Alfred Hilsberg to write about him. Hilsberg counted him as one of the "doers" ("Macher") within the early West German punk scenes. The journalist gives a clear insight into the organization of the record store and the affiliated undertakings. The punk Eugen Honold, publisher of the fanzine *Pretty Vacant*, sometimes helped in the shop and Hilsberg characterizes Maeck as a communicative individual:

> His store and the distribution of homemade and imported badges walk the thin line between commerce and communication. At the moment the agile side prevails: "Even when you mail order you are corresponding with people, and in the shop, it is even more direct." The shop evolved into a punk-department store that is probably unique in West Germany: from US new wave magazines to t-shirts and import records from the US and England to two dozen fanzines you will find almost everything that you need for the outfit and for information.²²

What Hilsberg describes as communication between punks and Maeck are simple transactions that fall under capitalist practices.

Soon after the installment of the shop, however, problems arose. The Hanover-based band *Rotkotz* recorded an LP with the help of the English band *Pop Rivets* and wanted to distribute it in West Germany. Hollow Skai negotiated a deal for the band with Maeck and Rip Off to sell the LP cheaper. Normally, the band would have sold the disc for DM 10 but Rip Off got it for DM 8 under the condition that Maeck would sell the album for DM 12 in his record store instead of DM 16. When the band found out that the record was sold for the higher price, they were pretty pissed off with Rip Off. Hollow Skai documented the whole incident in an issue of his fanzine *No Fun*. There, he even provides an ad where the *Rotkotz* LP is advertised as costing DM 16.²³ Furthermore, Skai was fair enough to let Maeck have a word in this incident as well. In a letter to Skai, he explains the situation: "[A]nd your clamor because of the high price came a bit early: it was a total misunderstanding, meaning it was erroneously stated and adopted in the mail-order list. There are more people working at RIP OFF and it probably went under that a special price was agreed, SORRY!"²⁴ Even if this was just a lousy excuse, Maeck admitted that the shop was thriving, and he needed more people to run it. He also states that he needs good business relations because he has to make a living from the store: "Anyway, the bare facts are: I live off of Rip Off, others (Sigi, Jacky, Alfred) make some extra money by taking care of the shipping or running the shop [. . .]. I have to stress that we are proud of this store, which is not just a consumer's shop but also a meeting point for punks."²⁵ Hollow Skai did not fully buy Klaus Maeck's reaction, but he also stresses that he did not want to call for a general boycott of Rip Off. However, he still found the business practices to be shady and thought that the band literally was ripped off by Maeck.²⁶

By 1980, such allegations of greed for profit became louder. Hilsberg was accused of profiting from punk music and his status was questioned. Thomas Buch, who published the fanzine *Limited Edition*, had "the feeling that it's all about money" and the "Hilsberg clan" runs all the activities concerning punk in Hamburg.²⁷ He also cleverly reminds his readers that Hilsberg forgot to mention one important "Macher" in his three-piece article series: "himself."²⁸ Others, like Lorenz Lorenz, called him a "monopolist" and condemned Rip Off in the same breath.²⁹ The record store had serious opponents and the punks who sensed capitalist practices were furious. For them, selling those vinyl records was not only about offer and demand. They identified themselves with those musical products and that is why they accused Rip Off of using this situation and in effect charging too much money because they were the only ones who provided the records.

The Experience of Hanging Out and Shopping in Rip Off Records

While Rip Off Records was part of controversial discussions about how profitable punk should be, there still were plenty of punks who welcomed shopping in the record store. As already mentioned, Maeck employed more people to run the shop and already saw it as a meeting point for punks shortly after opening. Rip Off became a fixed place on punk travel itineraries to Hamburg. The city was generally seen as a sort of punk hot spot and among the favorite travel destinations of punks. Other favorite destinations were West Berlin and the Ruhr area. West German punks liked to travel to these areas because you could experience a lot of things in one place.³⁰ Rip Off Records offered a hangout for punks from all over the country and was as such frequented a lot. One Hamburg punk even complained because he held the record store responsible for not selling enough fanzines. He had a hard time in selling the first issue of his fanzine *Ich und mein Spiegelbild* because "the Hamburg scene almost only consists of freeloaders who only read the fanzines at rip off without even thinking about buying one [. . .] soon I will only produce one copy and display it in rip off."³¹ For him, the record store was a central hangout for the Hamburg scene that unfortunately led to problems in distributing his own fanzine.

Rip Off Records invited punks to browse and socialize in the Hamburg city center. Many of them vividly recall how important the record store was for them. Ale Sexfeind, who later played with the German fun punk band *Die Goldenen Zitronen*, characterized the shop as an important place where he could obtain all the things he needed: "Records, fanzines, badges—they had everything that you could buy nowhere else."³² Kid P. alias Andreas Banaski, who published the fanzine *Preiserhöhung* and wrote for the music magazine *Sounds*, frequented the shop as well: "There, you were hanging around with some friends."³³ A special feature of the record store was a coffin that functioned as a decorative element, as a table, or as a record crate. Hilsberg even mentions that people sometimes were lying in there.³⁴ Besides Klaus Maeck, there were other people who helped in the shop and created the special atmosphere punks liked so much. One of them was Jacky Eldorado, who hailed from West Berlin and became a staple at Rip Off. Eldorado is sometimes referred to as "Germany's first punk."³⁵ He branded himself in that way by appearing in some of the early articles in the two well-known weekly magazines *Der Spiegel* and *Stern*. In *Der Spiegel* he appeared as a bouncer in the West Berlin club *Punkhouse*.³⁶ However, by far more important was a photo of him and Iggy Pop at a gig at the *Akademie der Künste* in West Berlin in the *Stern*. This snapshot shows Eldorado who tries to lick the thigh of Iggy Pop. The journalist and organizer of *Europe's Only Iggy Pop Fan Club* Harald Inhülsen took the photo and sold it to the magazine.³⁷ This photograph became a part of punk culture in West Germany and many punks knew who Eldorado was because of his media appearance.³⁸ He even recalls incidents where former customers remembered him from the record store and shamelessly admitted that it was easy to shoplift at Rip Off.³⁹

While that was not good for the business in general, the staff did their share so that the store was not just for shopping but also to generally hang around. Eldorado, for example, played the first single "Wir wollen keine Bullenschweine" of the Hamburg band *Slime* so often that punks got annoyed: "According to Jäckie Eldorado, he played 'wir wollen keine . . .' 23 times in a row the other day at Rip Off so that even the last punk left the store."⁴⁰ Maeck recalls in an interview that they sometimes did things like that to annoy punks. When they wanted to hear certain records in the shop, the staff played something completely different on purpose just to piss off the punks.⁴¹ They made fun of their possible customers because they knew that some of them only came to hang around. Nevertheless, other punks decidedly went there to buy records and badges. When Munich punks traveled to Hamburg for the *Geräusche für die 80er* festival in the *Markthalle*, they also paid a visit to the record stores in town: "On Friday we went to [Unterm, K.S.] Durchschnitt and to Rip Off. There's nothing much to say about Durchschnitt, [. . .] it is better to go to Rip Off a few streets away. They don't stock many records (at most 100-150 items!) but they have many fanzines, and the shop is a kind of punk meeting point."⁴² This statement shows that besides shopping the space itself played a crucial role for punks. It functioned as a kind of contact point to get to know the city's punk landscape. When visiting Hamburg for gigs it was part of the ritual to go to Rip Off Records and buy goods: "After I bought various consumer products at Rip Off, I went to the Markthalle."⁴³ Rip Off Records became a fixed place to go to when in Hamburg. Punks could gather what they searched for in the shop and meet other like-minded people.

New Endeavors and Rebranding the Record Store

While by 1980 punks made pilgrimages to Rip Off to buy records, badges, and fanzines and some of them highly criticized the business practices of the store, Hilsberg, Maeck, and Frank Z(iegert) of the band *Abwärts* ventured into new territory. Hilsberg intensified his relationship to new music from Germany by founding the label *ZickZack* Records. Klaus Maeck clarified in a letter to a fanzine in Hanover how those three people and the record store were connected: "ZICKAZACK is a label that we run, mainly alfred and then me and frank from abwärts, at the moment we finance the project through the sale revenues from the shop."⁴⁴ In fact, Rip Off Records and the mail-order business were crucial for starting the label. For promoting *ZickZack*, they used the network Rip Off already had in the punk community and sent free copies of the EP by *Abwärts*, which was the second release on the label, to punks in West Germany. Bobby Blitzkrieg from *Alles tot* fanzine in Rinteln near Hanover thanked Rip Off for the free record: "rip off sent free of charge advertising copies, exemplary idea, I got one of the samples. for me, they are the best group in Hamburg."⁴⁵ From his statement, we can deduce that this was a general practice to promote the record and the band. Indeed, another copy was sent to Munich in the south of Germany: "Received it free of charge from Rip Off—nice service to send a free copy! That is why I should write a review (And who thinks free records are a bad thing anyway???)"⁴⁶ Hilsberg and Maeck utilized their existing network to promote the label and the band.

All records by *ZickZack* were in distribution by Rip Off Records as an ad in a fanzine for the band *Abwärts* clearly shows.⁴⁷ When taking all the actions together that were linked to the name Rip Off, it is no surprise that people thought that this was a serious enterprise as Hilsberg states in the introductory quote. Badges, mail order, record store, and now a label suggested that these people were making a fortune with punks in West Germany. There were so many people who thought that Maeck got rich from running these initiatives that he had the urge to explain how Rip Off was ran and what the motives were. In the already cited letter to a fanzine he lets off a longer rant:

> people often call us (rip off) pigs, money pullers, führers etc. and that sucks because even people like you, who are active as well, don't seem to be in the know. we are several people here, who set up an enterprise within the free market economy without any big experience. that means that some people get work who otherwise would have to get a job at the docks or in a supermarket. [. . .] we also try to live off of it and that's hard. but all of the people involved have a personal interest in the new music. [. . .] the records in our store are that expensive because we have to calculate differently than govi or other chains.⁴⁸

Maeck sounded rather desperate that all his hard work was dismissed as being capitalist and that he was only in it for the money. In a way, all these undertakings were DIY in their own respect but came across as soulless mass-cultural capitalist ventures.

People got so angry with Alfred Hilsberg and the accompanying business practices that furious letters arrived at the *Sounds* editorial office. The editors answered that they were not in charge for what freelancers like Hilsberg were doing besides writing articles for the magazine. Interestingly enough, they also mention Rip Off in their answer although the writer did not explicitly refer to the record store. However, Hilsberg and Rip Off seemed to be so closely connected that *Sounds* felt the urge to justify themselves in advance: "'Rip Off' pays as much for ads as other customers do. Besides that, Alfred recently stopped working for the Rip Off store and the distribution and only concentrates on ZickZack and the work as a tour operator."⁴⁹ Maeck and Hilsberg tried to separate the different branches and organize the chaotic structure they had created. Part of the restructuring was that the record store moved house and got a new name. In 1981, the new store opened under the name *Aus lauter Liebe* ("Out of pure love") at Pilatuspool 11, which was a short walk away from the first location at Feldstraße 48. Maeck also announced the new shop in advertisements with drawings of the directions from the old to the new address. *Aus lauter Liebe* was open from Monday until Friday from 3.00 to 7.00 p.m. and from 11.00 a.m. to 2.00 p.m. on Saturdays. The record distribution stayed at Feldstraße 48 where Klaus Maeck also lived.[50] The new record store was under the supervision of Jacky Eldorado who also put out music on *ZickZack* under the name *Aus lauter Liebe*. Before opening the shop in early 1981, Maeck and Eldorado faced problems. The landlord at the new location tried to get out of the signed contract once he heard that they wanted to sell punk records. After a legal dispute, the shop could eventually be opened.[51] The shop owners got to encounter how difficult running a business could be.

Despite the new name of the record store, connections with Rip Off stayed intact. West German punks communicated that Rip Off moved:

> As a first record store Rip-Off has to be mentioned. After they got kicked out of their old shop at the end of last year (1980) [. . .], they hastily opened a new store, only about 300 meters away from the old one. [. . .] The new shop looks rather nondescript, and I personally do not like it that much. A sterile atmosphere prevails. Anyway, maybe it'll get better.[52]

The punks who wrote this quote also gave directions how to get to *Aus lauter Liebe*.[53] It is interesting that they do not mention the new name. Rip Off was famous enough at that point that the new name was not as important. Everybody knew what Rip Off stood for. Others got the name change and also thought that *Aus lauter Liebe* was inferior to the old Rip Off.[54] However, there still were punks who thought that the new record store had a good range of records and cassettes.[55] Visiting punks compared the record store to what was going on in their hometowns and raved about *Aus lauter Liebe* because they had really good products in stock: "I also went to the shop 'AUS LAUTER LIEBE'—we needed something like that here, a real punkshop: Good German and English records and most of all really cheap cassettes and a lot of fanzines."[56] This punk from Munich was really thrilled with the record store. After all, *Aus lauter Liebe* also had customers from all over West Germany but that did not go down well with others. The images of a money-counting Klaus Maeck were still circulating.[57] The accusation that Rip Off and the associated *Aus lauter Liebe* were making large profits off punks prevailed.

In the end, *Aus lauter Liebe* was not profitable enough.[58] Maeck and Eldorado moved back to the old store at Feldstraße, renovated the place, and installed a new counter.[59] One punk noted that but thought that the new counter was too big. He also states that Maeck and Eldorado tried to sell videotapes and concludes that they were really becoming like big consumerist department store chains.[60] Maeck did not save up money for taxes and amassed a big amount of debt. He was liable for the record store and the other enterprises under the name Rip Off. In 1983, he declared bankruptcy and that is how the story of Rip Off Records ends, with a simple economic failure.[61]

Conclusion

It is ironic that Rip Off Records failed because of financial problems because the punks always suspected that they made a fortune with the shop and mail order. Rip Off was an important place for the development of punk in West Germany and provided the actors with records and badges that were not readily available at that time. By utilizing a microhistorical perspective, we can see how complicated the relationship between punks and the record store was. On the one hand, it was a hangout and punks loved to go there to shop or shoplift. Some came a long way to buy records, badges, and fanzines there. On the other hand, Rip Off was seen as a capitalist or mass-cultural venture from the outset. That the people involved in Rip Off had to somehow make a living was not considered. On this micro level capitalism becomes a cultural formation that made an impression on all the people involved, those who worked to sell records and those who had trouble in obtaining these goods. Punk in West Germany and beyond must be analyzed in connection to the setting of a mass cultural and capitalist mode of thinking to fully uncover how punk functioned in societies. The record store Rip Off in Hamburg is a good example to test out such a microhistory of capitalism and see how people negotiate with themselves and others what living should look like.

Notes

1 Boris Penth and Günter Franzen, *Last Exit: Punk: Leben im toten Herz der Städte* (Reinbek bei Hamburg: Rowohlt Taschenbuch Verlag, 1982), 221.
2 Alf Burchardt and Bernd Jonkmanns, *Hamburg Calling: Punk, Underground & Avantgarde 1977-1985* (Hamburg: Junius Verlag, 2020), 58-9.
3 Jeff Hayton, *Culture from the Slums: Punk Rock in East and West Germany* (Oxford: Oxford University Press, 2022), 105.
4 Burchardt and Jonkmanns, *Hamburg Calling*, 56.
5 Ibid.
6 Christian A. Müller: "'Die nicht-kreativen Hintergründe liefern': Tonträgerindustrien in Ost- und Westdeutschland im Strukturwandel der 1950er bis 1980er Jahre," in *Der Mythos von der postindustriellen Welt: Wirtschaftlicher Strukturwandel in Deutschland 1960-1990*, ed. Werner Plumpe and André Steiner (Göttingen: Wallstein Verlag, 2016), 165-6.
7 Thomas Welskopp, "Einleitung und begriffliche Klärungen: Vom Kapitalismus reden, über den Kapitalismus forschen," in *Unternehmen Praxisgeschichte: Historische Perspektiven auf Kapitalismus, Arbeit und Klassengesellschaft*, ed. Thomas Welskopp (Tübingen: Mohr Siebeck, 2014), 2; 20.
8 Giovanni Levi, "On Microhistory," in *New Perspectives on Historical Writing*, 2nd ed., ed. Peter Burke (Cambridge and Malden: Polity Press, 2001), 99-100.
9 Ibid., 99.
10 Burchardt and Jonkmanns, *Hamburg Calling*, 57.
11 "Small ads," *Sounds*, August 1978, 17.

12 Thomas Hecken, "Punk-Rezeption in der BRD 1976/77 und ihre teilweise Auflösung 1979," in *Punk in Deutschland: Sozial- und kulturwissenschaftliche Perspektiven*, ed. Philipp Meinert and Martin Seeliger (Bielefeld: transcript, 2013), 248–53.
13 *Der Schunt* 1, August 1978, 11.
14 *No Fun* 11, c. 1978, no pagination.
15 *Preiserhöhung* 1, January 1979, no pagination.
16 *No Fun* 21/22, c. 1979, no pagination.
17 Ibid.
18 *Alles tot* 2, c. 1979, no pagination.
19 *Pretty Vacant* 6, c. 1979, no pagination.
20 "Rip Off ad," *Sounds*, March 1980, 70.
21 *Ants* 3, January 1980, 5.
22 Alfred Hilsberg, "Macher? Macht? Moneten? Aus grauer Städte Mauern (Teil 3)," *Sounds*, December 1979, 47.
23 *No Fun* 31, c. 1979, no pagination.
24 Ibid.
25 Ibid., Maeck sees the shop as a "Treff-Punk." The right German word for meeting point would be "Treffpunkt." I suspect that this is not a typo but an intended play on words to stress that mainly punks were hanging out at Rip Off Records.
26 Ibid.
27 *Limited Edition* 5, c. 1980, no pagination.
28 Ibid.
29 *Die Einsamkeit des Amokläufers* 2, c. 1980, 8–9.
30 *Krrz - sinnentleert*, c. 1980, no pagination.
31 *Ich und mein Spiegelbild* 2, c. 1980, no pagination.
32 Michele Avantario, "Von Krawall bis Totenschiff: Punk, New Wave und Hardcore," in *Läden, Schuppen, Kaschemmen: Eine Hamburger Popkulturgeschichte*, ed. Christoph Twickel (Hamburg: Edition Nautilus, 2003), 49.
33 Ibid.
34 Ibid.
35 Hayton, *Culture from the Slums*, 96.
36 Anonymous, "Punk: Nadel im Ohr, Klinge am Hals," *Der Spiegel*, January 22, 1978, 141.
37 Christine Brinck, "Karriere mit Knieschützern," *Stern*, January 19, 1978, 129. Harald Inhülsen also printed the photo of Iggy Pop and Jacky Eldorado in the final edition of his Iggy Pop fanzine; *Honey, that ain't no romance* 3, May 1978, 3.
38 See, for example, *Shit-Bolzen* 7, c. 1979, no pagination; *Langweil* 110, c. 1980, 31; *Schmier* 9, c. 1981, no pagination.
39 Christof Meueler, *Das ZickZack Prinzip: Alfred Hilsberg—Ein Leben für den Untergrund* (München: Wilhelm Heyne Verlag, 2016), 114.
40 *Bericht der U.N. Menschenrechtskommission über Menschenrechtsverletzungen in der Bundesrepublik Deutschland* 4 A, October 1980, no pagination.
41 Meueler, *Das ZickZack Prinzip*, 113.
42 *Langweil* 111B, c. 1980, no pagination.
43 *Bericht der U.N. Menschenrechtskommission über Menschenrechtsverletzungen in der Bundesrepublik Deutschland* 1, April 1980, no pagination.
44 *Bericht der U.N. Menschenrechtskommission über Menschenrechtsverletzungen in der Bundesrepublik Deutschland* 4 A, October 1980, no pagination.

45 *Alles tot* 6, c. 1980, no pagination.
46 *Langweil* 109, c. 1980, 4.
47 *Abwärts*, c. 1980, no pagination.
48 *Bericht der U.N. Menschenrechtskommission über Menschenrechtsverletzungen in der Bundesrepublik Deutschland* 4 A, October 1980, no pagination.
49 "Letters to the editor," *Sounds*, December 1980, 4.
50 Burchardt and Jonkmanns, *Hamburg Calling*, 82; Meueler, *Das ZickZack Prinzip*, 113.
51 Jürgen Stark and Michael Kurzawa, *Der große Schwindel??? Punk - New Wave - Neue Welle* (Frankfurt am Main: Verlag Freie Gesellschaft, 1981), 194–5, 202.
52 *A.d.S.W.* 1, c. 1981, no pagination.
53 Ibid.
54 *Hirntumor* 1, c. 1981, 6.
55 *Die Konzentration* 3, c. 1982, no pagination.
56 *Gehirnerschütterung*, c. 1981, no pagination.
57 *Sounds* 5, c. 1981, no pagination.
58 Meueler, *Das ZickZack Prinzip*, 195.
59 Burchardt and Jonkmanns, *Hamburg Calling*, 59.
60 *A.d.S.W.* 2, c. 1982, no pagination.
61 Meueler, *Das ZickZack Prinzip*, 194–7.

17

Soul Bowl

Rare Soul Uncovered

Christopher Spinks

Introduction

This chapter considers the intricate and complex link between the first wave of the northern soul subcultural movement of the 1970s and John Anderson. The founder of Groove City Records and Soul Bowl Records, Anderson is arguably acknowledged as "The World's Greatest Soul Record Dealer."[1] Musically, the northern soul scene depended upon the discovery and exposure of obscure Black American soul and rhythm and blues records that satisfied the demands of communal dancefloors. Anderson built his reputation for supplying rare studio-produced discs from the United States, which supported and established a community hierarchy revered and cherished by both fans and disc jockeys (DJs) alike.[2] This respect developed despite the geographical location of both Groove City Records and Soul Bowl Records, based in Glasgow and rural Norfolk respectively: both businesses were situated approximately 200 miles from the scene's epicenter of England's industrialized northwest and midlands. Moreover, this analysis allows for reflection upon Anderson's importance on the growth of the northern subculture and his reverence within that community as the tastemaker and prime record entrepreneur of the northern soul scene.

For the DJs embarking upon a lifelong crusade that relied upon a supply of new dancefloor fillers, Anderson became the guardian of northern soul's canon. Through Groove City Records and Soul Bowl Records, he supplied a constant source of rare and undiscovered soul tracks to an ever-demanding clientele of disc spinners, collectors, and fans. To understand this relationship, its formation, implementation, and subsequent development, this chapter evaluates the development of both the growth of northern soul and the part that Anderson played in that expansion during the 1970s. Endorsement of his role within this complex process comes through engagement with the key actors of the period, namely fans and DJs. Often dismissed by academic historians, those with firsthand knowledge of a subculture's emergence and development offer an untapped insight into its inner workings. This firsthand viewpoint often interpreted as lacking in a "more intellectual engagement" and is regularly bypassed from a purely academic viewpoint.[3] For a subculture that adopted and continues to promote a "you had to be there" attitude, embracing a collaborative approach within any research of this subject matter is essential to understanding the connection between Anderson and rare soul.

Literary Review

Participants of the northern soul scene traditionally consider any intrusion into their secret and close-knit faction with apprehension and remain cautious of outsider infiltration to this day. Consequently, there is very little in-depth academic research from the original period. Most of the contemporary literature issued comes from those with a vested interest in the scene. Stuart Cosgrove's autobiographical account, *Young Soul Rebels*, crosses the Rubicon between an in-depth personal recollection and full academic research. Like many commentators before and after him, Cosgrove identifies a religious fundamentalism within northern soul. In recalling its First Commandment: "there shall be no other music before soul." Cosgrove adopted this decree as a mantra for his life.[4]

From 2007, centered on his own experiences of amphetamine usage within the scene, criminologist Andrew Wilson's thought-provoking insight *Northern Soul: Music Drugs and Subcultural Identity* offers a vivid historical ethnography and delivers a compelling investigation into the scene's acceptance of drug usage, the social backgrounds of its users, and the relationship between crime and subcultural formation. *The Northern Soul Scene*, a 2019 collection coedited by Sarah Raine, Tim Wall, and Nicola Watchman Smith, is a series of cultural sociological journal articles. The book's main themes concentrate upon the recent twenty-first-century revival, including interviews with participants and the directors of northern soul-themed films. As the main academic anthology on the subject, it provides a platform for debate and discussion of the scene retrospectively and how it is seen now. Additionally, Raine traces both the modern themes of tribute and celebration proposed by newer members of the scene alongside the traditional viewpoint of maintaining a historical legitimacy by the original members of the scene in her 2020 monograph, *Authenticity and Belonging in the Northern Soul Scene*. Stephen Catterall and Keith Gildart's 2020 book *Keeping the Faith: A History of Northern Soul* offers its reader the most comprehensive academic account of the chronological progression of the scene's formation and its ultimate fall from popularity in 1981 with the closure of the Wigan Casino all-nighters. Through a meticulous reading of northern soul's portrayal within the popular Black music press of the 1970s, combined with an ethnographic analysis of a series of personal interviews with participants from the original movement, it delivers a "fascinating insight" into the original scene against a background of northern England's deindustrialization.[5]

Lamentably, academic literature confines Anderson's importance and his position as northern soul's musical gatekeeper to single sentences. This unique and untold story deserves detailed academic recognition. Aiming to bring to the forefront his role within the scene as the dominant supplier of rare soul, this chapter recognizes his tenacity and entrepreneurship that enabled northern soul's nationwide expansion during the 1970s.

John Anderson and Soul Bowl Records

John Anderson's (1949–2019) "insatiable interest" that remained with him throughout his life was his passion for Caribbean ska music produced by Prince Buster and released on British record labels such as Bluebeat.[6] Embracing the mod lifestyle of fashion and Black music during his formative years in Glasgow, Anderson became a collector of sought-after records, cherishing the thrill of seeking out and obtaining hard-to-find tracks. Ultimately, his veneration from within the rare soul record collecting community came from this built-in knack to hunt out the undiscovered and forgotten music of Black America.[7]

Together, with his mod girlfriend and future wife Merissa MacDonald, they embarked upon a journey of characteristically Scottish entrepreneurial vision, selling records by the day at Groove City Records and running underground Black music clubs at night at which Anderson would act as DJ. Visiting Philadelphia on an "Under 19" airline ticket, Anderson's first major shipment of soul imports was in 1968, with over 60,000 handpicked records being delivered to his mother and father's second-floor council flat in Pollock, Glasgow.[8] Reflecting upon his foray in soul record dealing, Anderson described himself as a "complete loser" citing soul music's evolution during the late 1960s into a raw and funk style reducing the demand for 45 rpm singles.[9] Small stores and independent record labels went out of business with the increased demand for 331/3 rpm long players.

In the 1970s, Groove City Records moved to Norfolk—initially, operating from a commercial outlet in the Victorian-styled town center at 15 Portland Street. Soon after Anderson changed the trading name to Soul Bowl Records. As a mail-order business offering exclusive and hard-to-get recordings, Soul Bowl "became the Harrods of soul collectors."[10] Always on the lookout for secure accommodation in which to store his ever-changing stock, Anderson utilized any spare buildings he could find, including an old disused church and a double garage. Finally, a neglected set of farm buildings at West Winch near King's Lynn were secured and became Soul Bowl's headquarters for the remainder of his time in Norfolk. For Anderson, this move to the rural wilderness reflected his inaccessible personality.[11] In order to maintain an adequate supply and constant turnover of new stock, Anderson regularly spent four or five months every year in the United States hunting down forgotten labels and overlooked artists.

For most collectors, their one endearing memory of the Groove City and Soul Bowl eras was the subscription weekly mail-order listings that were sent out. Merissa typed out the now-famous record lists, eagerly awaited in the post by hundreds of record collectors throughout the country. To verify a record's authenticity on the listing, the track carried the symbol (O) indicating it was an Original. DJ Richard Searling states that he felt an utter confidence in the validity of the singles offered on these lists.[12] One feature that appealed to many young soul fans was the one-pound Specials, which allowed the buyer to purchase a current in-demand northern soul track for one pound, well under its previous selling price.[13] The listings would also offer the famous Soul Bowl mixed packs of fifty records for ten pounds, allowing new converts to the scene to establish a collection without breaking the bank.[14]

Since the early days of the northern and rare soul scene, Anderson has played an idiosyncratic role. Elaine Constantine's gritty and realistic 2014 film, *Northern Soul*, acknowledges his importance in providing the music to its fans. The film is set during the first wave of northern soul in a fictional English northern town. Budding DJ Matt takes a raw and new convert to the scene, John into a high-street record shop. Picking up *Blues & Soul* magazine, Matt directs John to look at its back pages where he will find an advert for Soul Bowl, offering the authentic and rare tracks that the club DJs are currently playing.[15] In the movie, we see John making a phone call to order Lou Pride's "I'm Com'un Home in the Morn'un" (Suemi Records ST4567), referring to one of Soul Bowl's weekly lists. He later gives the record as a gift to Matt.[16] The weekly ritual process of receiving Soul Bowl's current availability listing, ringing up the office and being told by Anderson himself that their first choice had gone and having to revert to an alternative resonates with many record collectors.

Owning the latest dancefloor filler carried with it an esteemed status within the northern soul subculture. It rivalled the influence of an accomplished and formative dancer or even the drug dealer supplying a veritable cocktail of amphetamines enabling the dancers to keep going all night. With the DJs playing a seemingly never-ending selection of stompers, the scene's aficionados worshipped both the independent record labels and their little-known artists. Record departments of town center department stores only stocked popular chart records of the time, and the availability of imported American 45 rpm records came down to an exclusive collection of record and mail-order shops, the most renowned of these being John Anderson's Soul Bowl Records. In supplying the constant demand for new dancefloor fillers, the visionary Anderson became the musical curator to the establishment and progression of the northern soul subculture within its venerated first phase.

The Emergence of Northern Soul

During 1965, the adoption of "the sound of young America" within Britain's emerging soul club culture can be directly linked to two crucial events. In March 1965, the Motown Revue show brought their recording stars to tour the UK for the first time. The artists faced a challenging tour itinerary, playing twenty-one theaters over twenty-four nights, delivering two shows each night. Concurrently, the Tamla Motown Appreciation Society, with the influential Dave Godin at its helm, was in the process of persuading Berry Gordy that an exclusive label would further Motown's presence in the UK and Europe. Through EMI, Gordy found a distributor and at the start of the tour released the first six 45 rpm singles on the Tamla Motown label featuring artists on the Revue.[17] Despite the tour's poor attendance, both Andrew Flory and

Joe Street recognize the Motown Revue of 1965 alongside the 1967 Stax tour as key moments in the popularization of soul music within the UK.[18] R&B dancefloor tracks were part of the "Americanization" of the mod subculture, a teenage cultural division that had spread from its origins in London during the late 1950s into a national movement by the mid-1960s. An increasing number of mod-orientated clubs targeted this new audience, playing upbeat Motown and the rawer Memphis sound of Stax. Additionally, the clubs provided a platform for them to experience previously unseen Black musicians playing live.[19]

As the 1960s progressed, UK dance venues split into a North/South divide. London clubs such as the Scene and the Flamingo adopted the changes happening in the United States to incorporate the harder edge that was coming out of Memphis and Muscle Shoals. Prominent clubs in the north retained the distinctive popular commercially up-tempo Motown sound of the early 1960s. The crowd at Manchester's Twisted Wheel developed a passion for a frenetic dance style, incorporating and adapting steps and actions from past dance styles. The Wheel's dancers laid the genesis of a future musical movement, their actions copied and advanced on dance floors in clubs and halls throughout the northern soul music scene. Godin noticed this musical preference when football supporters of northern soccer clubs would visit his record shop to purchase oldies unavailable in northern record stores. After visiting the Twisted Wheel and other soul venues during the summer of 1970, Godin reporting for UK music magazine *Blues and Soul* penned his article "The Up-North Soul Groove."[20] In identifying a whole new subcultural environment centered around the playing, listening, dancing, and collection of rare soul, his commentary became an important foundational document, crucially definitive in its identification of a North/South divide and as the etymological source of the term "northern soul" within this context. Due to the codification of northern soul's distinctive sound in conjunction with their record collecting habits, Godin became known within this newly identified subcultural environment as the high priest of the rare soul movement.

The Northern Soul Scene

Originating in England's industrial heartland, the innovative northern soul scene of the late 1960s to the early 1980s remains an unprecedented period in the history of record collecting. Serious collectors competed with a new breed of enthusiasts to source and obtain rare up-tempo soul and rhythm and blues tracks made popular in the now-legendary clubs of northern and midland England. Uniquely, it was the first of the British postwar subcultures in which the appreciation of the scene's music was more important than relying upon previous subcultural mentalities of belonging to a gang or tribe to exist. According to Catterall and Gildart, northern soul "occupied a pivotal position in British youth culture."[21] Enthusiasts became part of a countrywide brotherhood, devoted totally to their mantra to "Keep the Faith" under the symbol of a clenched black fist. This situation led Stuart Cosgrove to remark that "no music in later life would ever touch its uniqueness."[22]

Encompassing an all-embracing fashion and music-orientated dogma centered on the dancefloor, the communal platform for northern soul's expansion were its clubs, with promoters and DJs organizing "all-dayers" and "all-nighters" to meet an ever-increasing demand. At each venue, DJs competed against one another, each looking to uncover the "next big dancefloor hit" to satisfy their audience's desire for obscure discs. A process made more complex by the fact that many of the tracks which eventually became emblematic of the northern soul scene were largely forgotten and undiscovered, a decade old, and languishing in American warehouses.

As the northern soul scene evolved during the mod-driven 1960s, the style of music progressed. To accommodate their chaotic but stylized dancing, the Twisted Wheel audiences preferred "the soul stomper." These tracks incorporated the "four-to-the-floor sound" which featured prominently within mid-1960s Motown songs originally written and produced by studio songwriters Eddie Holland, Lamont Dozier, and Brian Holland. Based in Detroit, Motown used a consistent band of musicians, known as the Funk Brothers, to produce a continuous series of up-tempo records that featured the driving bass of James Jamerson, howling saxophone breaks, and big instrumentation. Outside of Motown's agreement with EMI, very few independent American soul labels had distribution deals with British record companies, resulting in the limited availability of danceable tracks on British record labels. DJs and collectors of rare soul looked to the increasing availability of imported singles through dealers such as Anderson, seeking out the Motown sound from other Detroit recording studios such as Anna, Golden World, and Ric-Tic.

The urge to keep delivering dancefloor-filling tracks saw an increase in the popularity and availability of other labels outside of Detroit on Soul Bowl's weekly listings. Included among these discoveries were Okeh, Doré, and Arctic. Okeh is a long-standing R&B record label situated in New York, delivering a much more soulful approach through producer Carl Davis and songwriter Curtis Mayfield during the 1960s. Based in Los Angeles, Doré, in a similar vein to Okeh, expanded their musical scope and issued up-tempo soul records during the 1960s. Philadelphia's Arctic Records witnessed the emergence in prominence of the writers and producers Kenny Gamble and Leon Huff, the driving force behind the emerging "Sound of Philadelphia" in the 1970s. With the expansion of northern soul in the early 1970s, through the persistence of a select group of DJs, they formulated a specific sound and style and a portfolio of recognizable "oldies" established.

The expansion in popularity of the movement, both musically and subculturally, brought the problem of retaining exclusivity for its original members and DJs. In the early 1970s, tracks once unique to the northern soul dancefloors began to make regular appearances in the popular singles charts. Public interest in the northern soul scene increased. The media looked to impregnate a once underground secret society, seeking stories of culpability, particularly around the illegal use and supply of amphetamines. For many, this outside attention took away from the curatorship and connoisseurship that had been previously only available to those in the know.

From its establishment in 1973, the all-nighters at Wigan Casino became the most well-known component of the scene during the 1970s. Featuring original soul stompers previously played at venues such as the Twisted Wheel, and the Golden Torch in Tunstall, Stoke-on-Trent the early all-nighters attracted large audiences. At its height, the Casino audience would be over 2,000 strong and the club ultimately claimed a national membership of over 100,000. Due to its popularity, however, the Wigan bandwagon was subjected to unfavorable commercialization, which negatively affected its authenticity. One example being the creation of the Casino Classics record label. Truly awful commercially orientated pop songs that had no relationship to rare soul were played and actively promoted by some of the Casino's DJs. Releases on this label include puppet show TV themes and second-rate cover versions from the Casino's in-house band. These tunes had no connection to the original northern soul ethos, their only relationship to the scene being the appropriate beats per minute to satisfy new "tourists" to the dancefloor.

Some record companies took advantage of this idiosyncrasy by creating their own soul dance labels to break into the popular charts in tandem with the Wigan Casino phenomenon. Pye International created the Disco Demand label issuing classic soul originals alongside specific created novelties including the faked live mash-up track "Footsee" by Wigan's Chosen Few (Pye Disco Demand DDS111). The song was a commercial success reaching the Top Ten UK singles charts in 1975. For purists, worse was to come with the short-lived record labels Black Magic, Soul Galore, and Soul Fox which under the guidance of Simon Soussan "issued" in-demand records through retailer Selectadisc. These labels produced licensed reissues or even bootlegs, many of which, like those on the Casino Classics label, were substandard cover versions.[23] As far as the British public were concerned, for all things northern soul, all roads led to Wigan. This viewpoint gained further credence in 1977 with the national screening on prime-time television of *This England: The Wigan Casino*. The Granada TV documentary, directed by Tony Palmer, and filmed in the Casino Club during a Saturday all-nighter, intersperses images of the dancers and audience with representations of Wigan's past as an industrial town and its progressive state of urban decay.

This public intrusion into the mystery of northern soul became an issue for many on the scene and committed fans abandoned the venue in search of the original principles that had attracted them to northern soul in the first instance. Counteracting the detrimental impact of "Wiganization" upon the scene, a growing movement emerged among long-standing DJs.[24] They sought a return to the traditional approach of seeking out new and unheard rare soul sounds to deliver to their audience. The DJs turned to a tried and trusted source of original undiscovered and unknown sounds, Soul Bowl Records. The key to this progressive change would be Anderson's uncanny ability to know what sound would be appropriate for each of the venues. One influential catalyst to lead the revolution for change came from within the Casino roster of DJs, Richard Searling.

Disc Jockeys and the Introduction of Modern Soul

Searling's introduction to northern soul came from his wife Judith, a regular at the Twisted Wheel. Immediately before his eight-year residency at Wigan Casino, he had been the driving force behind the short-lived but influential all-nighters held at Va Va's in Bolton.²⁵ Working for Global Records, Searling first visited Philadelphia record hunting in July 1972. Over the next twelve months, he made regular visits unearthing hidden gems to feature during his DJ repertoire. One such track was the B side to Gloria Jones's 1965 single, "My Bad Boy's Comin' Home" (Champion 14003), the floor filling, "Tainted Love." This song became synonymous with the expansion of northern soul during the 1970s, crossing over into the pop world as a worldwide hit for Soft Cell in 1981. In mid-1973, Searling met John and Marisa Anderson at Global Records, where they struck up a friendship that lasted until John's death in 2019. Under the direction of owner Gerry Marshall, manager Mike Walker, and DJ Russ Winstanley, Wigan Casino opened for its first all-nighter on September 23, 1973. Aware of his reputation as one of the leading exponents in the northern soul scene for breaking previously unheard sounds, Kev Roberts recommended Searling to Winstanley as an addition to the resident DJ line-up. During this decade, the Casino became its high church and its DJs served as its evangelists preaching the "Keep the Faith" message through rare soul music. Northern soul generated a devout commitment from its congregation of dancers: Searling was at the forefront of this quasi-religious movement, paying homage to Anderson and his delivery of a continuous supply of undiscovered floor-fillers. Reflecting upon his time at Wigan, Searling observed how he felt challenged to source new sounds week in and week out, play them to a demanding crowd, and break them as regular spins on the Wigan playlist. He acknowledges that his reliance upon Anderson was immeasurable.²⁶

As the Wigan venture progressed, Searling, along with other DJs, sought to protect the rarity of their record discoveries through the process of "covering up." Handwritten labels concealed the track information on crowd-pleasing DJ promoted 45 rpm singles. Anderson worked with Searling to provide single-issue acetates with the artist and/or song title changed to protect its exclusivity. This process created a growing "mystique" around the track and/or artist identity.²⁷ Incorporating this custom also stalled the prevalent practice of bootlegging popular tracks. Rife throughout the northern soul scene, bootlegging flooded the record market making available inferior and poor-quality versions of rare records well under the prices being traded on the original singles. Once available to the public through either legal or illegal means, the tracks lost their uniqueness and association to either a DJ or venue and were subsequently dropped from their playlists.

With his reputation, prowess, and tenacity as a record dealer gaining prominence within American record dealerships, Anderson's expeditions stretched beyond the warehouses to the one-stops and small independent record labels. To give Soul Bowl access to limited-edition records, master pressings, and exclusive promotional copies, Anderson purchased as much of the surplus stock he could afford to ship back to the UK. This new influx of previously unheard gems attracted many prominent DJs, including Kev Roberts, Colin Curtis, and Blackpool Mecca's Ian Levine. They made the long journey from their homes in the north of England to rural Norfolk to trawl through these newly available discs seeking out the next big dancefloor filler. Within *Setting the Record Straight*, Searling recounts his visits to Soul Bowl's West Winch warehouse and highlights over 100 records that he first played during his period at the Casino. Many of which he covered up to protect their original identity and they were ultimately released in the UK via the Grapevine record label.

An underlying progressive groundswell resulted in a gradual introduction of modern soul to the dancefloor while the original scene formed around the 1960s produced stompers. Steeped in mystery, the mythology surrounding the breaking and subsequent popular development of this new style onto the DJ playlists is regularly contested on soul web forums such as Soul Source.[28] However, it is the Carstairs 1973 track "It Really Hurts Me Girl" (Red Coach RC802) which most commentators recognize as the breakthrough record in bringing about northern soul's evolution and acceptance of modern soul. Debate on the song's exposure within the northern soul scene has been embellished throughout the decades. Captivated by its "dangerous sound, a very throaty northern soul vocal . . . but with a slightly shuffly beat," Ian Levine reportedly heard the song on the radio while on a visit to Miami in 1973.[29] The legend has expanded on how Sherlock Holmes-like he tried to track down a copy of the single. The story of Levine's attempts to obtain this record progressively gained momentum through the internal grapevine that existed within the scene's record connoisseurs. Finally, Levine shattered this piece of northern soul folklore in 1999 and credited Anderson with bringing the single to the UK on one of his many visits to Philadelphia.[30] The acceptance of modern soul in the clubs opened up a new avenue for the DJs to develop their repertoire and enhance their now-legendary reputations. With his contacts and ever-changing stock of records, Anderson was able to meet the new demands placed upon Soul Bowl. His next move to meet the growing movement toward modern soul within the scene saw the formation of Grapevine Records.

Grapevine Records

In 1977, Anderson and Bernie Binnick established Grapevine Records. Based in Philadelphia, Binnick, who was the co-owner of Swan Records (one of the first American labels—along with Vee-Jay—to release Beatles records in the United States), had first met Anderson during his early record-hunting visits to the City of Brotherly Love. With their extensive contacts, Anderson and Binnick persuaded American label record owners to release previously unavailable master tapes for UK release. Particularly influential in the process were their past dealings with influential early-Motown producer Richard "Popcorn" Wylie. As an independent record label, its releases were in tune with what was happening on the northern soul turntables and dancefloors. Richard Searling preempted the release of many of the tracks released, playing them as exclusives at Wigan Casino. Their release on Grapevine allowed rare, coveted American originals popular on the northern soul scene to become readily available to eager British record buyers. Moreover, through Searling's prominent role at RCA records, Grapevine secured an arrangement that allowed the distribution of their records nationwide. In three short years, Grapevine issued forty-eight essential singles and three compilation long players, offering collectors of rare soul an unsurpassed opportunity to purchase affordable hard-to-get tracks.

Conclusion

Under John Anderson's leadership, Groove City Records and Soul Bowl Records became the principal supplier of rare soul that nourished and fed the northern soul DJs and record enthusiasts from its beginning through to the twenty-first-century revival. For Anderson, the thrill of the chase drove him on from the early days through to his later years. Soul music commentator Stuart Cosgrove penned Anderson's obituary in *The National*. Reflecting upon Anderson's "remarkable life," Cosgrove places him among the most renowned Scots, extolling his pioneering approach and willingness to take risks to enrich and undertake "a life few others could lead."[31] Without his entrepreneurship and doggedness millions of records from local independent record labels would never have seen the light of day.

Often passed over by researchers of 1970s subcultural development, the original northern soul scene occupied an integral position within British youth culture. The northern soul spirit is still very much alive today. Original participants and new converts of all ages are experiencing the complete subcultural aspect of the scene including dancing and record collecting not just in the UK but across the world. Despite the many changes in the format in which recorded music has been available since the first wave of northern soul in the late 1960s, collectors, scene participants, and, most importantly, DJs who continue to spin the discs in bars and clubs every weekend still prefer the seven-inch 45 rpm record. The opportunities that were offered to Anderson in the 1960s and 1970s of warehouses full of unplayed, forgotten, and unreleased singles have long gone, but his tenacity of hunting out the latest dancefloor fillers still lives on with today's followers. In the world of rare soul records, all collectors are beneficiaries of his legacy. Quite simply, it is difficult to overstate Anderson's contribution to the northern soul scene.

Notes

1 Michael Robinson, "John Anderson, World's Greatest Soul Record Dealer Is Dead," *The Daily Beast*, October 8, 2019. https://www.thedailybeast.com/john-anderson-worlds-greatest-soul-record-dealer-is-dead?ref=author (accessed February 2, 2022).
2 Stuart Cosgrove, *Young Soul Rebels: A Personal History of Northern Soul* (Edinburgh: Polygon, 2016), 121.
3 Keith Gildart and Steve Catterall, *Keeping the Faith: A History of Northern Soul* (Manchester: Manchester University Press, 2020), 2.
4 Cosgrove, *Young Soul Rebels,* 17.

5   Gildart and Catterall, *Keeping the Faith*, 6.
6   Soulmessenger, "John Anderson interviewed by Richard Searling in March 2014 at Prestatyn Northern Soul Weekender," October 16, 2019, YouTube Video, 35:25–36:30 https://www.youtube.com/watch?v=Mkls173rU48.
7   Stuart Cosgrove, "A Pitch-Perfect Funeral for Legendary John Anderson," *The National*, November 3, 2019. https://www.thenational.scot/news/18011168.pitch-perfect-funeral-scot-whose-soul-sang-music-black-america/ (accessed February 2, 2022).
8   Soulmessenger, "Anderson Interviewed," 6:30–7:10.
9   For a comprehensive deeply researched account of the social changes that revolutionized soul music during the late 1960s, refer to the trilogy of books by Stuart Cosgrove, *Detroit 67 The Year That Changed Soul, Memphis 68, The Tragedy of Southern Soul,* and *Harlem 69: The Future of Soul*.
10 Cosgrove, "A Pitch-Perfect Funeral."
11 Soulmessenger, "Anderson Interviewed," 16:30–16:40.
12 Soulmessenger, "Anderson Interviewed," 20:40–20:50.
13 Neckender, "Soulbowl the Lodestar in Discovering the Beauty and Diversity of . . ." February 19, 2015. https://www.soul-source.co.uk/articles/soul-articles/soulbowl-r3064/ (accessed February 2, 2022).
14 Allnightandy, "Soul Packs from Soulbowl Back in the 1970's," January 8, 2012. https://www.soul-source.co.uk/forums/topic/219311-soul-packs-from-soulbowl-back-in-the-1970s/ (accessed February 2, 2022).
15 Northern Soul Movie Script, https://www.scripts.com/script/northern_soul_14950 (accessed February 2, 2022).
16 Lou Pride's *I'm Com'un Home in the Morn'un'* was originally released in 1970 on the Suemi record label based in El Paso Texas. John Anderson unearthed the track on one of his regular Stateside visits. It became a popular track at the Cleethorpes Pier all-nighters in 1975 and is generally recognized by commentators within the scene as a key single bridging the gap between the stomping 1960s oldies and the more modern sounds from the 1970s. The single was heavily bootlegged but rare original copies still command upwards of £4000. Russ Winstanley's Northern Soul survivors—Various Artistis LP Vinyl (Outta Sight). https://www.northernsouldirect.co.uk/shop/lp/russ-winstanley-s-northern-soul-survivors-various-artists-lp-vinyl-outta-sight/ (accessed February 2, 2022).
17 To support the tour, the first six releases on the UK label were all issued on March 19, 1965. Only the Supremes *Stop in the Name of Love* (Tamla Motown TMG 501) broke into the UK singles charts. Russell Clarke, "Motown in Our Town: The 1965 Motortown Revue Hits the UK," April 8, 2015. https://rocknrollroutemaster.com/2015/04/08/motown-in-our-town-the-1965-motortown-revue-hits-the-uk/ (accessed February 2, 2022).
18 As a result of accepting and incorporating both the Motown and Stax sound, this reaffirms British youth subcultures' growing adherence to a transatlantic culture, see A. Flory, "Tamla Motown in the UK: Transatlantic Reception of American Rhythm and Blues," in *Sounds and the City. Leisure Studies in a Global Era*, ed. B. Lashua, K. Spracklen and S. Wagg (London: Palgrave Macmillan, 2014) and J. Street, "Stax, Subcultures, and Civil Rights: Young Britain and the Politics of Soul Music in the 1960s," in *The Other Special Relationship. Contemporary Black History* (New York: Palgrave Macmillan, 2015). https://doi.org/10.1057/9781137392701_8.

19. Street, "Stax, Subcultures, and Civil Rights."
20. Gildart and Catterall, *Keeping the Faith*, 41.
21. Ibid., 1.
22. Cosgrove, *Young Soul Rebels*, 3.
23. Kev Roberts acknowledges Soussan's influence in bringing a lot of original tracks to the scene; however, he considers him as "slippery" and "unreliable" in his commercial dealings. Northern Soul: An Oral History, Lauren Marten, January 14, 2016. https://daily.redbullmusicacademy.com/2016/01/northern-soul-an-oral-history.
24. Catterall and Gildart, *Keeping the Faith*, 92.
25. Va Va's was a small night club in Bolton holding around 500 people with two small dancefloors. The Friday all-nighters ran during 1973. In their advertising flyers, the club promoted DJ Richard "S" and the artists and sounds he regularly featured on his playlist. Both Searling and Va Va's reputation gained momentum within the northern scene. Coachloads of fans would descend on Va Va's from across the country, a scenario to be repeated when Wigan Casino opened later that year. Va Va's fell foul of the Greater Manchester Police when they found amphetamines being dealt at the venue, under police pressure, the local authorities curbed their music license to 2.00 a.m. Due to the change in timing, the last all-nighter occurred in August 1973, only five weeks before the launch of the Wigan Casino's all-nighters. Thomas Molloy, "The Empty and Unassuming Bolton Building That Was Once a Legendary Northern Soul Club," January 23, 2021. https://www.manchestereveningnews.co.uk/news/greater-manchester-news/empty-unassuming-bolton-building-once-18521793 (accessed February 2, 2021).
26. Richard Searling, *Setting the Record Straight: Music and Memories from Wigan Casino 1973-1981*, (Plymouth: Go Ahead, 2018), 10.
27. Ibid., 18.
28. Soulsearch, "The Carstairs—It Really Hurts Me Girl—Issue Vs Demo?" November 3, 2021. https://www.soul-source.co.uk/forums/topic/424348-the-carstairs-it-really-hurts-me-girl-issue-vs-demo/#comment-100150908 (accessed February 2, 2022).
29. Bill Brewster and Frank Broughton, *Last Night a DJ Saved My Life: The History of the Disc Jockey* (New York: Grove Press, 2014), 104–8.
30. Ibid., 101.
31. A Pitch-Perfect Funeral for Legendary John Anderson, November 3, 2019. https://www.thenational.scot//news/18011168.stuart-cosgrove-pitch-perfect-funeral-legendary-john-anderson/.

18

Lucky Records

Music Makes the People Come Together

Mariana Lins

In the era of streaming and e-commerce, it seems exotic that there are still bricks-and-mortar record stores in existence and even more so that there are people who continue to leave their house to buy their favorite albums. Moreover, it is also strange to think that these places that are so historically important for the collective spread of music and culture have almost completely disappeared, insofar as the shared in-person music experience continues to be a fundamental link between fans and idols, even after the rise of the internet. Print publications, online forums, and, later, social media platforms certainly define the fans' relation to the artists' work over time, but one thing that has not changed, for example, is the excitement of meeting together, usually at live performances, where emotions, memories, personal interactions, and collective identities are formed. This is why the record stores that still exist today continue to play a significant role in fan-artist relationships.

The French philosopher Edgar Morin affirms that "at the feet of each star rises, as if of its own accord, a chapel, i.e., a club," with the performance being the moment of climax he calls "Corpus Christi Days" for its in-person consummation of the cult of the idol.[1] Before the live encounter, however, the journey to the record store can be seen as part of the ritual that not only feeds the mythology of the music stars but also strengthens the sense of community among the fans, especially in niche establishments. Such is the case of Lucky Records, a small shop in the hip district of Le Marais in Paris. The store was founded in 1991 by three admirers of the American singer Madonna, and it soon cemented itself as a veritable temple of both worship and consumerism for her fans around the world.

In this chapter, I begin with my personal experience as a Brazilian Madonna fan and collector to consider how record stores, particularly Lucky Records, are situated in the process of mediation between the fans and the artist, encouraging a sense of belonging from the materiality of the emotions that emerge from albums, memorabilia, and a collecting culture that resists the digital transformation of the music industry. Frequenting the shop is part of a liturgy of interchange and sharing that draws fans nearer not only to their idols but, above all, to other fans, giving rise to affinity, support, and friendship. It is also in places like Lucky where practices of the cultural production of the fandom are developed in an alternative way, with their own structure of creation, distribution, and consumption.² One example is the magazine *Spotlight*, created by Corinne Plourde and Christophe Coatanoan (one of the founders of Lucky Records), which published photoshoots, sales information, album charts, and articles in French about the latest happenings in Madonna's career.

I also seek to discuss how all these aspects intersect with my experience as a fan in the Global South, where international shows and the idea of a pilgrimage to a shop in Paris nearly always sound like a utopian dream or at least one reserved for small groups of privileged fans. On two occasions, in 2013 and 2019, I was able to visit the shop and see part of the collection of over 10,000 items³ coveted by collectors around the world. Purchasing is, at times, of least concern when one is face-to-face with memorabilia that, for so many years, was seen only in photos or stories. There is something magical that turns the experience into much more than the acquisition of the object itself. I use my personal experience to reflect on how the culture of physical records, special editions, print publications, and "rarities" gains symbolic value from within the fandom and, also, examine Lucky Records from a generational point of view, considering that the business model used by Madonna throughout the last forty years is one in which the traditional album is still an important protagonist in her marketing strategies.

Get Together

Following an international pop music artist demands extra effort on the part of fans from outside the performer's country and culture. To start with, at least for me—as a Brazilian—learning to speak English was the first challenge. It was not enough to appreciate the melody of "Holiday" or feel the beat of "Material Girl" pulsing through my body—I needed to understand what Madonna was telling me in the songs, interviews, and shows that I watched on MTV or via VHS. At a time before Google or any type of instant translator, my only option was to collect English dictionaries, note down words, underline phrases, and lose myself in the indecipherable polysemy of a language that is so different from my daily Portuguese. The next bittersweet obstacle was to give in to my primary instincts as a fan and climb up another rung on the ladder of admiration by becoming a collector (much to the detriment of my bank account).

Equally pleasurable and frustrating, collecting albums and other rare items from an American singer in Brazil during an economic crisis and without many resources to import music or leave the country is, at the very least, dramatic. Not counting the national editions of records, books, magazines, and newspapers, further connection to content by or about Madonna nearly always depended on the goodwill of some acquaintance who traveled abroad or the network of fans that was reactivated with each new release in the record stores. Without the internet, going to physical stores was one of the most efficient ways of meeting or running into other aficionados of the artists we admired, apart from the rare times a show/tour made a stop in the country. Besides that, there was correspondence mediated by fan clubs, such as *Icon*, Madonna's official fan club created in 1990, which published an annual subscription magazine that shared the latest updates on her career. The publication had a classifieds section through which fans from various parts of the world placed ads seeking pen pals.

Beyond these mediated experiences, going in-person to the closest record store on the exact day of the release of a Madonna album was always, for me and my fan friends, a sacred ritual. Sometimes we arrived in time to see the boxes that carried the albums being opened, even before they were stocked on the shelves. Depending on the time of year, it was also possible to participate in events put on by the record label that included listening parties and potential drawings for promotional prizes such as posters, T-shirts, and badges. A release day was, thus, a party.

With the popularization of the internet, communication among fans grew more sophisticated, spreading through forums, sites, and social networks, drastically changing the dynamics of production, circulation, and consumption of the music industry.[4] For faithful collectors, however, the joy of visiting record shops appears not to have succumbed to the advance of technology and Lucky Records is proof. As a sonic temple for Madonna fans, the record store was founded over three decades ago by Georges Vidal, Maurice Robert, and Christophe Coatanoan (who passed away in 2004). At the time of writing, the shop continues to be a pilgrimage site for fans across the planet, even after Vidal and Robert decided to retire in 2021 and pass the store on to new owners.

Sanctuary

The story of Lucky is above all a story of friendship built upon a love of pop music. Vidal, Robert, and Coatanoan were three friends and collectors who held an affection for some of the most popular artists of the 1980s: Madonna, Michael Jackson, and Kylie Minogue. Their shared passion grew to the point of abandoning their respective careers and dedicating themselves exclusively to the record and memorabilia shop that they decided to create in the heart of Paris. On rue de la Verrerie, close to some of the most famous museums and monuments in the city, like the Centre Pompidou and the Place de la Bastille, they opened a small boutique that they christened Lucky Records, a reference to the song "Lucky Star" from Madonna's first album, released in 1983. According to Vidal, the venture took flight in large part due to the fact that Robert already worked in the recording industry and Coatanoan had worked in another store specializing in music. Vidal, for his part, had a background in finance and did not have any relation to the music market until entering into partnership with his friends.

In 1991, the three gave notice at their respective places of employment and set off together on what would become a project spanning decades. The objective, from the start, was that Lucky Records would be dedicated exclusively to pop music, positioning itself as an alternative to the majority of the record stores that put a heavier emphasis on rock. Georges Vidal told me in a February 2022 interview that "It was something really different at that time, because, as you know, most of the record stores focused on rock music and I think many, many people were waiting for a shop like this one [Lucky

Records]."⁵ According to him, the store came to meet a demand in the market, but it also fulfilled an emotional demand among fans of pop icons who sought a place where they could more freely share their admiration for their idols.

Lucky Records has always been a safe space, especially for LGBTQIA+ fans, as opposed to the "rocker" and heteronormative, male-dominated environments found in many other stores.⁶ The friendly atmosphere favored every type of "fangirling," independent of the gender or background of their customers. At Lucky, nobody was judged for "freaking out" over the innumerable posters displayed in the windows and on the walls or for feeling overwhelmed when encountering rare editions of nearly impossible-to-find albums. No customer would leave without spending some time at the small counter, tossing thoughts back and forth with one of the owners about Madonna's latest surprise, taking home whatever souvenir they could. Everyone had a story to tell and the store's customers always found listening ears. As Vidal recalls:

> I think we have always been very welcoming. People were not only customers, we were sharing something. They were not coming to the shop just to buy something. It was a story they were sharing with us, it was kind of friendship. People were coming to the shop along the years and some of them could not afford what we were selling, because some items are expensive, but they were still coming, we were talking and I think they were happy even without buying. For us, it made no difference. When you buy something, you are happy to buy it, if you pay €1 or €100. For us, if the people are happy, we are happy with them and for them.⁷

In the end, visiting Lucky had more to do with the experience than with the purchases, since it was about being with a community of people who shared the same emotions, the same happiness, and, not uncommonly, the same obstacles as you. My journey to the store in 2013 marked my first time abroad. Although my trip to France initially had no direct relation with my desire to visit Lucky, it is no surprise that it was the first place I visited in the city, even before the Eiffel Tower or the Louvre Museum. Such was my excitement that it seemed Madonna herself was waiting for me on rue de la Verrerie. Lucky was venerated on online forums, fan sites, social networks, and also among my friends, who had been there before and told me about the infinite albums, singles, magazines, books, and promo materials available on the shelves. In many respects, being there was as important an event as attending a Madonna concert.

Material Girl

Saving enough money to travel from Brazil to France took effort, without a doubt. Learning another language to communicate there did, too. But perhaps the most difficult part was coming face-to-face with around 10,000 items at Lucky Records and not having the slightest idea of what to take home, since crossing an ocean without buying a single thing was out of the question for me. This made me anxious, considering how tangled together emotions and consumer culture are. How one conforms to the web of desires produced by the cultural industry is as important as being a part of a fan community, feeling welcomed, and finding one's identity.[8] In a way, it is a tacit price to pay for not only being integrated in but also strategically positioned on the inside of the hierarchies of any fandom.

Since the store's opening, collecting has been and continues to be the flagship activity of Lucky Records. When I asked Vidal if the digital transformations and consequent crisis in the recording industry had impacted his business, he assured me it has not, in spite of noticing that some types of media have gradually grown scarcer or have challenged the position of others (like how CDs took over vinyl or how VHS was replaced by DVDs). He says that, together with Robert, he traveled the world to visit other record stores and unearth rarities for Lucky: "We used to go to Japan four times a year, same for the USA, Canada, UK, and all the record fairs in Europe." As the majority of the store's products are the fruit of transactions carried out among networks of collectors and traders across the world, they arouse customers' interest no matter the economic outlook or generation:

> We saw the CD sales going very low and the new generation buying vinyls again. It's very fashionable now. We have seen new young collectors arriving recently, but we can tell from the last show Madonna did in Paris, in 2020, that most of the people coming were over 40 years old. You know, the artist is getting older and the audience and the fans are getting older, too, that is life. But we are seeing new young fans coming, who were not even born for the *Like a Virgin* album.[9]

The constant reinventions that became Madonna's brand also worked to garner new audiences and transit to other markets in recent decades, as can be seen in the singer's flirtation with Latino or Asian references at various moments in her career.[10] It is possible that her partnerships with younger artists like Nicki Minaj, Diplo, Maluma, or Dua Lipa have also contributed to a new generation of pop music fans being interested in her work, as Vidal observed in his later years at Lucky Records. It is notable, however, that even though the majority of aficionados are over age forty, what all of them have in common, independent of age, is the desire to see their love for the singer translated into records, posters, and other rare items they make an effort to collect.

In his renowned essay "The Collector," Walter Benjamin discusses the idea of "completeness" present in the accumulation of objects by aficionados, in an attempt to understand what is called the "devised historical system" of the collection. This is a private system that each collector creates for their own items, attributing to each of them a repertoire of memory and information that transforms them into an essential encyclopedic record of the time, space, origin, and social context in which they are inserted. Benjamin states, "The deepest enchantment of the collector [is] to enclose the particular item within a magic circle, where, as a last shudder runs through it (the shudder of being acquired), it turns to stone."[11] Everything becomes, thus, souvenir, frame, and pedestal of a manifestation of proximity and permanence that, in the absence of the idol, only memorabilia can supply.

Possessing a collectable object makes not only the object present and palpable but also its history and all the experiences related to it. It is not enough to admire it from afar or only as a memory, since collectors, like Benjamin points out, are people with a tactile instinct, beholders whose gaze has nothing in common with a secular owner's but who see their "object of worship" as something much more than its materiality.[12] There are lived stories, feelings, and symbologies that acquire a sense of urgency and "completeness" in the face of the inevitable withering away of life. The greatest effort of the collector's practice is perhaps in the use of emotion as a tool that bestows organicity on what is found scattered around the chaotic world we live in. The collection gives meaning to things, enunciates, produces a particular logic that is not necessarily linear but is coherent on some level.

There are many kinds of collectors, but it is certain that, for nearly all of them, accumulating memorabilia constitutes a practice of sociability that sparks interchanges and dialogues that are very dear to any fan community. When sharing his impressions of Lucky Records' customers, Vidal, who stopped collecting when he began working at the store, reflects on a process that he has watched up close for thirty years—one that has gained new meanings over time:

> When there is something you would like, when you are collecting whatever it is, books, magazines, CDs, or vinyl, it is something you think you cannot live without. You have to own it. That is the point. For me, as I get older, I am a little bit over this. I still love to buy, because I buy things, but maybe now, for me, it is better to wait for something than to have it.[13]

It is possible that the collecting practice acquires new meanings as the fans get older, like Vidal seems to suggest. In any case, memorabilia is likely a source of one's own biography, passions, memories, and friendships gathered along the years.

Spotlight

An important aspect of the collecting practice is how it can represent fairly complex knowledge-sharing and social structures[14] while being, at the same time, a communication tool. In their study about hierarchies in music fandoms, Jessica Edlom and Jenny Karlsson analyze fan engagement in the creation of brands, products, and content surrounding the promotion of their idols. They discuss how these activities grant a certain status elevation within the community.[15] They are seen as important reference points of fandom by those who are younger or less "initiated" in the community.

Since its inception, Lucky Records has been a meeting place and reference for fans of various nationalities, motivated by the robust collection of records and collectable items sold by the store. In addition, Lucky's owners began publishing the magazine *Spotlight*[16] in 1998. It was filled with photos, interviews, information on the charts, and opinion articles about Madonna's career, similar to what *Icon*, the fanzine of the singer's official fan club, was already doing.¹⁷ At the forefront of its publishing were Coatanoan, Robert, and another friend of theirs, Corinne Plourde. Plourde was responsible for the fanzine's visual concept and for writing some of the articles. The idea, explains Vidal, was to produce material for the francophone audience that did not speak English and that, consequently, had difficulty accessing the content in *Icon*:

> I can say it started locally, then it grew and even people who were not really French were buying the magazine because we sold it to all Madonna's collectors. Some of them were complaining saying, "We need an English edition, a Spanish edition," but it was very difficult. We were not making money out of the magazine, because the printing was very expensive and the cost of it was paid by the subscriptions, but I think it was made with love. We did our job with love. That is the word in Lucky Records, that is the main thing, the strongest thing. It was love. Love to the artists, love to Madonna, love to the fans. I think they were feeling this and they gave us love too.¹⁸

With time and popularity among fans, *Spotlight* improved its editorial quality and caught the attention of the French recording subsidiaries that Madonna had contracts with (first Warner Music and later Universal Music). I asked Vidal if they received any investment in the project or if they were compensated in some way for the many years of promoting the singer's career. He assures me they were not, and everything was paid for by the money coming from subscriptions, though sometimes the record labels placed ads in the magazine. *Spotlight* stopped publication in 2017, in large part because of the speed that social networks imposed on news sharing. It stopped making sense to spend so much effort creating a publication that would arrive already out-of-date in subscribers' mailboxes.

In one of the editions that I have, an 82-page double edition from December 2012, there are three ads: one by the French film distributor Studiocanal, promoting the DVD and Blu-ray release of the feature film *W.E.* (2011), directed by Madonna; one by Rhino, Warner Music's French subsidiary, that had just rereleased a special boxed edition of the artist's discography; and one by Universal Music, announcing the album *MDNA* (2012). Apart from the ads in the magazine, some commemorative events organized by Lucky Records got official authorization from the label, albeit without financial support. Vidal also remembers that he and cofounder Maurice Robert even got invitations to certain presentations, like the sold-out pocket show that Madonna did at Olympia de Paris, but this was not a rule. Normally the two continued buying tickets to see her concerts.

While publishing *Spotlight*, Lucky also supported other literary releases related to Madonna's career: books in sophisticated hardcover editions, filled with detailed information about her tours, discography, awards, and so on. It was not uncommon for Vidal and Robert to be involved in the research and development of these materials, becoming resources not just for the fans but also for the pop music studios. Such is the case of the trilogy *The Eighties* (2008), *The Nineties* (2009), and *Millennium* (2010) by Frédéric Guilloteau, which have a detailed overview of the singer's discography divided by decades and are now a rare collector's item since they have been out of print for years, according to Vidal. A copy on eBay does not sell for less than 300 USD. Another book by Guilloteau released with Lucky's support is *Madonna on Stage* (2008), a compilation of data, photos, and articles about the artist's tours.

The content production within the fandom, as shown by Edlom and Clarkson (2021), guarantees not only a privileged status among other fans but also a certain discourse of authority. Both authors indicate the existence of an internal hierarchy in fandoms whose values are based on the different levels of knowledge and engagement of fans in promoting their favorite artists. When it comes to senior/expert fans, it is true that in addition to the prestige of having access to certain information and contacts, the content produced by them on their idols can usually guide practices, topics of discussions, and also enrich the education of others. This is why Matt Hills defends that fan social capital goes side by side with fan cultural capital.[19] What Lucky Records (and its owners) achieved is a great example of this.

Take a Bow

Upon completing thirty years at the helm of Lucky Records, Georges Vidal and Maurice Robert decided that their retirement was overdue. The announcement, published on February 28, 2021, via the store's social networks, coincided with the Covid-19 pandemic, though the decision had already been made long before coronavirus completely transformed the world. With consideration for their long-term customers, the partners gave advance notice, leaving the destiny of the venture open by saying in a post that Lucky's story "may continue without us."[20] Reactions across the fan community were mournful, and many sent messages expressing dissatisfaction with the possibility of the store's closure.

A few days after the announcement, Vidal and Robert received at least five different offers to pass on the store and "continue the story." The two suddenly found themselves in the middle of a selection process—not to choose the best financial offer but rather to find the one that would best adjust to Lucky's principles and its most important function: the facilitation of community and friendship. They chose to sell the business to five fans who were regular customers of the store and longtime friends. This, for Vidal, seemed to be the perfect combination: "Friendship in the shop is something very important. We felt that with them the story does not end."[21]

In June 2021, Lucky closed for renovation and reopened weeks later with completely new décor. Its look was minimalist, lacking the famous posters of Madonna stuck to the walls and windows, and sporting an updated logo. The new store proposed to be devoted to pop music without keeping the singer at its core—although it continues selling her albums and memorabilia like usual. The shop's loss of visual character left some fans disappointed with its new direction but for Vidal, this attitude conflicts with one of the strengths of Madonna herself: change. As he shared with me:

I think it was important to show that it was going to be different. It is not a museum, it was not something that was going to be stuck in the past because we were leaving. The shop had to change, to increase, to be different, to live. We can say, as Madonna's fans, that she is always changing, always being different. Madonna has always been working, she is still alive and she has been changing all through her career, and so is the shop. Lucky Records was not a museum.[22]

Vidal's statement underscores the dynamic character that Lucky has built over time, consolidating its vocation to be more than just a place to buy and sell collectable objects. It continues to be a space of interchange, experience sharing, welcoming, and, above all, emotion. The magic of possessing memorabilia or a rare edition of a record comes from this process of cultural and social inclusion carried in memories, stories, biographies, and experiences that only make sense if they are shared, if there is dialogue, if there is someone with whom to exchange ideas and feelings about fandom. This is perhaps the greatest legacy that Vidal, Robert, and Coatanoan have left with Lucky Records.

Notes

1. Edgar Morin, *The Stars: An Account of the Star-system in Motion Pictures* (New York: Evergreen Profile Books, 1961), 74.
2. Henry Jenkins, *Textual Poachers: Television Fans & Participatory Culture* (New York: Routledge, 1992).
3. Eric Bureau, "Ce disquaire parisien est presque un musée à la gloire de Madonna," *Le Parisien*, March 11, 2020. https://www.leparisien.fr/culture-loisirs/musique/ce-disquaire-parisien-est-presque-un-musee-a-la-gloire-de-madonna-11-03-2020-8277159.php.
4. Fabián Arango Archila, "El impacto de la tecnología digital en la industria discográfica," *Dixit* 1, no. 24 (2016): 36–50.
5. Georges Vidal, Interview by author, Zoom Interview, February 18, 2022.
6. Sophia Maalsen and Jessica McLean, "Record Collections as Musical Archives: Gender, Record Collecting, and Whose Music Is Heard," *Journal of Material Culture* 23, no. 1 (2018): 41.
7. Vidal, Interview.
8. Henrik Linden and Sara Linden, *Fans and Fan Culture: Tourism, Consumerism and Social Media* (London: Palgrave Macmillan, 2017), 3.
9. Ibid.
10. Mariana Lins, *A estetização da política na performance de Madonna* (Recife: Universidade Federal de Pernambuco, 2017).
11. Walter Benjamin, *The Arcades Project* (Cambridge, MA: Harvard University Press, 1999), 205.
12. Wagner Alexandre Silva, *Comunicação, consumo e colecionismo: produção de memórias e práticas identitárias do fã-colecionador de estátuas e dioramas bishoujo* (São Paulo: ESPM, 2015).

13 Georges Vidal, Interview by the author, Zoom Interview, February 18, 2022.
14 Arjun Appadurai, *A vida social das coisas: as mercadorias sob uma perspectiva cultural* (Niteroi: Editora da Universidade Federal Fluminense, 2008).
15 Jessica Edlom and Jenny Karlsson, "Keep the Fire Burning: Exploring the Hierarchies of Music Fandom and the Motivations of Superfans," *Media and Communication* 9, no. 3 (2021): 123–32.
16 The name of the magazine is a reference to Madonna's song of the same name, released in 1987.
17 "New Issue of Spotlight Magazine Out," *MadonnaTribe*, October 9, 2006. http://www.madonnatribe.com/decade/2006/new-issue-of-spotlight-magazine-out/.
18 Vidal, Interview.
19 Matt Hills, *Fan Culture* (London: Routledge, 2002).
20 Lucky Records (@luckyrecordsparis), "www.lucky-records.com Bonjour à toutes et à tous, Nous venons vous annoncer que nous envisageons dorénavant de prendre notre retraite…" Instagram, February 28, 2021. https://www.instagram.com/p/CL2MZqFFVCH/ (accessed February 24, 2022).
21 Vidal, Interview.
22 Ibid.

19

Rough Trade Paris, 1992–9

The History of a Scene

Jean Foubert

Upon looking back on my student years in 1990s Paris, a few beloved places stand out in my memory. One of them, the Village Voice, was an independent bookstore named after the famous New York City-based weekly of the same name. Situated on the rue Princesse, the shop was entrenched in the heart of Saint-Germain-des-Prés, the historically famed bohemian area on the southern bank of the Seine (*la Rive gauche*). Located on the opposite side of the river (*la Rive droite*), another one of my favorite hangouts was a record store, the French branch of Rough Trade London, which had opened its doors in October 1992 on the rue de Charonne in a then ungentrified, definitely not-so-fancy, and still inchoate district of Bastille that would soon become the new epicenter of Paris' underground music scene. Today, I still fully sense the breadth of impact both stores had on my cultural upbringing and can accurately remember the trivia and the seriousness of those days, the long hours spent tracking down rare, sometimes unheard-of, imported books, magazines, and records which, for the most part, were simply not available anywhere else in France at that time. In my eyes, the Village Voice and Rough Trade Paris were not just shops in the strict, mercantile sense of the word. They were focal points or locales vibrating with the pulse of a seething, artistic, and intellectual scene which provided me with a sense of community and belonging. Sadly, both stores went out of business a long time ago. The Village Voice closed in July 2012, Rough Trade Paris more than a decade earlier in February 1999. Coincidentally, both of them knew their heyday at a time predating the advent of the global digital market and the rule of the algorithm. If the Village Voice's closure must be largely attributed to Amazon's spectacular and fateful, hegemonic growth, Rough Trade Paris shut its doors shortly before digital processes and technologies began to dramatically reshape the twenty-first-century landscape of music industry. When, at the end of his interview, Stéphane David, one of the two Rough Trade Paris' founders, muses about the shutting down of the store, he points out to the difficulty of the challenge facing today's record store owners in the context of a globalized digital economy:

It might eventually not have been such a bad thing that the whole adventure came to an end at that very moment, right before MP3s and digital music took over and totally changed the world of music as I had known it for years. Sometimes I wonder how independent record store owners currently manage to cope with the harsh and complicated realities of today's market. It must be so difficult to find a way.[1]

As a teenager in the late 1960s and early 1970s, David grew up in the surroundings of the French septentrional metropolis of Lille listening to the glitter and pop-rock sounds broadcasted by Radio Caroline and other British offshore pirate radio stations. As the 1970s were drifting away, he began to become deeply enamored with the then booming, disco-flavored, music scene before forming an interest in post-punk and new wave acts at the onset of the 1980s. By that time, David had moved to Paris where he and two friends teamed up to start *Mea culpa*, a cult fanzine subtitled *Autopsies sonores* (*Sound Autopsies*). *Mea culpa* consisted of record reviews aimed at reflecting the editors' distinctive, uncompromising taste in bands ranging from the Pastels, and Sonic Youth, to the Durutti Column, Front 242, and, more generally, groups influenced by Captain Beefheart, the Fall, or Joy Division:

The idea behind *Mea culpa* was basically about writing on and standing up for bands and records that virtually no one else in France, the press as well as the record industry, cared about: Creation Records and The Jesus and Mary Chain, Shop Assistants and the Scottish Anorak Pop scene, or Factory Records releases like Section 25. It was difficult to get information on these bands but it was even harder to get copies of the records we reviewed especially if you didn't live in Paris. So, another interesting thing about the fanzine was that we had a mail order page that allowed our readers to order records straight from an English Plymouth-based distributor we had struck a deal with.[2]

Mea culpa's experience was short-lived and, between 1985 and 1987, only nine issues of the fanzine were released. As an underground statement, *Mea culpa*'s reading audience may have been limited in scope yet David's fanzine was a forceful, influential undertaking which uncovered a collection of new, fast-emerging trends branded as indie, largely dismissed as irrelevant or insignificant by the French, pop-rock criticism establishment. In that sense, *Mea Culpa* stood up against the dominant discourse of totemic institutions such as the magazine *Rock&folk* which, self-assertively, posed as the depositor of a then somehow already old-fashioned, clichéd, rock and drug mythology. All things considered, *Mea culpa* most certainly broke the ice and paved the way for a new, innovative, indie scene that would eventually culminate with the openings of record stores like David's first venture into the record-selling business Dancetaria (1988–93), named after the New York rock club or, later, Rough Trade Paris. As a seminal act, David's fanzine ushered a period which saw the launching of major indie music magazines like *Les Inrockuptibles* (1986) and *Magic* (1995) or the daily airing of Bernard Lenoir's *L'Inrockuptible* on French nationwide radio stations like

Europe 1 (1988–90) and France Inter (1990–2011), a program featuring live sessions which would earn Lenoir the nickname of "the French John Peel."

Soon after the end of *Mea culpa*, David was proposed the opportunity of managing a record store in Paris. "I still had a few friends in my hometown, Lille. They had a shop Boucherie moderne which essentially sold indie pop-rock releases and a few records coming from the Belgian electronic scene, stuff from Front 242, Neon Judgement, or Siglo XX."[3] Under the moniker Dancetaria, Boucherie moderne also operated as the French record distributor for the British indie label Creation Records. Dancetaria was thinking of opening a new shop in Paris and offered Stéphane a position of manager which he readily accepted. Instead of Boucherie moderne, they decided to name their Paris branch after their distribution organization and, in spring 1988, Dancetaria finally opened its doors on the Rue Cardinal Lemoine in the fifth arrondissement of Paris. As a new underground entity, the store had to face the competition from New Rose which, as a legendary shop and record label started by Patrick Mathé and Louis Thevenon in 1980, had come to be recognized as one of the true, unsoiled emblems of the 1980s French independent rock culture. Both located on the left bank of the Seine, the stores were located just a ten-minute walk from each other and had to share relatively mutual and intercrossing sections of the record market. David recalls:

> Obviously, there were people that would go to both Dancetaria and New Rose. So, we had records in common and it was occasionally something of a tough contest between the two shops. Yet there were a few significant differences in taste. New Rose was into kinds of music which I had never been interested in: things like Southern blues rock, artists like Johnny Thunders, bands like The Gun Club . . . On the contrary, Dancetaria was open to the kind of music that New Rose, at least part of its staff, may have less favored: synth pop, industrial and part of the recent, indie pop, and post punk stuff. As time went by, we eventually managed to attract a specific audience to the shop.[4]

Within the broader, sociological context of conflicting countercultural trends in music, Dancetaria and New Rose made up a dyad which crystalized the opposition of an older, independent, punk and rock generation with its emerging, British, indie kid counterpart. By the end of the 1980s, *Les Inrockuptibles* had become the main magazine for the new wave of indie cultures and values in France. Jean-Daniel Beauvallet, *Les Inrockuptibles*' former editor-in-chief, smiles back insightfully at that time of cultural changes:

> I had and still have a true passion for New Rose. As a kid in the early 1980s, I was living in Tours, a city located two hundred kilometers south of Paris. Every Wednesday, my parents would go to Paris and, every week, I asked my mum to stop by at New Rose to get the records I'd read about in the press. Still the shop was sort of intimidating and probably not the kind of place a late 1980s indie kid would feel comfortable with. Punks and skinheads hung around in front of the store and part of the regulars clearly despised everything that was associated with new indie

trends. Ironically, we used to call them *les crédibles*, (the rock-credible guys). I loved New Rose but I felt more at home hanging out in stores like Dancetaria and Rough Trade.⁵

An incipient locale of the Rough Trade Paris scene, Dancetaria soon evolved into a meeting point of the younger, French, indie generation. Most importantly, it turned out to be the place where David met his future Rough Trade associate, Jérôme Mestre. Born in Paris in 1967, Mestre started to forge an interest in music when he was in middle school at the age of twelve:

> At that time, I was not into one kind of music in particular. I just loved it. There were no social networks, no internet. Finding out about music was basically about exchanging tapes and records with friends. Gradually, though, I began to attend live acts, to hang around in record shops like New Rose and to read the English musical press as well as lots of French fanzines. Later, when Dancetaria opened in 1988, I was listening to bands like Butthole Surfers, Silverfish, Sonic Youth, or Terminal Cheesecake. Stéphane was highly supportive of that scene. We rapidly befriended and, as Dancetaria worked pretty well, he asked me if I'd be interested in helping him with the management of the shop.⁶

The story sounds casual, almost clichéd. Still, it needs to be taken much seriously as it sharply characterizes the emergence of an important social and cultural network consolidating in and around Dancetaria. Old pals showed up. New friendships were struck. Connections were made. Journalists from *Les Inrockuptibles* frequently dropped by to chat and dig up news and records selected by David and Mestre. A lunchtime regular, Lenoir, the radio presenter, did the same and regularly invited the Dancetaria staff to come and play a selection of new tracks in his radio program *L'Inrockuptible*. As Mestre recounts:

> During the whole of the Dancetaria and Rough Trade years, we always got along really well with Lenoir and the guys from *Les Inrockuptibles*. We shared common tastes and were eager to support artists we truly liked as much as we could. It was also kind of a virtuous circle. We provided them with the stuff they would play or review. People would then come to our shop to buy the records we all stood up for.⁷

When Dancetaria was about to close in 1993, Daniel Dauxerre, a former New Rose seller who would quickly join Rough Trade Paris, had been working there for a few months. Dauxerre had fallen in love with the music of a young, noisy, French pop combo called Darlin'—the band's name a tribute to the 1967 Beach Boys' song from their album *Wild Honey*. After becoming their manager, Dauxerre worked hard to secure the release of the band's first songs on the compilation *Shimmies in Super 8* published by Stereolab's label Duophonic that same year. Their sound was described as "daft punky thrash" by the UK music weekly *Melody Maker*. Soon after, Darlin' disbanded and two of its members, Thomas Bangalter and Guy-Manuel de Homem-

Christo, initiated a new project which they named Daft Punk after that somewhat derogatory review. The French Touch was already in the air.

Unlike Rough Trade Paris which would rapidly offer a very large array of electronic and dance music releases, "Dancetaria's DNA was deeply rooted in the 1980s-post punk and indie pop rock movements. Something like 90% of the records that we sold came from that scene."[8] David is assuredly right. Yet, as Mestre also points out:

> Right from the start, some EBM and early acid house stuff was available in the shop. Meanwhile I don't think we never really had any Chicago house or Detroit techno sound in Dancetaria. But the situation was quickly evolving. There was a growing sense that an era was coming to end. In the late 1980s and at the beginning of the 1990s, it simply felt like we were going in circles. After the grunge tsunami, there was nothing truly new or exciting about new indie rock releases.[9]

Simultaneously the rave movement was popping up practically everywhere in Paris and in the rest of France. Mestre recalls that he "instantly got into it. There was something really cool about the underground, rebellious, illicit spirit of a rave party. That was also a time when we both began to listen to things like early Ozone and Warp recordings."[10] To a large extent, David and Mestre's curiosity and open-mindedness as Dancetaria record sellers most surely contributed to the blurring of the frontier between rock and electronic cultures which defined the aesthetic and sociological landmark of Rough Trade Paris. One of Mestre's old friends, the musician and producer Marc Collin, coleader of the band Nouvelle Vague, remembers:

> One day, I stepped into Dancetaria shortly after it had opened. Back in those days, I was involved in a new wave group Spleen Ideal and was fully immersed into the pop and post-punk scene. Stéphane was playing a new EP by 808 State. And, for me, it was really something of a blast as if a new world of unknown possibilities had suddenly opened up. The surge of house and techno was assuredly a decisive turning point in my artistic career.[11]

Subsequently, Collin initiated and took part in a multitude of impressively eclectic projects, including bossa nova covers of new wave classics with Nouvelle Vague, a lounge, easy listening LP with Ollano (1996) and its honey-sweet pop hit "Latitudes" or various film soundtracks like Graham Guit's *Les Kidnappeurs* (1998), or Luis Javier's *Tiempo Felices* (2014). However, from the beginning of the 1990s till the mid-2000s, he also remained actively involved in the French electronic music landscape and started an acid jazz combo with Mestre under the moniker Indurain (1992–4) before producing house EPs with Erik Rug (Dirty Jesus, 1996–7) and new wave inspired tracks and remixes with Ivan Smagghe (Volga Dub, 1999–2005), both of whom worked for a while at Rough Trade Paris.

Life moves pretty fast and relatively soon after their local branch in Paris was established, Dancetaria as a distributor organization faced financial difficulties which, in turn, began to seriously affect the dynamic of the shop. As Mestre details:

Starting from January 1992, our distributor was confronted with severe financial and marketing issues that were just too difficult to deal with. The whole thing was alarmingly going awry and the shop quickly ran out of new releases. We couldn't simply go on living on our back catalogue. During his *Mea culpa* years, Stéphane had sympathized with the guys from Rough Trade London. He found an arrangement with them and every two weeks he'd take his car and cross the channel all the way to London to bring back records we sold on the following day. It was basically a question of survival. But we had also built a strong business connection. Parallelly they certainly already had the idea of opening a shop in Paris. In the meanwhile, Dancetaria was going bankrupt. So, we seized the opportunity and went to London to negotiate a deal with them. We were allowed to open a record shop operating under the name of Rough Trade Paris as independent managers. In exchange, they'd be our exclusive suppliers and they'd be retributed on the basis of a percentage calculated on our turnover.[12]

The deal was struck. David was fired in June 1992 and Mestre resigned at the end of the summer. "The problem was that we lacked money and were then looking for a cheap affordable place to rent. Setting up the Rough Trade shop in the Bastille district was a question of pure chance,"[13] Mestre recounts. Nonetheless, David somehow instinctively sensed that starting their new venture in that neighborhood would prove a good strategic choice. "We spotted that small place on the rue de Charonne and I remember having the feeling that there was something going on in the whole area. The headquarters of Paris's main countercultural radio station Radio Nova and BPM, the pioneering techno shop, were close. So, it didn't seem like a bad idea at all to open up the store there."[14] David and Mestre had the correct insight as the clientele rapidly exploded and Rough Trade Paris promptly proved a remarkably successful enterprise. Two years after the inaugural opening of their shop in September 1994, Rough Trade Paris had thus to move to larger premises which, fortunately enough, were located on the same street, a two-minute walk from the original store. Along with the closure of Dancetaria and New Rose in the first half of the 1990s, the creation of Rough Trade Paris partook of a wider movement: the massive migration of a large part of Paris' underground music scene from the left bank of the Seine to the right bank of the river and the Bastille district. With its buzzing concentration of record stores and cafés, the area gradually came to be known as "the Golden Triangle" among fellows of the older techno generation such as Laurent Garnier.

"Clearly, with the signing of the deal with Rough Trade London, there was a complete change in terms of structure and dimension,"[15] explains Christophe Basterra, cofounder and former editor-in-chief of the French pop magazine *Magic*. Originally more of a New Rose guy, Basterra recalls:

> I started working at Dancetaria with Daniel when Jérôme and Stéphane both quit. When Rough Trade Paris opened, it definitely changed a lot of things. In 1992, they did that festival featuring the likes of Colm, Moonshake and Pulp which was totally amazing even though there were not that many people. They simply had

practically everything, French house tracks and Warp intelligent techno, The Verve and Vincent Gallo. Lots of records we reviewed in our fanzine *Magic Mushroom* and later in *Magic* came from the shop. More importantly for us, they had received promo copies of the first Tindersticks single through their connection with Stuart Staples who worked at Rough Trade London. They gave us one. We loved it and it turned out we'd be amongst the first publications to issue something on that band. That helped us a lot and there was this nice interaction with the guys. Daniel, Ivan, or Jérôme wrote a few papers for us.[16]

If Rough Trade Paris benefited from Rough Trade London's aura and network, the shop retained much of Dancetaria's identity. Still a political science student at that time, the DJ, producer, and co-instigator of the label Les Disques de la mort Smagghe joined the Rough Trade Paris staff in its early days to finance, as he puts it, "a record addiction." When asked about the identity of the place, he explains that Dancetaria's indie pop-rock legacy and the Brit grain were the main defining aspects of Rough Trade's politics and spirit: "This is clearly what distinguishes the shop from a Bastille store like BPM which was a pure emanation of the club and rave culture with a distinct focus on the US house and techno scene while we, at Rough Trade, mostly came from a Brit indie background." Running parallel with Smagghe's take on the Rough Trade scene, the long-standing BPM seller, future resident DJ of Paris' club Le Silencio, designed by filmmaker David Lynch, and now Radio Nova programmer, Romain BNO, remembers getting into house music in 1988 after reading an article written by gay activist Didier Lestrade for the French daily *Libération*. BNO then found a job at Sal Russo's BPM which was still known as Bonus Beat. A friend of Marco Marco Pannella, the leader of the Italian Radical Party, the erudite and charismatic Russo had extensive knowledge in different kinds of music from hip-hop to prog rock. A well-known story has it that BPM's owner had been to the United States to play new wave in clubs and that he had brought back American techno to Europe. In BNO's words:

> We were purists in the sense that we were clearly mostly drawn to the form of radical statement offered by US, underground, electronic music. Yet we also liked Brit stuff like Warp's bleep sounds. But we were not interested in some of the crossover, pop, house things Rough Trade sold. Meanwhile Ivan came to our store to buy records and, reciprocally, I went to Rough Trade to buy records too. Both shops had a cool, most friendly relationship.[17]

In contradistinction with BPM's orthodox nature, Rough Trade Paris stood as a heterodox entity, a topic Smagghe further elaborates on: "Originally there were just a few electronic records available at the shop. Jérôme and Stéphane were obviously aware of the growing importance of the dance scene and they rapidly decided to open the basement of the shop to that kind of music."[18] When the second shop opened, Rough Trade Paris stuck to that two-story architectural configuration with the major difference that, from then on, "the vinyl dance section was at street level while the CD and indie rock section was located upstairs."[19] For Mestre, "it was important to separate the two sections of the shop. To put it bluntly, you can't play a garage house track and a Field Mice record in the same room."[20] To a large extent, Smagghe shares Mestre's position: "Truly, the shop accreted two different groups of people that didn't always necessarily mix with each other: DJs and clubbers on the one hand, indie guys on the other." Yet, as he also rightly points out, "there was that sort of floating, amorphous, in-between space where cultures kind of blend and which we amusingly used to refer to as *les disques de l'entresol* (the mezzanine records): Daft Punk, Mo'Wax, Warp, and other affiliated labels. That's to me the most distinctive feature of Rough Trade Paris's identity."[21] Here Smagghe echoes Mestre's oft-heard statement: "We basically pushed for things that were barrier breaking."[22] A then Radio Nova programmer, the DJ producer Gilb'R recounts:

> I came from the hip hop, rare groove scene and had never been into the new wave and rock thing. In those days, I was fully immersed into the breakbeat and drum and bass scene which was heavily present in Rough Trade Paris. One day I realized I'd never been upstairs and I was curious to know what was going on there. I went up to chat with Daniel and Stéphane and, starting from that very moment, I started to become interested in indie stuff like Stereolab.[23]

Ranging from I:Cube's filtered house through Chateau Flight's new jazz breakbeat to Zombie Zombie's haunting, jazz-infused, electronic psychedelia, Gilb'R artistic management of his aptly named label Versatile continues to resonate with the crossbreeding economy that was characteristic of the Rough Trade scene.

The Rough Trade Paris years were a time of intense creativity and genre-blending experimentations that, according to Dauxerre, was also influenced by the late Andrew Weatherall's groundbreaking fusion of dub, indie pop, new wave, and post-punk elements into dance music.[24] Rough Trade staff launched labels characteristic of their eclectic taste. Dauxerre's Kung Fu Fighting released Tommy Hools's trip hop sound as well as P. Jack's electro pop while Mestre's Artefact issued Beth Hirsch's folk music along with Rug's house productions under the moniker Daphreephunkateerz. Meanwhile, then Rough Trade Paris seller and DJ producer Arnaud Rebotini—known for his critically acclaimed Zend Avesta vanguard electronic LP *Organique* (2000) or Cannes award-winning movie *120 Battements par minute*'s soundtrack (2017)—produced drum and bass tracks and teamed up with Smagghe to launch a new wave-flavored dance project named Black Strobe. What largely contributed to the emergence of a myriad of new artists and labels at that time was, according to Collin, "the relative affordability of home studio material. It wasn't just cheaper to do music. More importantly, it allowed you to do music on your own. So far, records had been associated with bands, singers, and songwriters. In the 1990s, with things like Air or Daft Punk or more underground projects, the focus of interest shifted to producers."[25] The DJ producer Chloé, who had begun to hang around in the Bastille record shops and would in the early 2000s regularly play with Smagghe at the Pulp *Kill the DJ* parties in Paris, agrees with Collin's analysis and adds: "The other thing is that the vinyl record market used to be much more flourishing. Today, as an artist, you don't earn any money with a platform like Spotify."[26]

In those days, the vitality and the energy which could be felt in and around the shop were amazing. David Blot, Nova Club's radio presenter and scriptwriter of the comic book *Le Chant de la machine* (2001), cocreated the *Respect is burning* party concept which helped build up the worldwide fame of the French Touch scene. He recalls:

> It wasn't just about Rough Trade. It was about the whole of the Bastille area which resembled an immense laboratory. Seen from today's perspective, with the impact of social networks on our lives, it may be hard to imagine what it was really like. There was Radio Nova of course and, apart from Rough Trade, there were lots of exciting record shops everywhere like BPM, Techno Import, or Vibe Station.... We all hung around in the same stores and cafés, met at the radio, and went to the few rare events we might have enjoyed. Some of the people I saw at Rough Trade would come to our *Respect* parties. Ivan played for us and came up with the *Respect* heading. It really gave the sense of a social continuum.[27]

Rough Trade Paris was a place where music buffs, DJs, producers, journalists, or sound designers like Karl Lagerfeld's longtime collaborator Michel Gaubert met and socialized, discussed and bought records, exchanged ideas and mixes. It was a place of opportunity too. Beauvallet remembers meeting there a young Joseph Ghosn who later worked as editor-in-chief of *Les Inrockuptibles*: "We were simply impressed by his vast knowledge in Lee Hazlewood and asked him to write papers for us."[28] A Rough Trade Paris regular, Ghosn would go to the shop every week. He and his friends had a radio show *Mondial Twist* which was "in a way, a satellite of the store. We were used to pinning our playlist onto a board that displayed all kinds of news in the downstairs part of the shop. There was something truly emotional about those weekly routines, especially when it came to record digging." One day, an old friend called him. "He was in the phone booth opposite Rough Trade. He thought I had bought the sole available copy of a Polish rock band. We quarreled. It was like I actually stole the record. We didn't speak to each other for years."[29] For Sébastien Boyer Chammard who succeeded Gilb'R as Radio Nova programmer, "there was indeed something of Nick Hornby's novel *High Fidelity* about Rough Trade. But the shop had a soul, something larger stores never had. Whatever might have been said of the guys—that some of them were rude and unpleasant—, they were true persons genuinely dedicated to the love of music."[30]

What partly accounted for the original success of Rough Trade Paris was that, suddenly, "records were available at exactly the same time as they were in London. We were completely in touch with what went on and were so ahead of our competitors. In some cases, we were practically in a situation of monopoly," Mestre recounts. "When the first Tindersticks LP was released and had positive reviews in *Les Inrockuptibles* and *Magic*, we sold tons of copies. The same occurred with Oasis."[31] Yet, starting from the mid-to-late 1990s, Rough Trade Paris began to suffer from the increasingly aggressive competition of a syndicated store like the Fnac which began to sell identical records with a discounted price. Mestre explains:

> Right from the start, there was some kind of a punk DIY philosophy behind the whole thing. We had no money, no real business plan, and were in no position to

negotiate a good deal with Rough Trade London. It quickly turned out that the conditions were financially untenable for us. Then everything simply happened too quickly. We just didn't have the time to calmly sit down at a table to discuss the future or to develop complementary activities for the shop. A sense of urgency prevailed on all decisions. I think that people don't realize how time-consuming running a record store like Rough Trade Paris was. There were no computers, no software to help you with. Most everything was done by hand.[32]

In parallel, David experienced a form of exhaustion and weariness: "The first three years at Rough Trade Paris had been the most exciting part of my life as a record seller. Yet I might have been in that business for too long and I started to lose interest in music. I was the oldest of the staff and I didn't feel like going on selling singles to twenty years old kids." He then adds: "Over a decade, customers had changed. They used to come not just for the records but also to chat and get some advice. Then they came with a specific idea and if you didn't have the record, they left. After Rough Trade's closure, I stopped listening to music for almost two years."[33] Ghosn was there the day the shop closed: "Rebotini played a track by Plastikman and we all went to a restaurant. Then we parted and it was all over."[34]

Notes

1. Stéphane David. Zoom interview with the author. August 6, 2021.
2. Ibid.
3. Ibid.
4. Ibid.
5. Jean-Daniel Beauvallet. Zoom interview with the author. March 5, 2021.
6. Jérôme Mestre. Zoom interview with the author. June 24, 2021.
7. Ibid.
8. David.
9. Mestre.
10. Ibid.
11. Marc Collin. Zoom interview with the author. January 29, 2021.
12. Mestre.
13. Ibid.
14. David.
15. Christophe Basterra. Zoom interview with the author. January 21, 2021.
16. Ibid.
17. Romain BNO. Zoom interview with the author. June 13, 2021.
18. Ivan Smagghe. Zoom interview with the author. July 8, 2021.
19. Ibid.
20. Mestre.
21. Smagghe.
22. Mestre.
23. Gilb'R. Zoom interview with the author. March 31, 2021.

24 Daniel Dauxerre. Zoom interview with the author. June 16, 2021.
25 Collin.
26 Chloé. Zoom interview with the author. June 29, 2021.
27 David Blot. Zoom interview with the author. October 31, 2021.
28 Beauvallet.
29 Joseph Ghosn. Zoom interview with the author. January 19, 2021.
30 Sébastien Boyer Chammard. Telephone interview with the author. January 11, 2021.
31 Mestre.
32 Ibid.
33 David.
34 Ghosn.

20

Musicians in the Record Store

Celebrity Encounters through Amoeba Music's *What's In My Bag?*

Christine Feldman-Barrett

During the mid-to-late twentieth century, an era which witnessed the increasing popularity of brick-and-mortar record stores in the United States, numerous social functions became significant to such retail spaces. Whether a local, independent shop or one that was part of a city- or nationwide chain, record stores provided their customers an ambient refuge from other aspects of everyday life. While the main imperative of retail space is the selling of products, the mid-1950s' emergence of rock 'n' roll—and the youth culture it created—meant that record stores evolved into popular meeting places. They became crucial sites for musical exploration and socializing among their young clientele.[1] They were also spaces that allowed for daydreaming about one's musical heroes. As teenagers in Peoria or Pittsburgh became fans of nationally or internationally renowned performers, local record stores evolved into one of the few spaces where far-flung music celebrities seemed a touch closer to home.

In Cameron Crowe's film *Almost Famous* (2000)—which depicts the passionate mindsets of teenage rock fans during the early 1970s—devout music lover and self-proclaimed "Band Aid" Penny Lane describes record stores as places where fans can "visit" their famous rock-star friends.[2] Since part of the film's plot is about Penny meeting, befriending, and accompanying touring rock bands, her statement speaks to how albums both seen and heard within record stores' walls become stand-ins for the people who created them. Moreover, her comment recognizes that most people "meet" popular musicians through their recordings. However, Penny Lane's mention of this particular "celebrity encounter" within record stores also evokes another. Record stores, alongside radio stations, have been important promotional spaces for musicians since the postwar period. Unlike radio airplay, "in-store appearances" provided the opportunity for artists and fans to meet and were normally booked into performers' touring schedules while supporting new albums. The record store became an intimate venue for musicians to play a few songs from their latest release, autograph records, pose for photos, or have brief conversations with their fans.[3] Significantly, in-store appearances have been one of the few opportunities (especially prior to online engagement between celebrities and fans), where "visiting with one's 'rock-star friends'" was an interactive, two-way reality rather than a parasocial daydream prompted by recordings and their packaging.[4]

During the 2000s, the in-store appearance was transfigured and redefined in virtual space by Amoeba Music, a California-based independent record store established in 1990. Its initial Berkeley location was eventually joined by stores in San Francisco (1997) and Hollywood (2001).[5] Since 2008, the store's online video series, *What's In My Bag?* (from here: *WIMB?*), has featured musical artists speaking about favorite albums, singles, and other potential purchases from Amoeba. The videos simulate the traditional in-store appearance, with some early episodes including brief musical performances and, oftentimes, references made to the artists' latest releases.[6] Additionally—and not unimportantly for this discussion—the series also evokes the occurrence of unplanned meetings between musicians and their fans in record stores. After all, as fans and consumers of music themselves, professional musicians often shop at stores like Amoeba. Such non-planned encounters, as discussed by Kerry O. Ferry, can destabilize the celebrity-fan dichotomy because fans are meeting celebrities while both parties are engaged in an "ordinary," everyday activity.[7]

By asking each episode's guests which items have been placed in their Amoeba bag, the interviewees are portrayed in two ways: first, as a professional musician/celebrity, and second, as highly knowledgeable music fan/consumer. The following discussion examines this dual role as presented within the *WIMB?* series. Moreover, attention is paid to how the "celebrity" versus "fan" aspects of the videos were affected by the Covid-19 pandemic, whereby health concerns and restrictions meant that musicians began recording episodes at home. Thus, this chapter's focus is not only on how the dual celebrity/fan identity is prompted by the usual record-store setting of the series but also on how the series suggests increasingly blurred boundaries between these two subject positions—particularly when the record store space itself (both physical and virtual) is absent.

What's In My Bag?: Introducing the Series

As noted earlier, Amoeba Music's *What's In My Bag?* online series began in 2008. All episodes are available through Amoeba's website and the store's official YouTube channel, which has a substantial number of subscribers. The YouTube channel offers playlists for every season. At time of writing, the Amoeba Music website stated that there were 800 episodes of this series, with 730 featuring "artists." This category mainly comprises musical artists (whether bands, singer-songwriters, or DJs), though some actors, comedians, visual artists, and music industry professionals feature as well.[8] In its early seasons, Amoeba employees and customers also shared their favorite music, but this seems to have been phased out of later seasons, with a total of only seventy such videos. Episodes are sometimes as brief as three minutes and generally not much longer than fifteen. Earlier seasons also capture more interviews conducted across all three Amoeba locations, whereas later seasons tend to feature the Hollywood store.[9]

Another change early in the series is a set introduction to every episode. The introduction begins with establishing shots of the featured artists browsing the store's merchandise while a clip of their music is typically played. It concludes with a close-up of the band's or artist's name on a record rack divider—referencing their section of vinyl in the store. This shot then cuts to the artists introducing themselves, usually adding: "and this is what's in my bag." After a brief cutaway to the *WIMB?* title image, the guests begin talking about what they have found in Amoeba (placed in an actual Amoeba bag or not). While many of the featured musicians select vinyl albums, it is often the case that singles, CDs, and cassettes are produced and discussed as well—usually with a reason provided for the choice of (non-vinyl) media. Books, DVDs, and collectables are also sometimes included. Overall, the content of each episode remains the same across the seasons—with the interviewees discussing the professional and/or personal significance of their selections.

Given both the professional and personal contexts of musicians appearing in record stores, I only viewed the episodes featuring band members, singer-songwriters, or DJs who perform onstage. No matter their level of fame, it was important that these were people who would be recognizable to some of Amoeba's customers. Since this project was undertaken during the Covid-19 pandemic, the Season 12 playlist (2019), alongside the episodes recorded privately after stay-at-home measures were advised in 2020, were the ones selected. The videos created by artists at home during quarantine and various lockdowns worldwide were compiled into a playlist called *What's In My Bag? [Home Edition]*, available through the official Amoeba Music YouTube channel. In total, 118 videos were examined, with 47 from Season 12 and 71 from the *Home Edition* playlist, which in May 2022 spanned from May 1, 2020 to April 26, 2022.[10] Notes were taken for each video, citing which items were discussed and why—and, specifically, how the artists' discussion of them prompted references to aspects of their professional, celebrity identity or to a more personal, music-fan perspective. For the latter videos, observations continued to be made regarding how the artists' dual identities played out given the absence of the record store itself. Since musicians are usually considered VIPs of such stores to varying degrees, I was curious how the wide-ranging and equalizing effects of a public health emergency might reconfigure their positionality and the performative aspect of celebrity.

What's In My Bag? Season 12 (2019)

With this season shot on location at Amoeba Music's Hollywood store,[11] these episodes visually simulate both the planned and unexpected encounters between fans and music celebrities in record stores. Given Amoeba's Hollywood location—with Los Angeles as a long-standing hub for the American music industry—the latter kind of encounter is likely more commonplace there than in record shops elsewhere across the United States.[12] That said, though the episodes' establishing shots position the musicians browsing through the store's goods like "everyone else," most do not show them interacting with other customers/fans. In this season, for example, only one video—

featuring Fat Mike from the band NOFX—shows such interaction.[13] Nonetheless, these opening shots underscore both the "everyday practice" of a musician scouting for records alongside other consumers *and* their difference from them.

In Season 12, the main part of the episodes most always takes place in what appears to be a private and quiet "employee only" space beyond the shop floor. The featured artists are seated on a red couch in front of a multicolored, painted wall. This location, while surely practical in terms of avoiding background noise and so on, also further emphasizes the musician's professional status in an otherwise public, consumer space. Taking them away from the shop floor indicates they have been specially chosen to recount their selected Amoeba merchandise, unlike others in the store. The things the featured artists choose to talk about are as varied as their chosen items, but their commentary still unfailingly positions them as both celebrity (and potential influencer) *and* a fan with a long history of enjoying and engaging with music.

In reviewing this season's episodes, the ways in which musicians self-identify come across through comments made about touring; their association and/or friendship with other musicians; their knowledge of music production and/or songwriting; their record label; other music-related work (i.e., when band members mention DJing at clubs or musicians talk about managing bands); and, unsurprisingly, those that directly promote their recent musical releases—as would be standard for traditional in-store appearances. In one deviation from the record-store setting of this season, Argentine musician Juana Molina is not interviewed at Amoeba, but "Live at Abelton Loop," an event that has both the musician and an interviewer in a theater onstage discussing Molina's choices and playing excerpts from songs on music equipment set up for them.[14] Here the performative, professional role of musician-as-celebrity is visually more pronounced. However, in reviewing artists' comments from Season 12, it was interesting to discover that more commentary foregrounded their position as longtime and well-versed music fans. Their music selections were mapped onto their individual biographies—what Lauren Istvandity would call their "lifetime soundtrack."[15]

Indeed, throughout Season 12, many musicians talk about the personal resonances the musical artifacts hold for them as a fan. These remarks linked albums or songs to pivotal biographical moments, often referring to a legacy of music appreciation within their families. Other comments ostensibly align their music consumption practices with those of the *WIMB?* audience. Remarks regarding collecting vinyl or an artist's complete discography, for example, were not uncommon. Additionally, some interviewees described memorable first purchases and/or favorite artists, albums, or songs. Fan identity was also evident when artists included rock biographies, memoirs, cultural histories, and other nonfiction books about music among their choices. Finally, some musicians positioned themselves as music consumers by stating their love of record stores and/or Amoeba Music in particular. Rock musician Duff McKagan opens his episode mentioning how much he has always loved record stores, while singer-songwriter Jessica Pratt fondly describes Amoeba as "an institution."[16] While establishing that their appreciation of music and record stores is similar to that of the *WIMB?* audience, the featured musicians nonetheless also come across as especially knowledgeable music connoisseurs. This is clear from how they provide detailed information about the albums, artists, or genres they are discussing—or make selections showcasing a wide palate of musical tastes that viewers may not be expecting from genre-specific artists. In these ways, they are also acting as a tastemaker and potential influencer. The former is on display when, for example, a member of Superorganism knows that the 2018 fiftieth anniversary reissue of the Beatles' *White Album* was remixed by Giles Martin or when country singer Orville Peck states his disbelief that many people are still unaware of Bobbie Gentry and her song catalog. The latter comes to the foreground when, for instance, singer-songwriter Emma Ruth Rundle states her interest in Indonesian Gamelan music or when a member of post-hardcore band Touché Amoré enthuses about jazz.[17] In sharing this information, the *WIMB?* featured artists distinguish themselves as heavily invested music fans—similar to the friends viewers might turn to for music tips within their own social circles.

In sum, Season 12 of *WIMB?* provides a good sample of what the series offers overall. The way the videos are constructed, and the knowledgeable comments shared by featured artists, never fully subvert the interviewees' special status within the store. Nonetheless, their subject position as "music celebrity" is tempered by and often immersed within discourse that celebrates music fandom—something that is shared between the *WIMB?* guests and the series' viewers.

What's In My Bag? [Home Edition] (2020–2)

The global spread of a novel coronavirus, soon named Covid-19, was declared a pandemic by the World Health Organization in March 2020. Early on, the United States was one of the countries most impacted by severe illness and fatalities caused by the disease.[18] Varying levels of restrictions came into play worldwide, with many people fearful and reluctant to frequent public spaces. The decline of record stores in the early twenty-first century notwithstanding (with stalwarts like Amoeba Music one of the few independent record emporiums remaining), it was fascinating to see how the record store space of the *WIMB?* series—and the dual celebrity/fan identity of its guests—was transformed when, in May 2020, musicians started creating videos for the series at home. In this situation, the celebrity/fan subject position played out differently—with the performative nature of celebrity both emphasized and de-emphasized due to the equalizing nature of a public health emergency.[19] Numerous news stories during the initial wave of the pandemic pointed out that it was not only restaurants and retail spaces that were high-risk places but performance venues as well. Playing concerts and touring, which are significant to professional musicians' careers, were also suddenly suspended for an unknown period of time.[20]

The new reality of the pandemic, however, did not mean that the *Home Edition* episodes (also tagged *WIMB? @Home*) covered mostly new ground in their content. Just as in Season 12 (and previous seasons), the musicians' sharing of favorites remained the focus. Moreover, the featured artists continued to come across as both music professionals and knowledgeable music lovers. As in the 2019 episodes, much talk is devoted to autobiographical connections to certain artists or albums (e.g., my mother used to play this album all the time; this is the first single I bought with my own money). The biggest noticeable change, besides the varying private locations of the episodes, is how both the artists' professional status and the record store as an influential, physical space and place are reaffirmed despite Amoeba Music's visual absence from the episodes. Moreover, as the artists are not depicted browsing through the store, chosen items come from their home collections. Without the visual structure of the record store and the often public-facing professional activities of musicians there, the episodes must somehow reaffirm the artists' celebrity position amid a humbling world event.

Indeed, the shock of the pandemic—and the worldwide physical and social isolation that swiftly followed—is evident not just through the *WIMB?* videos' home settings but by comments made by the featured artists. This is particularly true in the episodes from 2020, though some from 2021 and 2022 also reference the new realities and terminology specific to the pandemic. For instance, Sean Solomon, a member of the indie rock band Moaning—in the very first *Home Edition* video from May 1, 2020—links the pandemic and its fatalities to an album by Broadcast, a band whose young lead singer unexpectedly died of the H1N1 flu in 2011. This tie-in of a previous pandemic and its tragic outcomes is thoughtfully spoken about by Solomon, especially in his comment that music helps "people live on in everyone's memories."[21] In another 2020 episode, the recently adopted term "social distancing" is mentioned by rock singer Alison Mosshart when introducing Neil Young's album *On the Beach*, which depicts a lone Young on the cover. Neil Young is also part of country singer Margo Price's 2020 selections, noting a song on his then recently released *Homegrown* album has lyrics about washing one's hands—timely given the push for heightened awareness of basic hygiene amid the new illness.[22] Another episode, featuring the US Hip Hop act Blue & Exile, situates the duo visually within the pandemic by showing them wearing masks despite the private location of their video. This suggests that one is a guest at the other's home with proper safety precautions being taken.[23] The historical context of the *Home Edition* episodes is also foregrounded by many of the artists' sign-offs, which often end with messages along the lines of "stay safe" or "stay healthy"—acknowledging the anxieties and health risks facing artists and fans alike.

Despite the equalizing effect of the pandemic, which made everyone's primary identity that of human being susceptible to a novel and deadly virus, the videos still found ways to underscore the featured artists' professional identity. One way this was achieved was through comments made related to touring—or, rather, a lack thereof. This makes sense, especially during the early days of the pandemic, when it was unclear when concerts and touring might resume. Such tour-related statements span the melancholy to the hopeful. A member of indie band Cloud Nothings expresses that they "miss touring a lot." The Mountain Goats' John Darnielle, an American musician, in speaking about a live album recorded at a Belgian venue mentions he would love to perform there "when we're allowed to play in clubs again."[24] US rapper and singer-songwriter Dessa says that "with any luck I'll see you on an actual, real tour soon." Similarly, rock drummer Clem Burke shares—when referring to a recent "supergroup" project of which he is a member—"Hopefully, one day, we'll get to play when this is all over," adding "and hopefully I'll get to play with my band Blondie as well again."[25] As in-store appearances have been normally linked to touring, the forced absence of this core activity among professional musicians cannot help but be mentioned in several of the *Home Edition* videos.

Another way the *Home Edition* episodes try to emphasize the professional status of the featured artists is through their choice of backdrop for the video. Some try to maintain a neutral background that betrays little of their homelife, with only white screens or walls visible. This choice, of course, suggests the boundaries celebrities often want to keep between their private and public lives. Others situate themselves in front of stereo equipment or record collections—environments which do not necessarily mark them out as someone other than a music connoisseur—but nonetheless emphasize the centrality of music in their lives. More on the nose are the episodes that allow viewers to see music equipment behind or beside the artist—reminding audiences that they are more than just consumers of music. Several of the videos, such as those by American bands Khruangbin and Black Pumas, depict the musicians sitting in recording studios rather than at home—overtly signaling their professional credentials to everyone watching.[26]

Certainly, the nature of the pandemic, and the way that it forced all manner of entertainers back into "home studios" (*The Tonight Show*, for example, was soon broadcast from host Jimmy Fallon's Hamptons home), could not help but soften the performer/fan divide.[27] Not only was everyone experiencing long periods of geographic confinement and physical distance from loved ones, but the pandemic meant that public spaces such as record stores—spaces that can demarcate musicians from their fans and other consumers—were temporarily absent from everyday life.[28] In the *Home Edition* episodes, redrawing some lines between celebrity and fan is also attempted through additional post-production flourishes. This is done, for example, by keeping the title card shown during the *WIMB?* introduction the same, reminding viewers that this is still an Amoeba Music production. To the same end, cut-away shots to music video clips or separate, professionally filmed close-ups of the items mentioned by the musicians remain from the previous *WIMB?* seasons—adding a familiar slickness to the otherwise more lo-fi productions.

Some featured artists also evoke aspects of the pre-pandemic, in-store episodes by showing viewers their personal Amoeba bags or other branded paraphernalia from the store. However, supportive commentary of record stores—notably Amoeba—is something undoubtedly shared between music celebrities and fans. And, while some artists' statements also imply an eventual return to the shop in tandem with touring, that aspect of their future visits is not always mentioned per se. In introducing *Led Zeppelin II* as part of her collection, for instance, folk-rock singer Jewel tells viewers, "Everywhere I go, I like to go into record stores," adding that she has "vinyl collected from all over the world." British electronic artist Leon Vynehall specifically refers to Amoeba, stating: "I really hope I can come across the pond and be in one of the shops sometime soon whilst on tour."[29] While most viewers at home are not fellow musicians missing activities like touring (though fans are missing attending concerts), the assumption in these videos is that both artist and viewer share the desire to support record stores. In an episode from July 2021, US guitarist Matt Sweeney thanks Amoeba for "all the years of providing so many people with great, great, records," saying that he looks forward to visiting the new Hollywood location (which opened in April 2021) "as soon as it's safe."[30] His is a sentiment likely shared by countless others, whether they are musicians or fans.

Music Celebrity in the Record Store Revisited

An examination of *What's In My Bag?* episodes created both prior and during the Covid-19 pandemic invites consideration as to how music celebrity functions under traditional versus novel conditions—when in-store appearances at Amoeba Music could not happen, let alone be captured and uploaded onto the internet. While the early twenty-first century has already changed the nature of celebrity given the advent of reality TV, YouTube stars, influencers, and interactive social media platforms like Twitter—which allow for real-time engagement between celebrities and "regular people"—the pandemic underscored the staged and performative aspects of fame.[31] Instead of a celebrity encounter that is a "wow moment" for fans, the *WIMB? Home Edition* episodes produced between May 2020 and April 2022 mostly de-emphasized the professional difference between musicians and fans. While the chosen, private settings sometimes drew attention to the person's musical identity (e.g., a studio location, recording equipment, instruments in the background)—as did mention of tours that could not take place—the record store, in this case Amoeba Music, was a space they missed visiting from the perspective of a knowledgeable music consumer. Given the leveling power of Covid-19, the record store came to symbolize a lost, communal pleasure among all music enthusiasts. In discussing beloved musical items and other collectables from their homes, the musicians featured in the pandemic-era episodes of *WIMB?* were yearning not necessarily for a space to reassert their unique professional status but for a place where musical biographies and sensibilities are forged and shaped—a place of endless discovery—and one offering the shared pleasure of sonic delights.

Notes

1 For emergence of youth culture via rock and roll, see Mitchell K. Hall, *The Emergence of Rock 'n' Roll: Music and the Rise of American Youth Culture* (New York: Routledge, 2014). For record stores as teen haunts in the mid-twentieth century, see the example of Tommy Edward's Hillbilly Heaven in Deanna R. Adams, *Cleveland's Rock 'n' Roll Roots* (Charleston, SC: Arcadia, 2010), 30, and Los Angeles's Tower Records on the Sunset Strip in Robert Landau, "Excerpt: Live on the Sunset Strip: The Street That Made Music History," *Boom: A Journal of California* 2, no. 4 (2012): 85.

2 *Almost Famous*, directed by Cameron Crowe (2000); Universal City, CA: DreamWorks Home Entertainment, 2001), DVD.

3   A 1947 *Billboard* article addresses the importance of "personal appearances" at record stores alongside radio promotion. "Booming Wax Promotion—Boon to Dealers," *Billboard*, May 31, 1947, 10–11. For a more recent discussion: Frank DiCostanzo, "In-Store Tours Offer Alternative; Appearances Benefit both Artists and Merchants," *Billboard 108*, no. 27 (1996): 71.
4   David Herrera, "Parasocial Engagement for Musicians and Artists: A Systemic Review of Theoretical Foundations with Applications," *Meiea* 17, no. 1 (2017): 14.
5   K. Santos, "Amoeba Music: California's Most Iconic Record Store That's Still Kicking," *The Culture Trip*, June 14, 2019. https://theculturetrip.com/north-america/usa/california/articles/amoeba-music-californias-most-iconic-record-store-thats-still-kicking/ (accessed June 3, 2022).
6   Drew Costley, "Amoeba Music Exposes Famous Artists' Tastes with 'What's In My Bag'?" *SFGate*, September 4, 2017. https://www.sfgate.com/music/article/Amoeba-Music-famous-artists-Whats-in-my-bag-11988789.php (accessed June 3, 2022).
7   Kerry O. Ferry, "Through a Glass Darkly: The Dynamics of Fan-Celebrity Encounters," *Symbolic Interaction* 24, no. 1 (2001): 26, 36.
8   "What's In My Bag?" Amoeba Music. https://www.amoeba.com/whats-in-my-bag/#/grid/1 (accessed May 28, 2022).
9   A sampling of episodes was viewed from the 2008, 2009, and 2014 seasons to capture the general structure and content of the clips prior to carefully examining Season 12 (2019) and the *Home Edition* playlist (from May 2020 to April 2022).
10 Amoeba, "What's In My Bag?—Season 12," YouTube Videos. https://www.youtube.com/playlist?list=PLjtg-TKQBIhuBdr2PgXso6ijX9qSEfJHQ (accessed May 28, 2022) and Amoeba, "What's In My Bag? [Home Edition]," YouTube Videos. https://www.youtube.com/playlist?list=PLjtg-TKQBIhugq3DVnG0UuA7U6IjbNoYm (accessed May 28, 2022).
11 The Hollywood store moved to a new location and opened on April 1, 2021. Chris Willman, "Amoeba Music: A Look Inside the Sprawling New Hollywood Store," *Variety*, April 1, 2021. https://variety.com/2021/music/news/amoeba-music-new-store-hollywood-location-move-photos-video-1234942264/ (accessed June 4, 2022).
12 For LA as a music industry hub, see Richard Florida and Scott Jackson, "Sonic City: The Evolving Economic Geography of the Music Industry," *Journal of Planning Education and Research* 29, no. 3 (2010): 314.
13 Amoeba, "Fat Mike (NOFX)—What's In My Bag?," YouTube Video, 8:29, October 15, 2019. https://www.youtube.com/watch?v=8sehPbf8l90&list=PLjtg-TKQBIhuBdr2PgXso6ijX9qSEfJHQ&index=11.
14 Amoeba, "Juana Molina—What's In My Bag? (Live at Abelton Loop)," YouTube Video, 11:29, May 14, 2019. https://www.youtube.com/watch?v=q7xeXEGJx1g&list=PLjtg-TKQBIhuBdr2PgXso6ijX9qSEfJHQ&index=32.
15 Lauren Istvandity, "The Lifetime Soundtrack: Music as an Archive for Autobiographical Memory," *Popular Music History* 9, no. 2 (2014): 136–54.
16 Amoeba, "Duff McKagan—What's In My Bag?," YouTube Video, 10:12, May 29, 2019. https://www.youtube.com/watch?v=xFwpPFFGPt8&list=PLjtg-TKQBIhuBdr2PgXso6ijX9qSEfJHQ&index=30; Amoeba, "Jessica Pratt—What's In My Bag?," YouTube Video, 8:52, February 12, 2019. https://www.youtube.com/watch?v=du9Mnfg-wcM&list=PLjtg-TKQBIhuBdr2PgXso6ijX9qSEfJHQ&index=45.
17 Amoeba, "Superorganism—What's In My Bag?," YouTube Video, 12:05, August 17, 2019. https://www.youtube.com/watch?v=Ls3MmYpvM_s&list=PLjtg-TKQ

BIhuBdr2PgXso6ijX9qSEfJHQ&index=18; Amoeba, "Orville Peck—What's In My Bag?," YouTube Video, 11:45, November 5, 2019. https://www.youtube.com/watch?v=727Ygm3iXVY&list=PLjtg-TKQBIhuBdr2PgXso6ijX9qSEfJHQ&index=8; Amoeba, "Emma Ruth Rundle—What's In My Bag?," YouTube Video, 7:41, September 4, 2019. https://www.youtube.com/watch?v=OkRMDC_Fb0Q&list=PLjtg-TKQBIhuBdr2PgXso6ijX9qSEfJHQ&index=39; Amoeba, "Touché Amoré—What's In My Bag?," YouTube Video, 6:10, July 16, 2019. https://www.youtube.com/watch?v=cu-7EDU9DjA&list=PLjtg-TKQBIhuBdr2PgXso6ijX9qSEfJHQ&index=24.

18 Elisabeth Mahase, "Covid-19: WHO Declares Pandemic because of 'Alarming Levels' of Spread, Severity, and Inaction," *BMJ* (online) 368 (2020): m1036; Joe Murphy, Jiachuan Wu, Nigel Chiwaya, and Robin Muccari, "Graphic: Coronavirus Deaths in the US, Per Day," *NBC News*, April 7, 2020 (Updated June 3, 2022). https://www.nbcnews.com/health/health-news/coronavirus-deaths-united-states-each-day-2020-n1177936 (accessed June 3, 2022).

19 This aspect of celebrity's communicative "performance" as impacted by Covid-19's social isolation is discussed in Spring Duvall, "Quiet Celebrity in the Time of Pandemic: Stripping Away Artifice in Performances of Self, Cultures of Citizenship, and Community Care," *Continuum* 36, no. 2 (2022; 2021): 229–43.

20 "How Coronavirus Is Wreaking Havoc on Music," *Rolling Stone*, April 27, 2020. https://www.rollingstone.com/pro/lists/coronavirus-music-business-latest-974262/ (accessed June 3, 2022).

21 Amoeba, "Moaning's Sean Solomon—What's In My Bag? [Home Edition]," YouTube Video, 4:52, May 1, 2020. https://www.youtube.com/watch?v=ZIRCwNErFpg&list=PLjtg-TKQBIhugq3DVnG0UuA7U6IjbNoYm; "Broadcast's Trish Keenan Dies after Getting Swine Flu," BBC, January 14, 2011. https://www.bbc.com/news/uk-england-birmingham-12194530 (accessed June 3, 2022).

22 Amoeba, "Alison Mosshart—What's In My Bag? [Home Edition]," YouTube Video, 8:17, May 14, 2020. https://www.youtube.com/watch?v=6Uh97kZkqUs&list=PLjtg-TKQBIhugq3DVnG0UuA7U6IjbNoYm&index=3; Amoeba, "Margo Price—What's In My Bag? [Home Edition]," YouTube Video, 5:16, July 14, 2020. https://www.youtube.com/watch?v=NK8HSLda5eQ&list=PLjtg-TKQBIhugq3DVnG0UuA7U6IjbNoYm&index=8.

23 Amoeba, "Blu & Exile—What's In My Bag? [Home Edition]," YouTube Video, 7:25, July 21, 2020. https://www.youtube.com/watch?v=qcZqje3-tDw&list=PLjtg-TKQBIhugq3DVnG0UuA7U6IjbNoYm&index=9.

24 Amoeba, "Cloud Nothings—What's In My Bag? [Home Edition]," YouTube Video, 7:26, March 23, 2021. https://www.youtube.com/watch?v=dfkFArJWnHQ&list=PLjtg-TKQBIhugq3DVnG0UuA7U6IjbNoYm&index=38; Amoeba, "The Mountain Goats—What's In My Bag? [Home Edition]," YouTube Video, 7:53, December 1, 2020. https://www.youtube.com/watch?v=pB_3QD4U4XY&list=PLjtg-TKQBIhugq3DVnG0UuA7U6IjbNoYm&index =26.

25 Amoeba, "Dessa—What's In My Bag? [Home Edition]," YouTube Video, 6:17, July 13, 2021. https://www.youtube.com/watch?v=JBHJR4dEKlw&list=PLjtg-TKQBIhugq3DVnG0UuA7U6IjbNoYm&index=53; Amoeba, "Clem Burke (Blondie)—What's In My Bag? [Home Edition]," YouTube Video, 7:59, January 20, 2021. https://www.youtube.com/watch?v=GwLnadyeJgM&list=PLjtg-TKQBIhugq3DVnG0UuA7U6IjbNoYm&index=30.

26 Amoeba, "Khruangbin—What's In My Bag? [Home Edition]," YouTube Video, 6:04, February 23, 2021. https://www.youtube.com/watch?v=V8RPpOlvChk&list=PLjtg-TKQBIhugq3DVnG0UuA7U6IjbNoYm&index=34; Amoeba, "Black Pumas—What's In My Bag? [Home Edition]," YouTube Video, 6:04, August 3, 2021. https://www.youtube.com/watch?v=ntlA1Eo72Fc&list=PLjtg-TKQBIhugq3DVnG0UuA7U6IjbNoYm&index=56.
27 Duvall, "Quiet Celebrity in the Time of Pandemic."
28 Ibid.
29 Amoeba, "Jewel—What's In My Bag? [Home Edition]," YouTube Video, 3:29, December 8, 2020. https://www.youtube.com/watch?v=wMho8lKvDtQ&list=PLjtg-TKQBIhugq 3DVnG0UuA7U6IjbNoYm &index=27; Amoeba, "Leon Vynehall—What's In My Bag? [Home Edition]," YouTube Video, 8:17, May 25, 2021. https://www.youtube.com/watch?v=wK7MMXfl4YA&list=PLjtg-TKQBIhugq3DVnG0UuA7U6IjbNoYm &index=46.
30 Amoeba, "Matt Sweeney—What's In My Bag? [Home Edition]," YouTube Video, 5:49, July 20, 2021. https://www.youtube.com/watch?v=1A03tfNblMU&list=PLjtg-TKQBIhugq3DVnG0UuA7U6IjbNoYm&index=54.
31 For brilliant discussion of these ideas, see Philip Auslander, "Everybody's in Show Biz: Performing Star Identity in Popular Music," in *The Sage Handbook of Popular Music,* ed. Andy Bennett and Steve Waksman (London: Sage, 2015), 323–5.

21

"Contents Expected to Speak for Themselves"
A Preliminary Understanding of North American Self-Service Record Retail

Tim J. Anderson

The aim of this chapter is simple: to provide an opening for an undiscussed history. In this case, the history of self-service record retail, a mode that is so common that it treads into the banal. While this research does not pretend to offer a definitive history of self-service—even in the United States self-service stores have varied from location to location and, as one might intuit, so have their histories—it is important to recognize that self-service is not and has never been a universally embraced form of retail shopping.[1] Still, this mode is dominant not only for record stores but for supermarkets, department stores, and other sites of retail. I draw primarily from secondary literature that addresses the concept of self-service in twentieth-century retail and post–Second World War reports from *Billboard* from the late 1940s until the 1960s. These *Billboard* articles provide useful discussions of how to provide self-service stores and the advantages of moving to this model of record retail, as well as addressing issues of record stock, music segregation, and sales staff. Each of these issues engages the record store as a stage for a specific self-performance where taste can be best displayed for and satisfied by the record shopper. As such, the chapter argues that the self-service record store is purposely staged for specific customer performances and displays that continue today in both brick and mortar and digital sites of record retail.

As I write this at the beginning of the third decade of the twenty-first century, the concept of self-service retail feels like one of the great givens of the new millennium. In a world of hyper-personalization, address, and presentation, the screens that dominate our lives are ready to service our online profiles and avatars every second, every day. This persistent monitoring is the result of a logic that promises us bespoke premade playlists and artist suggestions as long as we allow these streaming platforms to view our choices, rejections, and streams to create our performative profiles. The better the interface, the longer the user performs in it, thus allowing each gesture to underscore and detail potential artists and releases for that specific user. The embarrassment of a particular choice may be hidden from peers through the advantages of "privacy settings," but the interfaces never forget as once youthful choices may be later repurposed in a "remember when" playlist.

While the surveillant screens of music-streaming platforms may be a twenty first century expression of self-service, when it comes to record retail the desire to have the consumer display their tastes as they search for experiential goods is not new and, in fact, is the source of conflict and connection in the narrative of *High Fidelity*, a text which exists as a novel, feature film, and limited-run television series.[2] For the most part, the customary take on record stores has been that they are places filled with people devoted to aesthetic judgments that they are often willing to vocalize. Despite their reputation as hyper-masculine spaces in dire need of feminist intervention,[3] the record store's place in US history is significant. Joshua Clark Davis claims that "in the postwar United States, record stores were perhaps *the* place where consumers most commonly interacted with individuals who made their living from popular culture."[4] Davis points out that for African Americans in particular, and at one point for Black merchandisers, the record trade could be envisioned as "an arena in which African Americans could pursue a broader strategy of bolstering economic self-sufficiency and sustaining black public life."[5] As such, one can envision the record store as a kind of retail-oriented public sphere where the display and debates surrounding music and media develop or hinder specific taste publics and the identities they curry.

A Brief History of Self-Service Retail

Like all retail, the US record store is a site among other American retailers in aggregate and draws from the techniques and technologies of department stores and supermarkets. Beginning in the twentieth century, both supermarkets and department stores began to investigate a self-oriented performance of consumption.[6] The need to connect record stores and their evolution to these other sites of consumption is of concern for popular music scholars: sites of purchase and consumption can never be removed from concerns of music distribution and reception. Part of taking music seriously is studying how its modes of consumption provide significant contexts for a record's social position and meaning. A record's display in a store may, for example, signal expectations of popularity, importance, or genre. Because engaging in music is always a social act, the logistical maneuvers necessary to place a record in front of a listener includes any number of possible intermediaries and propositions including its staging at the retail level. Commercial recordings are always produced and distributed through specific cultural material, imaginations, and practices focused on generating specific acts of association. The record store is composed and supported by distinct material cultures that make it a significant site for meaning-making.

Will Straw reminds us that "music arrives in our lives propped up by multiple forms of material culture."[7] Two of those material cultures—the cultures of music distribution and promotion—are severely understudied by popular music scholars. However, promotion and distribution are always directly connected to the points of sale and, perhaps more so, with the rise of streaming platforms. As important as these are, all too often these acts of display and promotion are forgotten through their ephemerality. From a store's rack jobbers—those vendors who rent space to display goods for sale—to distributors armed with posters and "promo copies" designed with specific spaces and audiences in mind, a record's initial retail placement is influenced by numerous forces and actors. Among those forces influencing record stores are those within the historical context of twentieth-century "modernization" as it pertained to retail consumption.

First and foremost, the history of all self-service retail is directly connected to twentieth-century issues of modern production and distribution. Reporting in 1928, Evans Clark noted in a *New York Times* article addressing the rapid growth of supermarket and retail chains that while "yesterday was the age of mass production," the installation of burgeoning networks and cost-saving protocols in the 1920s had allowed these chains to grow in an "age of mass distribution." For Clark, the problem was that the address of scientific management that had revolutionized production had "yet to be applied as thoroughly to selling."[8] Nonetheless, the seeds of self-service as a more rational and effective sales modality were planted twelve years before Clark's article. It is conventionally understood that the origins of self-service in the United States first appeared with Clarence Saunders's 1916 opening of Piggly Wiggly, a Memphis-Tennessee-based grocery store. The store soon became a model for other retailers. The first aspect that made it unique was its open display of goods on the floor. Unlike general stores of the nineteenth and early twentieth centuries that kept goods behind the counter and housed away from customers, Piggly Wiggly placed the goods on the floor for customers to peruse for themselves. At the time most retailers, including grocers, employed a "small army of clerks" that "would assemble your order for you . . . Even chain stores used clerks."[9] The second innovation was the "pen system." Just as it sounds, a pen system is a system of physical guidance that corrals and guides customers in and out of the store, with aisles assisting the purchaser's flow past potential choices. Both of these innovations were seen as cost-saving measures. By allowing customers to choose and collect their goods, "Saunders's model cut costs by cutting out the clerks."[10]

The open display of goods and penning concept would spread quickly with retailers during the First World War, and, by the end of the war, speakers would begin to tout this style of service in conventions and seminars.[11] While self-service saved costs vis-à-vis labor, it also allowed stores to expand and grow out of the restraints of product specialization.[12] But perhaps most importantly, according to John Stanton, a professor of food marketing at Saint Joseph's University, the method of pen and display offered a way for consumers to "make decisions as to what it was they wanted to buy, [which then would eventually lead to] companies trying to catch consumers' attention, . . . [and would lead to] the origin of branding."[13] By the end of the 1940s, pen and display for self-service would be considered as a set of standard operating procedures for supermarket chains and other retailers throughout the United States.

The "pen and display" method, which is still used today, is only one among many self-service systems that also include "food automats," automatic tellers, self-checkout, and online recommendation algorithms. As Paul du Gay points out, there has never been one singular "self-service" system. Instead, self-service retail comprises "a loosely connected set of technologies, many of which [have] been in existence for some time in various parts of the world, and which [can] be lashed together, more or less coherently, to adapt to local demands and circumstances."[14] Writing about grocers, Franck Cochoy argues that their movement to self-service was dependent on "two semi-simultaneous, semi-consecutive movements" that allowed them to open their stores up, making them accessible and legible to consumers and staff. Cochoy also notes that the generation of open display techniques (and their physical arrangements) led customers and allowed them to easily identify goods that were key. However, he also ascribes improvements in sanitation, training, and new technologies—such as motor transport and telephony—for the human productivity necessary to make self-service possible. For example, he notes that the addition of grocery cart trolleys in the 1940s and 1950s altered the self-service grocery business and sales capacity in manners that Clarence Saunders simply could not conceive at the time of his innovations.[15]

While some techniques such as pen and display were eventually adopted and adapted by record retailers, examining articles in *Billboard* between 1944 and 1963 suggests that self-service centered on five principles as a set of working principles for record retail. These principles are as follows:

(1) The consumer has an "inner desire" that can be sparked.
(2) There are professional consultation services to assist any store in making the transition to self-service.
(3) Management needs to understand that record store clerks and personnel are different than those working at supermarkets.
(4) The effective display of goods is vital to the success of self-service.
(5) Key to this process is the proper spatial organization of "flow" and "pens."

Accordingly, in the remaining sections of this chapter, I will address principles 1, 2, 4, and 5. I already have written elsewhere about point three as it pertains to questions of gender[16] and believe it is a topic that merits its own dedicated discussion. Instead, I want to note here that the distinction between record and superstore clerks rests with the fact that supermarkets trade in commodities where issues of difference are purposely flattened despite branding and promotional efforts. Recorded music, on the other hand, exists as an experiential good where differentiation is always stressed, no matter how bound to categories of genre or utility it is. This is a distinction that can never be ignored by record retailers or music platforms.

Acknowledging the Customer's "Inner Desires"

The focus on a customer's desires and their sense of self is a fundamental principle of self-service. In a 1925 *New York Times* article on the emergence of the "dresseteria," a self-service shop where racks and spaces would allow customers to try on clothing for themselves, the newspaper pointed out that this technology was part of a particularly atavistic call for "twentieth century men and women [to return] to a time when no service could be had" and shoppers could peruse among racks whose "contents are expected to speak for themselves."[17] Locating a clothing store alongside both the "grocerteria" and the "cafeteria" as yet another self-service space encouraged the stimulation of desire through the presentation of multiple choices.[18] What made these "terias" modern was their emphasis on choice and the expressions of desire that customers generated when they confronted a variety of goods.

This modern creation of variety is key. In the nineteenth and early twentieth centuries, a significant portion of the United States lived in nonurban areas where department stores did not yet exist and choices of consumer goods were limited to what local general stores could access and provide through local and regional contexts.[19] However, the growth of national distribution networks combined with technologies of display allowed for a broadening of markets where goods could be better accessed by the average consumer through the innovations of catalog printing, distribution, and mail-order retail.[20] The result was that the possibility of what we call "personal choice" could be provided by these modern developments in consumer culture. As du Gay points out "a focus on the practical development of self-service allows us to undertake a more nuanced understanding of how 'consumer identities' are formed and how they relate, or not, to work-based identities."[21] Du Gay's analysis suggests a Foucauldian exploration where we begin to research "how consumers came to see themselves as persons of a certain sort—as creatures of freedom, of liberty, of personal powers of choice and of self-realisation, in relation to their everyday practices of shopping—through their immersion in that range of technologies we have come to know as 'self-service.'"[22] To make the move to self-service meant more than visibly offering "choices" to consumers. The move demanded coordinating and presenting to consumers a view of self-service as something emancipatory, something which involved not only retailers but every aspect of manufacturing, distribution, and promotion. It also demanded new investments in display makers and even psychologists to connect with shoppers.[23] To convince the customer that they should feel happy to exercise their choice demanded a better understanding of the customer's psychology. As du Gay also notes, a 1955 *Shop Review* article argued that "the whole technique of 'Customer Self-Service' . . . depends on a Reclassification of the Merchandise *from the customer's point of view* . . . This is the *sine qua non* of Customer Self-Service" (italics in original). Merchandise had to speak to the consumers first so that they would be willing to labor as they served themselves.[24]

While the rhetoric of sparking an individual's desires is consistent with the appeal made to record retailers around this same time, the initial move to self-service retail among such stores was made to cut costs. For example, in 1944 *Billboard* reported that while record retailers first turned to self-service during the Second World War as a counter to manpower shortages, "today we have come to the realization that the self-service record department is no mere experiment to meet a temporary need."[25] Instead, it was clear that self-service was "studiously calculated to serve best the interests of both the store and the customer."[26] One of the implications of this move was that the ability to cater to individual choice was contingent on a store selling "all types of recorded music, regardless of types previously featured."[27] Furthermore, this meant that merchandisers would need to reorganize records in order to be better discovered and that retailers would need to understand that "records lend themselves naturally to innumerable groupings."[28] The editorial emphasized that the "fairly simple matter" of music segregation was key as it would guide the consumer to discover their choices as they perused "the various types of music on record in separate display racks, with signs to guide the purchaser from rack to rack."[29] Two years later, a 1946 *Billboard* editorial titled "Ultra Self-Service" claimed the importance of self-service. Its investment in displaying goods was that it "appeals to the inner desire of people to help themselves."[30] The same editorial noted that "self-service is being promoted by the manufacture of all kinds of display stands, tables and devices so that goods can be displayed for people to help themselves."[31] The development of appropriate display and presentation technologies, it turns out, would be key.

Display and the Need to "Buy with Your Eyes"

While it may be difficult to conceive of record stores without racks and bins, there was a time when these technologies had not been embraced let alone purchased and adopted by a majority of record retailers. These new technologies would require a significant investment.[32] To persuade record retailers to make these purchases, major labels, convinced that self-service would allow them to move more product, began to engineer educational campaigns. The first label to engage distributors and retailers alike in the United States was Capitol, who touted "self-service as the key to increased profits for record dealers."[33] In 1953 Capitol organized a fall educational campaign for retailers and distributors that included promoting merchandise such as "an album cover book and demonstration record album, album check list, broadsides, self-service brochure, self-service units, wrought iron stands for the units, children's records displays, [and] supplements."[34] *Billboard* reported in an August 1953 issue that "all dealers will receive a special brochure prepared by the firm's merchandising division [with a depiction] of self-service in record shops."[35] The brochure included instructions on how to begin self-service by answering questions of inventory control, catalog, and floor plans. Capitol also prepared a special twenty-minute Technicolor film featuring Mel Blanc, the voice of Bugs Bunny and other Warner Brother cartoon characters, that was titled *Self-Service—The Greatest!*.[36] Capitol had planned to offer "over 100 dealer showings of the film" that would be presented "around the country in key cities."[37] That same August, Capitol assembled 200 Midwestern record dealers in Chicago's Congress Hotel to announce the fall of 1953's release schedule and promote the utilization of self-service techniques.[38] To encourage adopting self-service techniques, Capitol's fall program offered

> no unusual financial deals via discounts or extended payment plans . . . However, the company has worked up three basic plans under which the dealer can qualify

for free racks making purchases of $450, $300 or $150. Dealers get, respectively, three, two or one rack for displaying and selling packaged merchandise.[39]

Within the next two years both RCA and Columbia followed Capitol's lead and provided retailers and distributors their own educational programs.

RCA, for example, began to raise its engagement in 1954 by merchandising a "special 45 rpm display,"[40] but it would soon pale when compared to the company's program for the following year. The 1955 program detailed a store layout service for both the new record dealer and the established store that wished to "change to self-service."[41] By this year the Recording Industry Association of America (RIAA) explained that retailers of all types were showing interest in store modernization programs.[42] RCA promoted their store modernization plans in the April 16, 1955, issue of *Billboard* with a two-page advertisement promoting how store modernization could be done via a "new low-cost Record Program," where fixtures, layout advice, and an "architectural store remodeling service!" would be offered with payment plan possibilities.[43] RCA worked closely with professionals in merchandise presentation, hiring W. L. Stenagaard and Associates who helped design and manufacture "the 'Face the Music' line of record display units for stores."[44] Meanwhile, as Capitol worked with Freedman Artcraft Engineering Corporation to assist retailers with possible floor-plan programs, Columbia turned to Holley Associates.[45] This Michigan-based store-design company also helped Columbia produce a self-service modernization program that offered "store planning" that could be somewhat individuated for each store and within a dealer's specific budget.[46]

While not explicit, the emphasis on store planning and record racks placed a notable emphasis on the issue of display. As Cochoy explains, part of the transformation of businesses to self-service was achieved "through the gradual introduction of 'open display'" of goods."[47] Indeed, without the development of appropriate display technologies, the possibility of a successful self-service supermarket would be difficult to envision.[48] Displays and brands provided retailers "a new way of selling," "whereby one capitalised on the statement that 'people buy food with their eyes.'"[49] At least one *Billboard* article noted that the expansion of self-service record rack technologies into supermarkets and department stores sparked the growth of the record rack jobber who would, by 1953, provide a new mode of record distribution that posed a threat to traditional "old line" record retailers.[50]

With this threat to older models of record retail in mind, Joel Freidman strongly advocated the acquisition of record displays in the 1954 *Billboard* article, "Smart Shops Put Records in Full View, Adopt 'Help Yourself' Policy."[51] Friedman declared that "when customers see more, they buy more!" He also admitted that you could not run a record store like a "supermarket operation, with no sales clerks and few clock clerks," but you could generate, "mass exposure of stock . . . increase sales in case after case" and bring "'impulse buying' to the record industry."[52] Furthermore, once the conversion to self-service occurred, retailers could add more stock and transform dead space into live display space through "the use of modern display racks." Because of these racks, "dealers can now carry far more stock than they previously could in far less space, and equip their stores in far more attractive fashion."[53] Again, presenting these products demanded that issues of segregation by genre, issues not paramount for retailers before self-service display, would come to the fore. One 1955 *Billboard* article suggested that "the more categories and the more individual artist listings, the more impressive the stock. Besides making the customers' selection simpler, this can establish an air of authority based on genuine knowledge. Even in self-service operations that is something the serious regular customer values."[54] The record retailer's display of genre categories was not simply an option but instead both a promotional and educational necessity. One display technology that American record retail would adapt from the world of groceries, cellophane, was "an innovation dating back to 1908, industrialised from 1917 and promoted by Dupont in [the trade magazine], Progressive Grocer."[55] The history of cellophane, a term that connects the now polypropylene product to its cellulose roots, allowed customers to "buy food with their eyes" and further disrupted food retail as customers no longer ordered from shop assistants but instead chose products off the store shelves.[56] This early twentieth-century invention would be adopted by record retailers in the 1960s. By 1963 a leading New England rack jobber by the name of Cecil Steen explained that he adopted one brand of cellophane for his stock and was spending $190 a day on "Cryovac." The resulting seal that Cryovac provided by wrapping records with a kind of skin tight polymer made them somewhat more "pilfer proof" while its tight, see-through fit provided a feeling of "lamination" and a "crystal-clear display of the jacket."[57] For Steen, this film was "of maximum importance of catching the impulse buyer."[58] Cellophane would provide both a means of preservation and the window for the self-serve customer. In this self-service environment the modern record, enshrined in see-through plastic, could be handled by anyone walking the aisles, inspected, placed back in its display, or purchased. In this configuration the customer could now buy with their eyes.

Coda—Staging and Extending Consumer Choice

As mentioned earlier, catching and penning a customer was fundamental to generating effective grocery self-service retail and would remain relevant when applied to record stores. However, extending times of contemplation was also key to self-service, a fact that demanded an understanding of both potential consumer movement and comfort. This fact would call for a reorganization of shops with their layouts assisting diversified browsing and "consumer purchases."[59] Ways that record stores could invest in assisting customer movement and comfort were offered as advice in a 1955 *Billboard* column titled "Approach to Floor Traffic, Set Sales." Howard G. Haas, the then vice president in charge of advertising for the Mitchell Manufacturing Company, noted that while genre organization and display were key, designed traffic flow would "help get the prospective phonograph customer into the store, where the salesman can confront him with his presentation."[60] If enclosing customers in one-way entrances and exits assisted in moving buyers to merchandise, Gordon Freedman, president of the Freedman Artcraft Engineering Corp that assisted in developing Capitol's self-service program, noted in a 1955 article to the record industry that penning for a considerable period of time often meant keeping customers comfortable. As such, one new technology, air conditioning, could be employed by retailers who wanted to attract people to a store and comfortably keep "them there to browse" in what used to be hot and humid spaces during the summer months.[61]

Echoing du Gay's note that self-service involved an amalgam of technologies, Freedman stated that self-service modernization demanded new paint, plate glass, cork for comfortable flooring, and new electrical and lighting fixtures to better stage wares. The ideal Freedman store layout included (1) inviting, sheltered exteriors, (2) large presentational windows, (3) complete self-service, (4) departmentalization, (5) checkout achieved unobtrusively by arrangement of counter and display, (6) open booths, (7) a children's department to pull traffic through the store and "down-to-the-floor" display for children's self-service, (8) use of playback equipment, and (9) versatile fixtures that can be changed about if desired, to re-departmentalize or for special promotions.[62] To be sure, Freedman's ideal would not become every record store. Yet the issues of display, traffic flow, containment, and recognizing the customer's point of view permeate these recommendations. If anything, in Freedman's ideal enclosed environment, customers of all ages and identities—not the records—would be such stores' most important items. In these and other record retailers, record customers could now be comfortable enough to wander about through aisles and make choices, content to speak for themselves.

Notes

1. At least one African American record retailer explained: "We didn't have self-service." George Bishop of the Mr. Entertainer store in Greensboro, North Carolina, stated that in the eyes of potential thieves, "self-service meant 'take one for free.'" See Joshua Clark Davis, "For the Records: How African American Consumers and Music Retailers Created Commercial Public Space in the 1960s and 1970s South," *Southern Cultures* 17, no. 4 (Winter 2011): 80.
2. See Nick Hornby, *High Fidelity* (New York: Riverhead Books, 1995); *High Fidelity* (2000), [Film]. Dir. Stephen Frears, USA: Touchstone Pictures; *High Fidelity* (2020), [TV series]. Hulu.
3. See, for example, Germaine Greer, "Why Don't Women Buy CDs?" *BBC Music Magazine: The Complete Monthly Guide to Classical Music*, September 1994, 35–7; Tim J. Anderson, "Female Treble: Gender, Record Retail and a Play for Space," in *Point of Sale: Analyzing Media Retail*, ed. Daniel Herbert and Derek Johnson (New Brunswick: Rutgers University Press, 2019),160–74.
4. Davis, "For the Records," 72.
5. Ibid., 73.
6. Stefan Niemer et al., *Reshaping Retail: Why Technology Is Transforming the Industry and How to Win in the New Consumer Driven World* (West Sussex, UK: Wiley, 2013), 15–18.
7. Will Straw, "Music and Material Culture," in *The Cultural Study of Music: A Critical Introduction*, ed. Martin Clayton, Trevor Herbert, and Richard Middleton (New York: Routledge, 2012), 227.

8   Evans Clark, "Big Business Now Sweeps Retail Trade: Huge Corporations, Serving the Nation through Country-Wide Chains, Are Displacing the Neighborhood Store—The New Age of Mass Distribution Is Working a Revolution in American Sales Methods," *New York Times*, July 8, 1928, 109.

9   Kat Eschner, "The Bizarre Story of Piggly Wiggly, the First Self-Service Grocery Store," *Smithsonian Magazine*, September 6, 2017. https://www.smithsonianmag.com/smart-news/bizarre-story-piggly-wiggly-first-self-service-grocery-store-180964708/#2BRlMscpbLR5mCCp.99.

10 After the first Piggly Wiggly opened, Saunders secured his concept with a series of patents belonging to his Piggly Wiggly Corporation. See previous note for source.

11 "Dry Goods Retailers Meet: Convention Here to Discuss Conditions Caused by the War," *New York Times*, February 11, 1919, 16.

12 Ashley Ross, "The Surprising Way a Supermarket Changed the World," *Time*, September 9 2016. http://time.com/4480303/supermarkets-history/.

13 Ibid.

14 Paul du Gay, "Self-Service: Retail, Shopping and Personhood," *Consumption, Markets and Culture*, no. 7 (2004): 149–63.

15 Franck Cochoy, *On the Origins of Self-Service*, trans. Jaciara Topley-Lira (New York: Routledge, 2016), 7.

16 Anderson, "Female Treble."

17 "We Now Behold the 'Dresseteria': Self-Service Idea Is Having a Try-out in the Dry Goods Zone," *New York Times*, February 8, 1925, 171.

18 Ibid.

19 See "Mr. Sears Catalog," *American Experience*, PBS-WGBH. November 21, 1989.

20 Ibid.

21 Du Gay, "Self-Service," 151.

22 Du Gay reminds us that early attempts to move Tesco's and Sainsbury's from counter to self-service did not come without stumbles. The High Street retailer, Tesco, converted one of their stores to self-service in 1947, only to have to return it to counter service when customers complained of the inconvenience of having to shop for themselves. The opening day of the first Sainsbury's self-service store in Croydon found a queue of only one person—the branch manager's wife! The idea that, once unveiled and put to work, self-service just took over the retail environment simply cannot be taken at face value. See du Gay, "Self-Service," 153.

23 Du Gay, "Self-Service," 155.

24 Ibid. For du Gay, the consumer of note in these discussions is typically the housewife. It may be that the most significant amount of effort devoted to creating the self-service consumer is this gendered subject; however, it is not clear that this is the case in terms of experiential goods subject to self-service regimes such as books and records. Further investigation is needed in this area to understand these subjects.

25 "Self-Service Set-up Sock Sales Stimulant: Technique Is No Mere Experiment to Meet a Temporary Need; It's Here to Stay and Will Grow," *The Billboard Music Year Book*, 1944, 158.

26 Ibid.

27 Ibid.

28 Ibid.

29 Ibid., 159.

30   Walter W. Hurd, "Editorial: Ultra Self-Service," *Billboard*, June 8, 1946, 100.
31   Ibid.
32   Du Gay, "Self-Service," 149.
33   "Big Disk Firms Plan All-out Push on Albums, Promotion," *Billboard*, August 1, 1953, 1.
34   "Capitol's Fall Line: Set Promotion on 89 Package Titles," *Billboard*, August 1, 1953, 14.
35   Ibid.
36   Ibid.
37   Ibid., 14, 30.
38   "Labels Announce Dating Discount Plans for Easy Dealer Payment," *Billboard*, August 29, 1953, 31.
39   Ibid.
40   Joel Friedman, "Smart Shops Put Records in Ful View, Adopt 'Help Yourself' Policy," *Billboard*, July 17, 1954, 30.
41   Joel Friedman, "Self-Help Helps Dealer: Self-Service Shoots Ahead; Record Firms Offer Designs," *Billboard*, February 26, 1955, 18.
42   "Conkling, All Incumbent Riaa Officers in Again," *Billboard*, April 9, 1955, 16.
43   Advertisement. "Rca Victor Introduces New Low-Cost Record Store Modernization," *Billboard*, 1955, 36–7.
44   "A Special Section . . . Moderning Your Record-Phono Store," *Billboard*, June 30, 1956, 30.
45   Ibid.
46   Bob Gardner, "Self-Service Benefits: Advantages Boom Dealer's Trade," *Billboard*, August 4, 1956, 38.
47   Cochoy, *On the Origins of Self-Service*, 7.
48   For example, one of the keys to successful self-service for supermarkets was the innovation of glass refrigeration units that were both "conservation devices" and "display tools" that acted as "equipment that sells goods." Ibid., 105.
49   Ibid.
50   Joe Martin, "Record Industry Foresees Quickened Pace in Changes: Evolution Focuses on Market Levels: Self-Service Tack Expansion Heightens Diskeries' Dealer Problems," *Billboard*, July 18, 1953, 1.
51   Friedman, "Smart Shops," 30.
52   Ibid.
53   Ibid.
54   Bill Simon, "Self-Service Proves Itself as Panacea for Dealers," *Billboard*, July 23, 1955, 46.
55   Cochoy, *On the Origins of Self-Service*, 144.
56   Barrett Axel, "History of Cellophane," *Bioplastics News*, July 23, 2019. https://bioplasticsnews.com/2019/07/23/history-of-cellophane/.
57   "The Pack & Wrap Make Boston Rack Healthy, Wealthy & Wise," *Billboard*, December 14, 1963, 3, 40.
58   Ibid.
59   Cochoy, *On the Origins of Self-Service*, 135.
60   Howard G. Haas, "Couple of 'How To's': Approach to Floor Traffic, Set Sales," *Billboard*, October 1, 1955, 30.
61   Gordon Freedman, "Fixtures Play Only a Part in Successful Disk Store Project," *Billboard*, June 30, 1956, 23, 30.
62   Ibid.

22

Lost in the Booth

British Record Store Listening Booths as Atmospheric Sites of Intimacy

Peter Hughes Jachimiak

I started buying records sometime late in 1978. At that time, we lived in a part of Birmingham called South Yardley in the shadow of this weird wavy office block, and on the other side of that block was a little row of shops that appeared to have been unchanged since the previous decade at least. One of them was a record shop called Discus . . . Discus had three disused listening booths, presumably a relic from the previous decade, when all record shops had to have listening booths.

—Pete Paphides, journalist and broadcaster[1]

With Paphides's evocative recalling of Discus's neglected booths, and the store's centrality amid a bustling high street which—in all probability—is now altered beyond recognition, this chapter is concerned with both record stores and particular places within them—namely, listening booths—as sonic microcosms. Thus, what follows will, first of all, examine such distinct sites within the record store from individuals' perspectives as very intimate, contemplative, spaces-within-spaces. Then, in an effort to acknowledge the socially enhancing aspects of record stores and their listening booths, whereby they are to be understood and appreciated as shared cultural spaces, both will be considered from a community standpoint. Moreover, employing the combined approaches of cultural history, urban geography, and the still-emerging study of atmospheres (see, for example, Tonino Griffero),[2] this work will consider both the aesthetic and atmospheric significance of the listening booth—that is, from the "space age," headphone-less, cell-like rows found in HMV, Oxford Street, London, during the mid-twentieth century to the more DIY-like, quirky places that were once found amid the late twentieth century high streets of the UK. In doing so, it will aim to connect deeply personal experiences with wider collective associations and the past with the present. As such, use of empirically gathered oral history testimonies[3] will be made throughout (along with similar material drawn from secondary sources), as a means by which to investigate the atmospheric dimensions of listening booths within record store environments. Indeed, ahead of doing so, we should bear in mind Shanti Sumartojo and Sarah Pink's assertions in establishing a "new agenda" as far as "thinking atmospherically" is concerned: "[w]e have argued that atmospheres must be anchored in the specific and contingent circumstances from which they emanate, and that this demands methodological approaches that attend closely to these conditions and their ongoing emergence."[4] So, with these "specific and contingent circumstances" in mind, an overview of the societal positioning of the record store and the historical context of the listening booth will be provided.

"Ancient," "Old," and "Vintage"—British Record Stores and Their Listening Booths

Perhaps rather predictably, among British publications documenting record stores and the so-called vinyl revival, there is often an emphasis on London's record stores, both high-street chains and local independents. Yet, despite their London-centric worldview, such texts offer invaluable insights into both the British and global resonances of record stores and record hunting. For example, Tom Greig[5] insists that paying a visit to an actual record store is something that cannot be replicated virtually—that is, by either shopping online or listening to music via streaming services. For him, the practice of searching, locating, and uncovering shop-based music cannot be bettered, whereby one is "interacting with people, an environment and a carefully selected spectrum of music."[6] Moreover, Garth Cartwright[7] argues that record stores do so much more than shift units to customers. In reality, they act as community-based sites of creative energy, where people meet, make friends, form bands, and embark on setting up record labels. Indeed, in moving from the physical aspects of record stores to their more esoteric qualities and highlighting the sensory-orientated experience of pouring over a shop's shelves, Marcus Barnes is perceptive. In his estimation, pleasures can be found "from the aroma of old acetate and the anticipation when on the hunt for a specific record to the buzz of the store itself and the physical act of flicking through shelf after shelf of records."[8] In short, the act of visiting a record store—fingertips flitting from record to record and chatting with staff and fellow customers—cannot be replicated when purchasing music online.

While such memories of London's record stores (both existing and now lost) focus upon the social vibrancy to be found within them, this chapter's examination of such environments strongly aligns with Marcus Barnes's work about the "buzz" of record stores. He explores how the inner self is experienced when visiting a community-orientated retail space that is also an emotionally intimate space. More specifically, I will examine, here, the places-within-places that were record store listening booths: as private spaces within these shops, they allowed listeners to lose themselves within their own feelings.

Cartwright[9] offers a detailed insight into the birth and rise of the Gramophone Company's very first HMV department store. Opening on July 20, 1921, the premises, situated at 363 Oxford Street, London, was marketed as "something beyond an ordinary merchandise emporium," whereby the very sonic impact of the store was phenomenal: "[u]pon entering and hearing Beethoven's 5th or Wagner's Valkyries slice through the air, certain customers felt faint, were left gasping for breath, so intense was the experience."[10] Indeed, this initial HMV site epitomized the controlled shopping environment, as—years before the stores' management would permit the public to actually handle the stock kept—customers were required to ask for a record of their choice, ahead of purchase, which a clerk would then take out of storage for it to be listened to inside one of twenty-five available "audition rooms" (i.e., soundproofed, stand-alone listening booths). However, by the mid-to-late 1950s, HMV, and other pristinely kept department stores that possessed record departments full of neatly ordered racks of classical music and Broadway show titles, began, rather begrudgingly, to stock a small number of rock and roll releases. Thus, almost overnight, they witnessed gangs of youths crammed into their listening booths who were scornful of all that HMV, and others similar, offered—all, that is, except the primal rock and roll that they yearned for.

Cliff White,[11] a member of HMV staff from 1964 until 1968, notes that by the mid-1960s, 363 Oxford Street, amid its basement-level record department, had a pair of "listening rooms" and seemingly endless rows of listening booths along both the left and back walls. Installation of the latter meant that dozens of shoppers could listen to countless records at any given time. As White makes clear, each of these wood-enclosed, numbered hubs was connected to its own record player. In fact, White's reminisces emphasize the utter liveliness of HMV's basement during these years—especially during lunchtimes and after the working day, when youths packed out the booths. Now, if White's account was typical of listening booths and their uses during the mid-to-late 1960s, and from an HMV assistant's perspective, then how did the record-buying public respond? An example of these views is provided by one such customer from the era, as recounted via the website *ST33*.[12] Indeed, by carrying out research into E. M. G. Handmade Gramophones Ltd. (EMG),[13] the online article's author inadvertently uncovered this customer's recollections of record buying in 1940s London. For, mentioned in this account are not only all-day trawlings through masses of secondhand and discontinued vinyl but the liberties taken by this customer listening to countless records before eventually buying only one or two at the end of each record-hunting day. In fact, a typical record-hunting day involved the customer arriving from Kent at Charing Cross station before 8.00 a.m. and then visiting the Gramophone Exchange on Shaftesbury Avenue. There, at the far end of the premises, were a number of listening booths—what the customer recalled as "musical retreats" that "we all loved so much."[14] Also according to this customer, while EMG possessed very similar booths to that of the Gramophone Exchange, EMG was not spacious, so—eventually—a far more accommodating record-buying respite was to be had amid "the more palatial HMV premises."[15] However, as—according to the customer—HMV's listening booths possessed "an odd, electrical sort of smell which made you feel ill,"[16] he felt so unwell sitting amid the booths that "[o]n every visit I was glad to get out and would occasionally commit the unforgivable, leaving before hearing all I wanted."[17]

This customer's account provides us with not only great insight into a record collector's search for classical music during 1940s London but a true sense of the physical-come-emotional atmospheres experienced when encased within record stores' listening booths. But, if that was London in the 1940s, what about the rest of the UK in the decades that followed? Via oral testimonies, my respondents provided a real sense of geographical scope of where their local record shops (and the listening booths within them) were to be found. For, while one of my respondents spoke about "L & H Cloakes, on Streatham High Road, south London,"[18] for others their local record stores were "The Sound of Music, Rotherham,"[19] and "a large-ish record shop around Swansea Market area."[20] Moreover, when attempting to recall such stores (which not only had listening booths but were ones they regularly visited), highly personal memories were evoked. Indeed, here's an account from one Cambridge-based customer:

> There was a record store with listening booths located in the heart of Cambridge when I was growing up there. The shop was called either Miller or Miller's. In fact, there were two shops [belonging to the same company] which faced each other on opposite sides of the street. They were glass-fronted places, but had an "old," "traditional" look even in the early 1980s. They were the kinds of places that had curved glass each side of the doors. Entering the shop with the listening booths, you were first greeted by pianos and classical instruments. The records in stock appeared to be only classical recordings and did not interest me at the time.[21]

Meanwhile, another respondent held memories, dating back to the mid-to-late 1970s, of two stores in particular:

> [A] store in Hayes, Middlesex called Rowleys Electrical Ltd. In 1978 I worked quite near this shop, and I would go in on my lunch hour. Record Centre in Slough also had one, but both shops are gone now. The shop in Hayes, Rowleys, was a throwback to the 1950s, with a fittingly vintage interior. They sold all electrical goods and the owner wore a work coat—the type a butcher or storeman wore.[22] Slough's set-up, meanwhile, was more contemporary in design.[23]

What is significant about these memories—especially regarding the record stores' unique, site-specific interiors—is that all was characterized by the coming together of the old (i.e., traditional) and the new (i.e., contemporary), where, for instance, classical music coexisted with popular music, and traditional architecture was to be found amid contemporary urban design and planning (curved glass fronts, New Towns,[24] etc.). But, if all of this highlights the stores' structural make-up—and, in turn, gives clear indications of the ways in which customers interfaced with such highly physical spaces—let us also consider what those structures and spaces sounded like: in short, what was listened to amid those spaces and what sounds emanated from those structures.

"I Was Like a Kid in a Toy Shop": Individuals' Experiences of Listening Booth Atmospheres

Graham Sharpe, when recalling his visits to a number of record stores deemed to be haunted, insists upon the nigh-on-paranormal nature of record stores (haunted or not) and, even, records themselves. In his view, record stores, due to their function, are populated with the "ghosts" of long-dead singers and musicians found amid the grooves of long-playing records, seven-inch singles, EPs, and the like. Moreover, such music-locked spirits are repeatedly resurrected when a customer "picks up, plays or discusses one of their records."[25] While "paranormal" considerations of recorded media and the mediumship of the listener (where we are in-between receptors of sound) have already been explored,[26] what is most relevant here are the feelings—that is, atmospheres—evoked when listening to music. According to Friedland Riedel, to engage with the notion of atmospheres is not, in any way, contradictory to the affective, two-way, bodily interaction when one engages with sonic stimuli. Rather, it is the offering of "a dynamic of mediation, namely between the environmental whole and individual bodies."[27] In short, then, sound and atmospheres, and atmospheres of sound, are things in which we are immersed—we are, in effect, conduits for them.

With the interwoven nature of sound and atmospheres, and our bodily in-betweenness in relation to both in mind, it is essential that we consider the ways in which we perceive atmospheres (sonically induced or otherwise). Moreover, it is vital for us to think about the body-centric perception of atmospheres with record store listening booths at the forefront of our minds. According to Griffero, one of the leading voices on this topic, it is the "perceiving" of atmospheres that is all important. Being open to our atmospheric environment is key to us being aware of, and being able to understand, our inner selves within our everyday surroundings.[28] However, rather than being a unique experience, Griffero insists that any perception of atmospheres is merely an extension of our everyday perceptions. That said, atmospheric perception is not a clumsy sensory-led grabbing of the world that immediately surrounds us. It is instead a more in-tune engagement with things and situations. In short, the true perceiving of atmospheres is "a holistic and emotional being-in-the-world."[29]

This is not to say that the perception of atmospheres is either straightforward or concerned with easily measurable environmental conditions. An academic appreciation and understanding of atmospheres are seemingly well-suited to the humanities, yet it is, perhaps, less suited to more quantifiable aspects of science-led human existence—where, for STEM-associated academics, such an approach to bodily felt atmospheres has yet to be pursued. However, when the phenomenon under scrutiny—such as the intimacy felt when inside a listening booth—is essentially immeasurable via, say, statistical analysis, the atmospherically driven approach would seem more fitting given what Griffero terms the "felt-body experience" of "quasi-things."[30] Regarding the latter, these are elements of life that are not wholly three-dimensional and/or fully sensed experiences, yet they nevertheless "exert on us a more direct and immediate power than that exerted by the object."[31]

Meanwhile, in relation to the former, the "felt-body experience," this is something which is not reliant upon, in absolute terms, the sensory organs. Griffero explains that the "felt body" possesses an absolute spatiality through its emotion-sensitive grounding in specific locations. For example, the feeling of warmth, in emotional terms, does not negate any perceptible surrounding drop in temperature, as, instead, "warmth, with its absolute spatiality, is perceived within the multiple felt-bodily isles."[32] As it is within our very own felt-bodily isles, Griffero insists, that we experience above and beyond what our five senses are able to perceive at any given time. Thus, in day-to-day terms, our "felt-bodily isles" are when our internal tissue semi-manifests itself as something that is (if only fleetingly) integral to our emotional condition. For example, this happens when we feel love as something that emanates from our inner chest (where and when the heart is responsible for more than pumping blood around the body) or when we have the feeling of "butterflies in our stomach," the organ represents much more than its bodily (digestive-based) function.[33]

The basis of Griffero's conceptualizing of the felt-bodily isles' perception of atmospheres comes from Hans Ulrich Gumbrecht's work—and, especially, a need to appreciate Gumbrecht's use of the German term *Stimmung* (mood). For Gumbrecht, *Stimmung* suggests both "mood" and "climate," with the former suggesting an indeterminate core sensation and the latter a tangible surrounding that holds a corporeal effect upon living things. However, Gumbrecht is at pains to point out that in German, *Stimmung* correlates directly with *Stimme* and *stimmen*, where *Stimme* means "voice" and *stimmen* is "to tune an instrument." Quite crucially for Gumbrecht, any tuning of an instrument evokes the notion that identifiable moods, atmospheres, and so on are engaged with on a musical scale-like continuum.[34] And it is this experiencing of moods and atmospheres as connected to music that is—certainly as far as this chapter is concerned—absolutely crucial here. As Gumbrecht elucidates:

> I am most interested in the component of meaning that connects *Stimmung* with music and the hearing of sounds. As is well known, we do not hear with our inner and outer ear alone. Hearing is a complex form of behavior that involves the entire body. Skin and haptic modalities of perception play an important role. Every tone we perceive is, of course, a form of physical reality (if an invisible one) that "happens" to our body and, at the same time, "surrounds" it.[35]

Empirically, Griffero's and Gumbrecht's theories are reflected and further articulated by the following testimonies, as these recollections speak to varying experiences of what "happens" when our bodies are "surrounded" by music. For example, one of my respondents, who was Rotherham-based, remembers the feelings that came with listening to punk singles at the Sound of Music record store. They and their friends would convene in the listening booths enacting their own "*Juke Box Jury*-styled thumbs up or down" scoring process before choosing which records to buy. Importantly, they never felt "anxious" or "intimidated" while there, appreciating fully how "the staff were very knowledgeable"—which speaks to how this music enthusiast gauged the store's overall *Stimmung* while shopping for records.[36] Moreover, another of my respondents, who "was a Skinhead (nice type!) from a very young age," was, at the store in Slough, "listening to anything like Roxy Music and Bowie, to Reggae, Soul, and Motown" and, at the Hayes store, "American Punk-New Wave stuff like Talking Heads or Television."[37] In these instances, such *Stimmung* was defined by multi-site musical and subcultural eclecticism. Elsewhere, a record store's *Stimmung* was, instead, not only characterized in negative terms but impacted by—quite negatively—a customer's subcultural "look" being at odds with the store's expectations of their clientele. At a classical-orientated store in Cambridge, another respondent commented that it was "not my kind of place at all." Dressed in his "mod uniform" he, when entering, "always felt like an undesirable type."[38] For others, though, the record store *Stimmung* was one of absolute comfort and almost child-like contentedness—for example, with a female respondent from Swansea:

> I remember listening to Scott McKenzie's "San Francisco (Be Sure to Wear Flowers in Your Hair)," and I bought it—in fact, still got it! I think I was 18? I didn't feel any need to rush to get out of there. It was a good experience, and I felt happy listening—good laid-back atmosphere, as it was in those times. I was like a kid in a toy shop.[39]

From this testimony, the atmospherically imbued interiors of the Sound of Music, Miller's, and all of the other stores were sites where any number of individuals, from various youth subcultures of the time, felt either welcomed or intimidated—where assistants would freely share their knowledge and expertise or keep a watchful, suspicious eye on patrons until they left the premises. Meanwhile, when hearing music within a listening booth, it was often a "laid-back" experience and akin to being "a kid in a toy shop." In essence (and with the latter analogy especially), this is a true encapsulation of the atmospheric felt-body being in harmony with its intimate, sonic-specific environment. But the individuals' experiences of both record shops and listening booths, at the level of the intimate, are, of course, only a partial account. The more collective interface is, now, something to which we turn—that is, record stores and their listening booths as communal spaces and places.

"From a Time Long Since Passed"—The Communities' Experiencing of Listening Booth Atmospheres

With a sense of where such record stores were to be found within the UK, an idea (structurally and aesthetically speaking) of what they looked like, and the ways in which their listening booths connected with people on an individual basis, it is also important to get a sense of where and how listening booths were positioned within such record stores. For, at Rotherham's the Sound of Music, the store's handful of listening booths were, according to one respondent, "past the main record sections and attached to a wall that was parallel to the opening shop door,"[40] while the two booths at Rowleys, Hayes, "were right-hand side,"[41] and, at Miller's, Cambridge, "a row of maybe three or four booths" were "situated at the back on the right of the cash desk as you walked in."[42] Meanwhile, at London's L & H Cloakes on Streatham High Road the booths "were near the front of the shop, along a side wall."[43]

The reason for sharing these descriptive positional recollections is that—especially in comparison to London's HMV on Oxford Street—such provincial record stores, and the number of listening booths they had, were all of a smaller scale. Furthermore, acknowledging the fact that space-age "commodity scientism" had characterized design and architecture during the late 1950s and early-to-mid 1960s,[44] one nonetheless gets the sense that—atmospherically and/or structurally—space-age-conscious design was virtually nonexistent at these stores. Instead, here were small-town, high-street record stores where space was at a premium. That meant everything—door openings, counters, cash registers, record racks, listening booths, etcetera—was often positioned exceptionally close together in a highly cramped manner. As such, more contemporary—and, thus, costly—design trends were not on display, whether that was down to unaffordability or simply not being a style that the proprietor felt comfortable with (or, even, was compelled to choose).

If the above was true for provincial record stores during this time, let us, once again, turn to the interviewees' testimonials, which describe the internal characteristics of their local record store listening booths. For, while occasionally having wearable headphones installed within them (i.e., when wall-mounted speakers seemed to be the norm), they were often devoid of seating and even doors. Moreover, the listening booths frequented by my respondents were almost totally lacking when it came to glass, plastic, or Formica-based materials, with those at the Hayes store being "quite dull in colour."[45] Those found at Rotherham's the Sound of Music "were more of a wooden structure,"[46] while those at London's L & H Cloakes "were mainly wood" with "what looked like peg board."[47] Cambridge's Miller's was much the same—where "each booth was lined with that board with hundreds of holes drilled into it."[48] Yet, despite their rather basic wooden design, some of the booths did allow for ease of movement once inside. At the Swansea premises, "there was plenty of room for two—maybe three at a push."[49] At other stores the listening booths offered little or no room for movement. The booths at the Sound of Music were "not really spacious,"[50] while the ones at L & H Cloakes "looked quite cramped"[51] and those at Miller's "were quite dark and small."[52] All in all, these record stores—and their listening booths especially—seemed somewhat antiquated. According to my respondents, those at the Sound of Music "could have been considered a little dated,"[53] while those at Miller's "looked as if they were left over from a time long since passed."[54] In fact, such an aura of outdatedness explains why the booths seemed to be hardly ever used. Also, as a customer of the Hayes store explained, their lack of a modern aesthetic meant not only an aversion to using them on a regular basis but a sense of emotional discomfort: "So, yes, I felt a little anxious using them—embarrassed really."[55]

With the previous testimonies, we get a real sense—certainly as far as record store listening booths of the UK during the late 1960s, 1970s, and very early 1980s are concerned—that a budget, almost "DIY"-like approach, and aesthetic predominated. While lip service was possibly paid to the basic principles of space-age ergonomics (in that the booths were spacious enough for any listener to be encased by their structure without feeling too cramped), we are left with an overarching sense of something—atmosphere- and mood-wise—quite drastically removed from cutting-edge, space-age aesthetics and mid-twentieth-century popular design. However, that is not to say that such low-tech listening booths—and the old-fashioned record stores within which they were found—totally lacked atmosphere, mood, or *Stimmung*. For this was not only the result of the felt-body isles internalizing both the record stores' and listening booths' antiquated and non-state-of-the-art environments but the externalizing of all in the manner of offering community-enhancing spaces-within-spaces.

Conclusion—*Stimmung*, Yearning, and the Cultural Past of the Record Store

According to Gumbrecht, *Stimmung* is, in many respects, a necessary conceptual means by which we can both make sense of and cope with life today.[56] But, it is far more than that: the longing for *Stimmung* has increased in recent years as many—and the older population especially—live in an era where there is a sense that our environment no longer cocoons us materially in any meaningful way. In that respect, any desire for atmosphere, mood, and the like is a hunger for a more visceral existence—and, quite possibly, a version of an existence "that presupposes a pleasure in dealing with the cultural past."[57]

Arguably, hunger for such an existence and experiences is vital—certainly from an empirical perspective—in that immersing oneself in, theorizing about, and projecting ourselves within and beyond atmospheres brings about useful rules of thumb for analysis in order to fully comprehend them and ourselves.[58] For, even as far as the most (seemingly) insignificant, day-to-day events are concerned, "thinking atmospherically" allows us to acknowledge that atmosphere is ever present and that the task at hand is to acclimatize ourselves to it, in order to "find the terms to best describe it and understand what work it is doing."[59]

This discussion recognizes that amid record store listening booths, characteristic atmospheres existed—*exist*, even, as they still live on in memory. Via testimonies from those who, in their youth, experienced such booths, letting their felt-bodies be enveloped by the music-imbued atmospheres, this chapter has been an attempt to better understand the effect such atmospheres had on them. Those music consumers who once sat or stood in a listening booth, losing themselves in a record, are now in middle or late age. Reflecting on their past, it might be said this cohort yearns for such an experience—an experience now long-lost and only attainable via their memories of the atmospheres and moods record stores, and their booths, once evoked and provided.

Notes

1 Pete Paphides cited in "Pete Paphides: The Record Collector," https://www.homeprotect.co.uk/discover/stories/pete-paphides-record-collector.html (accessed November 5, 2021).
2 Tonino Griffero, *Atmospheres—Aesthetics of Emotional Spaces*, trans. Sarah DeSanctis (London: Routledge, 2016).

3   The oral history testimonies were gathered by the author between November 17 and December 16, 2021. Respondents were selected from the author's already-existing social media contacts and the interviews were conducted utilizing Facebook Messenger. Suitability of selection was based upon age (they had to be in their mid-fifties or older to have experienced listening booths) and a high degree of interest in record buying and listening (certainly in their youth). A significant effort was made to achieve a reasonable level of diversity of race/ethnicity, class, and gender (in that one respondent was female, another was British Asian, and so on).
4   Shanti Sumartojo and Sarah Pink, *Atmospheres and the Experiential World—Theory and Methods* (London: Routledge, 2020), 119.
5   Tom Grieg, *Vinyl London* (Woodbridge: ACC Art Books Ltd, 2019).
6   Ibid., 8.
7   Garth Cartwright, "Introduction," in *London's Record Shops*, ed. Garth Catrwright and Quintina Valero (Cheltenham: The History Press, 2021), 11–20.
8   Marcus Barnes, *Around the World in 80 Record Stores—A Guide to the Best Vinyl Emporiums on the Planet* (London: Dog 'n' Bone Books, 2018), 7.
9   Garth Cartwright, *Going for a Song—A Chronicle of the UK Record Shop* (London: Flood Gallery Publishing, 2018).
10 Ibid., 21.
11 Cliff White, "HMV Revisited," in *Rock's Backpages Library*, November 2013. https://www.rocksbackpages.com/Library/Publication/rock's-backpage (accessed November 5, 2021).
12 "EMG—the shop," *ST33*. https://st33.wordpress.com/record-shops/closed/emg-the-shop/ (accessed November 5, 2021).
13 E. M. G. Handmade Gramophones Ltd. (EMG) was located on Grape Street, London.
14 "EMG—the shop."
15 Ibid.
16 Ibid.
17 Ibid.
18 P. G., Facebook message to author, December 12, 2021.
19 T. B., Facebook message to author, December 14–16, 2021.
20 M. B., Facebook message to author, November 17–18, 2021.
21 A. H., Facebook message to author, November 19–20, 2021.
22 The term "storeman" (who wore a "work coat"), in this instance, refers to a worker (usually a man—considering the era) who worked "behind the scenes" in a shop.
23 J. W., Facebook message to author, November 17–20, 2021.
24 "New Towns" was the term for post-1945, stand-alone urban developments that came about after the New Towns Act of 1946. Constructed in three waves, they were initially conceived in order to rehouse those following war-time bombing of city and town centers and slum-clearance programs thereafter. Later on, existing towns and peripheral areas of cities were added to in order to provide "over-spill" areas for an ever-expanding UK population.
25 Graham Sharpe, *Vinyl Countdown* (Harpenden: Oldcastle Books, 2019), 94.
26 See Jeffrey Sconce, *Haunted Media—Electronic Presence from Telegraphy to Television* (Durham: Duke University Press, 2000); David Toop, *Sinister Resonance—The Mediumship of the Listener* (London: Bloomsbury, 2010).

27 Friedlind Riedel, "Atmospheric Relations—Theorising Music and Sound as Atmosphere," in *Music as Atmosphere—Collective Feelings and Affective Sounds*, ed. Friedlind Riedel and Juha Torvinen (London: Routledge, 2020), 4.
28 Griffero, *Atmospheres*.
29 Ibid., 15.
30 Tonino Griffero, *Quasi-Things—The Paradigm of Atmospheres*, trans. Sarah de Sanctis (Albany: State University of New York Press, 2017), 11–14.
31 Tonino Griffero, *Atmosphere/Atmospheres—Testing a New Paradigm*, ed. Tonino Griffero and Giamiero Moretti (Sesto San Giovanni: Mimesis International, 2018), 13.
32 Griffero, *Quasi-Things*, 61.
33 Ibid., 62.
34 Hans Ulrich Gumbrecht, *Atmosphere, Mood, Stimmung—On a Hidden Potential of Literature*, trans. Erik Butler (Stanford: Stanford University Press, 2012).
35 Ibid., 4.
36 T. B., Facebook message.
37 J. W., Facebook message.
38 A. H., Facebook message.
39 M. F., Facebook message.
40 T. B., Facebook message.
41 J. W., Facebook message.
42 A. H., Facebook message.
43 P. G., Facebook message.
44 Timothy D. Taylor, *Strange Sounds—Music, Technology and Culture* (Routledge: New York, 2001).
45 J. W., Facebook message.
46 T. B., Facebook message.
47 P. G., Facebook message.
48 A. H., Facebook message.
49 M. F., Facebook message.
50 T. B., Facebook message.
51 P. G., Facebook message.
52 A. H., Facebook message.
53 T. B., Facebook message.
54 T. B., Facebook message.
55 J. W., Facebook message.
56 Gumbrecht, *Atmosphere*.
57 Ibid., 20.
58 Sumartojo and Pink, *Atmospheres*, 119.
59 Ibid.

Editors and Contributors

Editors

Gina Arnold is a professor, author, and music journalist. As a writer for *Rolling Stone*, *Spin*, the *Village Voice*, and many other publications, her work has been excerpted in many anthologies, including *Shake It Up: Great American Writing on Rock and Pop from Elvis to Jay-Z*, *The Rock History Reader*, *Rock She Wrote*. She is the author of four books, *Route 666: On the Road to Nirvana*, *Kiss This: Punk in the Present Tense*, *Liz Phair's Exile in Guyville* (2014), and most recently *Half a Million Strong: Crowds and Power from Woodstock to Coachella*. She is the coeditor of *Music/Video: Histories, Aesthetics, Media* (2017) and the *Oxford Handbook of Punk Rock*. She teaches courses in critical race studies at the University of San Francisco, United States.

John Dougan is Professor of Music Business and Popular Music Studies in the Department of Recording Industry at Middle Tennessee State University, United States. A former record store employee and music critic, he has contributed essays and reviews on music and popular culture to *Rolling Stone*, *Spin*, *American Music*, *Journal of Popular Music Studies*, *Popular Music & Society*, *Punk & Post-Punk*, *Minnesota History*, and online publications *Salon*, *Popmatters*, and *Perfect Sound Forever*. He has published dozens of artist biographies and discographies for the *All Music Guide* and is the author of two books: *The Who Sell Out* (2006) and *The Mistakes of Yesterday, The Hopes of Tomorrow: The Story of the Prisonaires* (2013).

Christine Feldman-Barrett is Senior Lecturer in Sociology at Griffith University in Australia and a youth culture historian and Beatles scholar. She has authored two monographs, *"We Are the Mods": A Transnational History of a Youth Subculture* (2009) and *A Women's History of the Beatles* (2021). She is also editor of *Lost Histories of Youth Culture* (2015), serves on the editorial board for the *Journal of Beatles Studies*, and is a member of the Subcultures Network. Further publications include articles in the *Journal of Youth Studies*, *Space and Culture*, *Popular Music and Society*, *Feminist Media Studies*, and other scholarly journals. Her scholarship and commentary have also been featured in *The Washington Post*, *The Guardian*, the *Conversation*, *CultureSonar*, and the ABC (Australia). Her favorite record store as a Chicago teenager was Wax Trax!

Matthew Worley is Professor of Modern History at the University of Reading, UK. He has written widely on British labor and political history, including books on the Communist Party of Great Britain, Labor Party, and Sir Oswald Mosley's New

Party. His more recent work has concentrated on the relationship between youth culture and politics in Britain, primarily in the 1970s and 1980s. He has published articles and essays in such journals as *History Workshop, Twentieth Century British History, Contemporary British History, Journal for the Study of Radicalism, Journalism, Media and Cultural Studies, Punk & Post-Punk* and chapters in collections such as the Subcultures Network's *Fight Back: Punk, Politics and Resistance* (2015). He is the author of *No Future: Punk, Politics and British Youth Culture, 1976–84*.

Contributors

Tim J. Anderson is Interim Chair and Professor of Communication and Theatre Arts at Old Dominion University, United States. Anderson studies the multiple cultural and material practices that make music popular and has published numerous book chapters, refereed journal articles, and two monographs: *Making Easy Listening: Material Culture and Postwar American Recording* (2006) and *Popular Music in a Digital Music Economy: Problems and Practices for an Emerging Service Industry* (2014). His latest research project focuses on records and the public sphere. His website is timjanderson.weebly.com.

Fernán del Val is Assistant Professor in the Department of Sociology I at Universidad Nacional de Educación a Distancia (UNED), Spain. He has been a postdoctoral researcher at Universidade do Porto (Portugal), professor of sociology of music at Universidad Alfonso X el Sabio (Spain) and of sociology of consumer at Universidad de Valladolid (Spain). He has published various articles and books on music, politics, media, and youth in Spain, and has been a visiting researcher at Newcastle University (UK), Universitat Pompeu Fabra (Spain) and Universidad de Barcelona (Spain). From 2015 to 2021 he was president of the Spanish branch of the International Association for the Study of Popular Music (IASPM).

Eromo Egbejule is Africa Editor at Al Jazeera English (Online) and was a church drummer before becoming a journalist after completing engineering school. Educated at the universities of Nigeria, Leicester (United Kingdom), and Columbia University (United States), he has reported across Africa, the Peruvian Amazon, and at the 2018 World Cup in Russia. His writing and photos have appeared in *The Guardian, Washington Post, Financial Times, New York Times, Frankfurter Allegemeine Teitung,* and as liner notes for Apple Music. He has also been a visiting lecturer at Malmö University in Sweden. His debut film, *Jesse: The Funeral That Never Ended* (2019), retells the story of Nigeria's most horrific oil explosion that went on for five days and killed at least 1,000 people.

Jean Foubert is an art and film historian, associate research fellow at Laboratoire de Recherche sur les Cultures Anglophones (LARCA Université Paris Cité, France), and teaches cultural and film studies at the École de management de Normandie in France. A former visiting scholar at the UC Berkeley (United States) film department, he has authored two books: *L'Art audio-visuel de David Lynch* and *Twin Peaks et ses Mondes*. He has published essays on Michelangelo Antonioni, Lee Friedlander, Alfred Hitchcock, Edward Hopper, Roman Polanski, and Slavoj Žižek. His forthcoming work focuses on Edward Dmytryk's *The Sniper* and Roman Viñoly Barreto's 1953 remake of *M, El vampiro negro*.

Lee Ann Fullington is a reference and instruction librarian/associate professor at Brooklyn College, City University of New York, United States. She holds an MPhil in popular music studies from the University of Liverpool and an MSLIS from Pratt Institute. Unsurprisingly, her record collection fights for dwindling shelf space with her book collection.

Holly Gleason was born in Cleveland, Ohio, came of age in Miami, and spent her key writing years in Los Angeles. A first call critic when she was often "the only girl," her expertise in rock, country, and singer/songwriters appeared frequently in *Rolling Stone, Los Angeles Times, New York Times, Musician, CREEM, Playboy, Miami Herald, Cleveland Plain Dealer, No Depression, Variety,* and *NPR*. Conceptualizer/editor of the Belmont Book Award-winning *Woman, Walk the Line: How the Women of Country Music Changed Our Lives*, she is the 2019 CMA Media Achievement honoree and editor of the forthcoming *Prine on Prine: Interviews and Encounters*.

Ben Green is a cultural sociologist with interests in popular music and youth studies. He is undertaking a Griffith University Postdoctoral Fellowship researching crisis and reinvention in Australia's live music sector. His previous work explores memory and heritage, cultural policy, youth, and well-being through ethnographic research in urban, regional, and trans-local music scenes. His first book *Peak Music Experiences: A New Perspective on Popular Music, Identity and Scenes* examines cultural memory in Brisbane music scenes based on interviews, observation, and documentary research.

Paula Guerra is Professor of Sociology and Researcher at the Institute of Sociology at the University of Porto (Portugal), and Adjunct Associate Professor in the Griffith Center for Social and Cultural Research in Australia. She is founder/coordinator of the network/journal *All the Arts* and of the project/conference KISMIF (kismifconference.com and kismifcommunity.com). She is a member of the Board of the *Research Network of Sociology of Art* of ESA and a leading international scholar on the topics of sociology of culture, youth, and arts. Currently, she is editor-in-chief (with Andy Bennett) of SAGE new journal *DIY, Alternative Cultures and Society*.

Peter Jachimiak is Senior Lecturer at the University of South Wales (Atrium, Cardiff campus, UK) and regular contributor to both *Subbaculture* fanzine and *Detail— The Magazine for Modernists*. His book *Remembering the Cultural Geographies of Home* (2014) is concerned with the spaces, places, and media forms that make up the childhood family home. More recently, his essay "'Curious Roots & Crafts'—Record Shops and Record Labels amid the British Reggae Diaspora" was included in the anthology *The System Is Sound: Narratives from Beyond the UK Reggae Bussline* (2021), and "Meadows, Relics, and Victorian Dolls' Houses—Places, Ephemera, and the Unreal Realities of Pink Floyd" in *The Routledge Handbook of Pink Floyd* (2022) considers the eerie, weird, and uncanny elements of Pink Floyd.

Jay Jolles is a PhD candidate in American Studies at the College of William & Mary, United States. He is an interdisciplinary scholar with interests in a wide range of fields including twentieth and twenty-first-century literature and culture, comparative media studies, critical theory, and musicology. His work has appeared in *The Los Angeles Review of Books*, *U.S. Studies Online*, and *Comparative American Studies*.

Ken Kato is a PhD candidate in the Department of Musicology, Osaka University, Japan, and a research fellow of the Japan Society for the Promotion of Science. He specializes in musicology and urban sociology, and his main research interests look at the interactions of globalization, public policy, and technology in popular music. His publications include *City Pop toha Nanika* (2022), coauthored with Moritz Sommet, Shibasaki Yuji, and others. Although the first ever music medium he owned was a Pokémon anime song CD in his childhood, for the past decade he has been fascinated by vinyl records.

Mariana Lins is a Brazilian journalist with an MA in communication for her dissertation "The Aestheticization of Politics in Madonna's Performance" (2017). She is currently a doctoral student in communication at the Federal University of Pernambuco (Brazil) and was part of a Doctoral Exchange Program at the Universidad de Oviedo (Spain) in 2019–20. Her research focuses on aging female artists and ageism in the music industry.

Lily Moayeri has been a music journalist since 1992. She has contributed to numerous publications, including *NPR*, *Los Angeles Times*, and *Rolling Stone*. Currently, her writing can be found in *Variety*, *Spin*, *Grammy*, *Billboard*, *Mix Magazine*, and *Flood Magazine*, among others. She is also a contributing editor to *The Guerilla Guide to the Music Business* and has penned an essay on electronic and dance music for *75 Years of Atlantic Records*. She hosts and coproduces the *Pictures of Lily Podcast*, a music commentary series about her interviewing experiences. Since 2004 she has served as teacher-librarian focusing on guiding students in navigating the intersection of technology and education.

Kenny Monrose is a researcher at the Cambridge University in the Department of Sociology and a fellow at Wolfson College Cambridge, UK. He is the co-chair of the University of Cambridge Race Equality Network and an affiliate of the Centre for Screen & Film within the Faculty of Modern & Medieval Languages & Linguistics. He is also a member of Center for the Study of Global Human Movement at the University of Cambridge. His interests lie in the areas of criminal justice and "race" identities, and is particularly interested in the impact of Afro-descendant music within and around the Black Atlantic.

Roy Montgomery was born in London, England, and currently lives in Christchurch, New Zealand. He lectures on environmental management and urban planning at Lincoln University, New Zealand. He worked in record shops in the 1980s and 1990s, plays electric guitar, and has made many home recordings since the early 1980s from which he has released several self-produced 45s, LPs, and CDs. He was involved in bands such as the Pin Group, Dadamah, Dissolve, and Hash Jar Tempo (with Bardo Pond), and he has collaborated with other artists such as Flying Saucer Attack, Liz Harris (Grouper), Julianna Barwick, Purple Pilgrims, and Haley Fohr (Circuit des Yeux).

Claudiu Oancea is currently a postdoctoral researcher at New Europe College, Romania, where he directs a research project on popular music in socialist Romania between ideology and entertainment. He has a PhD in history, and he has held several postdocs at New Europe College and at the University of Bucharest. He has also curated a series of records with Romanian progressive rock music from the 1970s, as part of the series "Romanian Sounds Unearthed," which he initiated in 2021. He is also a record collector and an avid traveler (and shopper) through record stores wherever he can find them.

Stephen Shearon is a musicologist, sometime-ethnomusicologist, and professor emeritus of music of Middle Tennessee State University, United States. A native of North Carolina, he came of age patronizing Record Bar as it began its rapid expansion. Most of his research has been on sacred, in particular Christian, music and culture. Since 2004 he has studied gospel convention singing and gospel music as an international phenomenon. In addition to articles such as "The Sacred in Country Music" in *The Oxford Handbook of Country Music* (2017), he coproduced the documentary film *"I'll Keep on Singing": The Southern Gospel Convention Tradition* (2010).

Karl Siebengartner studied history, English literature and language, educational studies, sociology, and politics at LMU Munich and history at the University of Sussex, UK. He is currently completing a research project on punk in West Germany funded by the German Research Council (DFG). His research interests include youth cultures after 1945, pop and mass cultures, media history, and transnational history.

Christopher Spinks is a PhD student at the University of East Anglia, Norwich, UK. He holds a first class BA (Hons) and first class MA in modern history, both from the UEA. He is currently researching the representation of individual ordinariness within the post-punk period of 1977–85 by commercially successful mainstream punk musicians. He grew up during the first wave of northern soul and was a regular at all-dayers and all-nighters, as well as a collector of soul records.

Jon Stratton is Adjunct Professor in UniSA Creative at the University of South Australia, where he is also a member of the Creative People, Products and Places Research Center. He has published widely in *Popular Music Studies*, *Cultural Studies*, *Australian Studies*, *Jewish Studies*, and *Media Studies*. His most recent books are *Multiculturalism, Whiteness and Otherness in Australia* (2020) and, edited with Jon Dale and Tony Mitchell, *An Anthology of Australian Albums: Critical Engagements* (2020). Currently, he is the series editor of 33 1/3 Oceania, published by Bloomsbury.

Paul Tarpey is Senior Lecturer at the Limerick School of Art and Design, Ireland, with an art practice exploring how legacies of resistance define place. Writings on place making and music include "Notes on an Irish Disco Landscape" (*Irish Times*), *Pungent Architecture* (on Manchester) in *Heart and Soul: Critical Essays on Joy Division*, and *Everybody Produce Their Own Bombs* (on Northern Ireland) in *Durty Words* (Durty Books Limerick). His research can be found at www.resistanceandplace.ie

Mark Trehus has spent a lifetime immersed in music and records. He has curated hobby record labels Treehouse and Nero's Neptune, managed and owned record shops Oar Folkjokeopus and Treehouse Records, and produced sessions for, among others, Koerner, Ray & Glover, and the Pagans. He has amassed a library of 50,000+ records, encompassing free jazz, blues, garage, psychedelia, reggae, and countless other areas of interest. He is contentedly married to Alice, his wife of seven years, with whom he shares two cats, Bernadette and Nadine. He still lives in the Twin Cities and enjoys traveling, fine-tuning his collection, and indulging his love of Bob Dylan.

Index

Note: *Italicized* and **bold** page numbers refer to figures and tables. Page numbers followed by "n" refer to notes.

4ZZZ 65, 68, 69
   Radio Times 62, 63
10cc
   "I'm Not in Love" 166
120 Battements par minute 226

ABBA 165
Abbey Discs 154
Able Label 62
Abwärts 191, 192
Adichie, C.
   Americanah 139
Adler, S.
   "On the Media" 165
affective curation, record stores as space for 110–13
Afrobeat 135
afro-descendant music 37
Aghasi 166
Aiken, L. 31
Airbnb 98
Akademie der Künste 191
Albinoni, T. 169
Aldea, D. A. 129
All City 158
All Things Must Pass 2
Almazora 26
A&M 179
Amazon 66, 68
"Ambivalence" 185 n.10
Amerika-mura 144
Amoeba Music
   What's In My Bag? (*WIMB?*) 230–7
Anderson, J. 34, 197, 199–200, 204, 207 n.16
Anderson, M. 204
Andjelic, A. 111, 115
Angus, R. 177

Anna 202
Anselmo 113, 117
ANZ, *see* Aotearoa New Zealand (ANZ)
Aotearoa New Zealand (ANZ) 175–85
   Crown Crystal Glass Factory 180–2
   EMI 182–4
   independent stores, rise of 180
   industry environment 179–80
   Otautahi Christchurch 176–7, 180
   record bars 177–9
   Small Wonder Records 180–2
Apple Music 135
Araba, O. A. L. 136
Armstrong, L. 42
Arnold, G. 7
Around the World in 80 Record Stores 134
artifact, displayed record as 156–9
Atlantic Records, New York 48
Atta, S. 139
aural public sphere 17–22
Aus lauter Liebe ("Out of pure love") 193
Australian Broadcasting Corporation 63
auto-ethnography 2, 175
Aztec Camera 147

bad (behavior) 89–90
Badabing 185
Baldwin, J. 26
Bangalter, T. 222–3
Barber, C. 74
Barcelona, record stores analysis 99–104, **101-2**, **104**
Bargman, B.
   Nice Guys Finish First: How to Succeed in Business and Life 52
Barnes, M. 253

Index

Around the World in 80 Record
    Stores 2
Barnes & Noble 170
Barthes, R. 79
Bartmanski, D. 116
    Vinyl: The Analogue Record in the
        Digital Age 75
Basterra, C. 224–5
Baxter, J. K. 177
BBC Radio 1
    Essential Mix 169
Beach Boys
    Shimmies in Super 222
    Wild Honey 222
Beatles 127, 129, 167, 184
    Ed Sullivan Show, The 8
    White Album 234
Beaumont, T. 65, 67
Beauvallet, J.-D.
    Les Inrockuptibles 221–2, 227
Bee Gees 165
Beethoven Music Center 167, 170
Beethoven's Symphony No. 9 143
Bell, S. 67–9
Benjamin, W.
    "Collector, The" 213
Bennett, A. 146
Bennett, T. 110
Bergman, A. 50
Bergman, B. K. 51
Bergman, H. 51–4
    ambitions 54–5
Berliner, E. 3
Bevan, A. 47, 49
    Grand River Lullabye 46
    Springboard 44
Beyoncé 19
Biafra, J. 63
Bichet, S. 22
Big Bopper
    "Purple People Eater Meets the Witch
        Doctor, The" 8
Big Freedia 19
Bigmoon 155
BIGSOUND program 65
Billboard 54, 143, 241, 244–8
Binnick, A. 205

Binnick, B. 205
biopolitics of resistance 110
Bjelke-Peterson, Sir J. 62
Black, D. 65
Black Pumas 236
Black Sabbath 167, 178
Black Spot Records 156
Black Strobe 120
Blanc, M. 246
Blockbuster Music 51
Blondie 127
Blot, D.
    Le Chant de la machine 227
    Respect is burning 227
bluebeat 31, 38 n.21
Blues and Soul 200, 201
blues dance 28–9
Bob Marley & the Wailers
    "Is This Love" 166
Bonomo, J.
    "Nostalgia" 5
Bonus Beat 225
Bonzo Dog Band 8
Boomplay 135
boss reggae 31, 38 n.20
bounce 18–19
Bourdieu, P. 110
Bovell, D. 33
Bowie, D. 42, 178
BPM 224, 225
Branson, R. 33
Bretton Woods system (known as "Nixon
        Shock") 141, 144
Bridge Over Troubled Water 160
Brisbane
    closures and cultural memory
        (2000s–2010s) 65–7
    in-stores and scene infrastructure
        (1990s–2000s) 63–5
    Pig City, imports and alternatives in
        (1970s–1980s) 61–3
    renewal and continuation
        (2010s–2020s) 67–9
British record store listening booths
    ancient 253–5
    as atmospheric sites of intimacy 252–
        60

communities' experiencing of 258–60
cultural past of 260
individuals' experiences of 255–8
old 253–5
vintage 253–5
yearning of 260
British War Museum, Manchester 157
"Brown" 18
Brown, C. "Gatemouth" 12
Brown, J. 42, 167
Buch, T.
   Limited Edition 189
Buckeye Biscuit 45
Buckingham, M. L. 47
Buckingham Nicks 47
Buffett, J. 47
Buhari, M. 137
Bunaciu, D. 126
Bundesrepublik 186
Burke, C. 235–6
Bush, K. 184
Business Day 139
Buster, P. 31, 199

Calle Tallers 105
capitalism 2, 137, 176, 181
   corporate 184
   disaster 19
   interwar 120
   microhistory of 186–94
Carolina Music Service 51, 53
Carry On 75
Carstairs
   "It Really Hurts Me Girl" 205
Carter, V. 44
Cartwright, G. 82, 253–4
   Going for a Song: A Chronicle of the UK Record Shop 2
Casino, W. 198
Cassidy, W. 160
Catterall, S. 201
   Keeping the Faith: A History of Northern Soul 198
Cavern Club 184
CBS Records 137, 175
CBS Sony 145

CDs 75–6, 81, 85, 88, 91
Centre Pompidou 211
Chammard, S. B. 227
Championship Vinyl 82
Champs
   "Do the Shag" 63
Chapin, H. 55
Chapin, S. 55
Chieftains 3 160
chikuonki (sound-storing machines) 142
Chloé 226
choral music 124
Chulainn, C. 160
Cisco 146
Claddagh Records 155
Clark, E. 243
Clark, J. P. 139
Clifford, C. 8
Cline, P. 4
"Coat" 185 n.10
Coatanoan, C. 210, 211, 215, 217
Cochoy, F. 244
Cold Chisel 64
Cole, T.
   Blind Spot 139
Collin, M. 223
Collins, C.
   Complicated Fun: The Birth of Minneapolis Punk and Indie Rock, 1974–1984 7
Collins, J. 44
Columbia 142
Comet Records 154
Comets 8
commodity scientism 259
communism 3
Compulsory Fun 181
Conefrey, P. 158
Cool Discs 155
Coolin' By Sound 68
Cooly Lully Review 187, 188
Cooper, K. 65
Cooper, R. P. 22
Cooperativa Radio TV 128, 129
Cosgrove, S. 198
Costello, E. 40
Courier Mail, The 66

Coxsone, L. 33
Cross, S. 185 n.11
Crowe, C.
    Almost Famous 41–2, 230
Crown Crystal Glass Factory 180–2
"Cult of the Record Bar, The" 50–7
cultural cosmopolitanism 64
cultural history 2, 60, 62, 120, 121, 154, 233, 252
Curtis, C. 204
Curtis Audiophile 103
Cusak, J. 41
Custard 63

Daft Punk 223, 226
Daily Records 105
Damon Cox of Intercooler 65
Dancetaria 222, 224, 225
dandyism 31
Daphreephunkateerz 226
Dariush 167
Darnielle, J. 235
Dauxerre, D. 222, 226
David, S. 219–20
Davidson, M. 54
Davies, G. 184
Davis, C. 202
Davis, J. C. 20, 21, 242
Dead Kennedys 62
    "Too Drunk to Fuck" 63
Deadly Earnest & the Honky Tonk Heroes 45
Debord, G. 159
Deezer 135
Degen, M. 98
Deller, J. 157
    Everybody in the Place: An Incomplete History of Britain 1984–1992 156
DeMain, B. 4
Denver, J. 44
Der Spiegel 191
devised historical system 213
Diawara, F. 134
digital streaming platforms (DSPs) 135
Dingle Record Shop 155
Diplo 213

Dire Straits
    Brothers in Arms 75
disaster capitalism 19
Disco Demand 203
Discogs 149
Discos Impacto 105
Discos Revolver 105
Discos Tesla 103, 105
Disc Road 146
Disc Union/Disk Union 146, 148
DJ MURO 148
DMR 148
Dock, T. 28
Dodd, C. "Coxsone" 30
Dogg, S.
    Love, Loss, and Auto-Tune 79–80
Dolphin Discs 155
Domino, F. 31
Domino Sound Record Shack 24 n.23
Donna Summer 165
Donny & Marie 166
Douglas, F. 184
Dozier, L. 202
Drag City 185
Drake 19
Dreamboat Annie 44
Dreaming My Dreams 46
dresseteria 244–5
Dromette, J. 10
Drunken Fish 185
DSPs, *see* digital streaming platforms (DSPs)
Dü, H. 9
Dua Lipa 213
Dublin's McCullough Pigott 154
du Gay, P. 243, 245, 249, 250 nn.22, 24
Dylan, B.
    Blonde on Blonde 78
    "Girl from the North Country" 10

Eagles, the 167, 178
Earth, Wind & Fire
    "Fantasy" 166
eBay 68, 149
EB Games 67
Echo Records 180
Edwards, V. 30

Egg Records   67
Eldorado, J.   190–1, 193
"Electrecord"   122, 123, *125*, 126, 127
   Electrecord Catalog (1958)   123–4
   Electrecord Catalog (1961)   124
Electric Fetus   10
elitism   76–8
Elvis   8
E. M. G. Handmade Gramophones Ltd. (EMG)   254
Empire Records   2
Empire Windrush   26, 28
Enescu, G.   126
Engine of Hell   158
Epic Records   48
Erlmann, V.   22
Ernest Tubb's Record Shop, Nashville   4
Euclid Beach Band   45
Europe's Only Iggy Pop Fan Club   191
Evans, E.   58 n.12
Evans, M.   58 n.12
Evil Dick   64
Évora, C.   134
Ewing, J. A.   141–2

Faaji Agba group   138–9
Fallaci, O.   165
Fallon, J.   236
Fari, Prince   33
fascism   3
Feldman-Barrett, C.   5
"felt-body experience" of "quasi-things"   256–7
"FEMA Fridays," Caesar's, West Bank   19
FEN (Far East Network)   143
Fender Telecaster Custom Deluxe guitar   180
Ferry, K. O.   231
Festival Records   179
Fifth Ward Webbie   18
First Programme of Economic Expansion, The   160
Fitzgerald, E.   42
Flack, R.
   "Closer I Get to You, The"   166
Flashdance   169

Fleetwood Mac   44
   Rumours   82
Flory, A.   200
Floyd, S.
   Power of Black Music, The   22
Flying Nun Records   183, 185
Foghat   44
Foley, D.   65
Footloose   169
Forbes, G. V. (aka Duke Vin the Tickler)   31
Forouhar, L.   166
Forster, R.   61
Frampton Comes Alive   2
Free Bird Records   154
Freedman, G.   248
Freedman Artcraft Engineering Corporation   247
Freidman, J.
   "Smart Shops Put Records in Full View, Adopt 'Help Yourself' Policy."   247
"Fuck Katrina"   18
Fullington, L. A.   76–7
Funk Brothers   202

Gagliardi, P.   167
Galaxy Records   180
Gamble, K.   202
Garage D'or   10
García, M.   98
Gates, T.   156–7
   Song for Frankie, A   156, 157
Gaubert, M.   227
Gautier, A. M. O.   17–22
Geffen, D.   54
Generation Z   169
gentrification   98
George, B.   3–4
George, E.   3–4
George's Song Shop, Johnstown, Pennsylvania   3–4
Geräusche für die 80er festival   189, 191
"Ghana Must Go"   137
Ghosn, J.   227
Giants of Science   65
"G.I. Blues"   53

Gilb'R 226
Gildart, K. 201
　Keeping the Faith: A History of Northern Soul 198
Gillies, A. 65
Glover, D. 177
Glover, T. 8
Go-Betweens 61, 62
Godin, D. 200, 201
Golden, B. 54
Golden Discs 154–5, 161
Golden Horde 158
Golden World 202
good (behavior) 86–8
Good Vibrations 154
Googoosh 166
Gordy, B. 200
gourmetization 97
Gracon, D. 98, 107
Graduate, The 160
Gramophile 143
Grand Ole Opry 4
Grapefruit Records 185
Grapevine Records 205–6
Green, T. 180, 184
Green Day & U2 18
Green Sleeves: The Irish Printed Record Cover 157–9
Greig, T. 253
Griffero, T. 252, 256–7
Grimes, M. 49
Grimey's, 12th Ave So, Nashville 49
Grimey's, E. Trinity Lane, E. Nashville, Tennessee 49
Groove City Records 197, 199
Groovy, W. 31
Guerra, P. 118 n.1
Guilloteau, F.
　Eighties, The 215
　Millennium 215
　Nineties, The 215
Guit, G.
　Les Kidnappeurs 223
Gumbrecht, H. U. 257
Guns n' Roses
　Appetite for Destruction 63
Gusman, I. 98

Haas, H. G. 248
Habermas, J. 2
Haley, B. 8
Hall, S. 29
Hall & Oates 46
Hard-Ons
　Dickcheese 63
Hardy, F. 167
Hariri, M. R. 170
Harmonia 143
Harper, R.
　Folkjokeopus 7–9
Harris, E. 44
Hathaway, D.
　"Closer I Get to You, The" 166
Hearts Beat Loud 2
Hendricks, J. M. 98
Hendrix, J. 86
Henry, J. 21
Hepworth, D. 66
High Fidelity 2, 72–82, 143, 183, 227, 242
　CDs and online marketplace 75–6
　elitism 76–8
　living popular music in 109–18
　morality for aesthetics, substituting 78–81
　pleasure 76–8
　specialist record shops 73–4
　taste 76–8
Hills, M. 216
Hilsberg, A. 186–9, 191, 192
hip-hop 135, 171
Hirsch, B. 226
HMV 66, 147, 148, 154, 177, 254, 258
Hô-ban 141
Hobbs, H. 18
"Holiday" 210
Holland, B. 202
Holland, E. 202
Holley Associates 247
Homem-Christo, G.-M. de 222–3
Homer, S. 65, 67, 68
Honold, E. 189
Hooley, T. 154
Hools, T. 226
Horn, F. W. 142
Hornby, N.

High Fidelity 2, 41, 72–82, 109–18, 143, 178, 183, 227, 242
Horne, L. 42
An Horse 66
Horses 44
Hot Press 154
Hracs, J. 98, 105
Hudson, H. 65
Humperdinck, E. 127
Hynde, C. 41

Ich und mein Spiegelbild 190
Icon 210, 214–15
Icona Pop 19
I:Cube 226
IM, *see* International Monetary Fund (IMF)
Independent, The 82
I Need That Record 2
Intercooler 66
International Monetary Fund (IMF) 137
Inventing David Geffen 54
Iran
    Ministry of Islamic Culture and Guidance 169–70
    pre- and post-revolution, western music acquisition in 164–71
    Vizarat-i-Farhang va Irshad-i Islami 169
Ireland
    artifact, displayed record as 156–9
    Black Spot Records 155–6
    Dance Hall Act 158
    psychogeographic drift, benefits of 159–61
    record stores, brief history of 153–5
Iron On 65, 66
I-Roy 33
Isaacs, G. 33
Ivory Music (formerly EMI Records) 137, 175, 177, 179, 180, 182–4, 201, 202

Jachimiak, P. 35
Jack, P. 226
Jackson, M. 211
    Off the Wall 80, 82

Thriller 82, 168
Jackson 5 165
Jansson, J. 98, 105
Japan
    Anti-Monopoly Law 144
    Dodge Line policy 143
    domestic and imported records in postwar Japan, frictions between 143–4
    Great Kanto Earthquake of September 1923 142
    imported records, influence of 141–9
    Import Trade Control Order 144, 145
    kokunai-ban 141–6
    Pied Piper House 145–7
    record industry, birth of 141–2
    Recording Industry Association of Japan (RIAJ) 141
    "Record Village," Shibuya, Tokyo 141, 147–8
    Saison culture 147–8
    Yunyû-ban 142–6, 148, 149
Jarre, J.-M. 169
Javier, L.
    Tiempo Felices 223
Jazzhole 134–9
Jeff, G. 76
Jesperson, P. 8–9
job insecurity 106–7
Jones, G.
    Last Shop Standing 2
    "My Bad Boy's Comin' Home" 203
Jones, K. 138
Jones, Q. 81
Jones, S. 35
Jordan, L. 31
Judas Priest 129
Jujiya 143
June, V. 49
JVC 142

"Kachūsha no Uta (Katyusha's song)" 142
Kamocsa, B. 127
Kaneyuki 143
Karlzen, M. 48

Karnofsky, M. 17
Karnofsky Tailor Shop 17
Katzman, T. 9, 10
Keazor, E. 136, 137
Kebra Disc 105
Keita, S. 139
Kerr, T. 9
Kersten, A. E. 168
Kesha 19
Keyser, P. 51, 53, 57 n.9
Khan, R. 165
Khomeini, A. R. 165, 167, 169
Khruangbin 236
Kid P. (alias Andreas Banaski) 190
Kill the DJ 226
King, B.B. 8
King Atupali 103
Kitchener, Lord
 "London Is the Place for Me" 28
Knuckles, F. 157
kokunai-ban 141–3, 146
 unique development of 144–5
Kornhaber, S. 78–9, 81
Kranky 185
Kritzler, J. 65
Kuti, F. 135
Kuti, J. R. 136

Lagerfeld, K. 227
Land That Time Forgot, The 48
Lane, C. 36
Last Shop Standing 2
Leavis, F. R. 80
LeBlanc, F. 4
Leckey, M. 157
 Fiorucci Made Me Hardcore 156
Led Zeppelin 86, 127
 Led Zeppelin II 236
Lee, B. "Striker" 30
Lee, P.
 Is That All There Is 42
Lehrer, T. 42
LeMenestrel, S. 21
Le Silencio 225
Les Inrockuptibles 220, 221
Let It Be 10
Levine, I. 204, 205

Lewis, J. 72
Libération 225
lieux 22
light music 132 n.26
Lilburn, D. 177
Limerick Record Fair (2021) 159
Lindsay, T. 9
L'Inrockuptible 220, 222
Little Lovers 66
"Live at Abelton Loop" 233
Live at the Button 48
location of record stores 105
Longhair 31
Longhorn Bar (aka Jay's Longhorn),
 Minneapoli 9
Lorenz, L. 190
"Lost and Found Sounds: Cultural,
 Artistic and Creative Scenes in
 Pandemic Times." 118 n.1
Loueke, L. 139
Louie Louie 115
Louisiana Music Factory 24 n.23
Lovett, L. 55
"Love Will Tear Us Apart" 183
Lucky Records 209–17
 bow 216–17
 get together 210–11
 "Material Girl" 210, 212–14
 sanctuary 211–12
 Spotlight 210, 214–16
Lugard, F. 135
Lunch, L. 63, 65
Lynch, D. 225
Lynn, L. 4

McCahon, C. 177
McCaslin, J. M. 135
McCormack, N. 157
McGee Repository Museum, Sligo Folk
 Park 157
MacGilivray, J. 36
McKagan, D. 233
Macklin, A. 53
Madonna 210–11, 216–17
 "Lucky Star" 211
 W.E. 215
Maeck, K. 186–9, 191–3

Magic   220, 224
Maher, S.   158
Majora   185
Maluma   213
Manhattan Records   148
"Mari"   181
Markthalle   188, 191
Marley, B.   46, 134
Marley, Z.   1
Marsh, M.   177
Marshall, G.   204
Martin, D.   42
Mary Trembles   66
material curation, record stores as space for   110–13
"Material Girl"   210, 212–14
Matéria Prima   105
Mayfield, C.   202
MDNA   215
Mea culpa   220, 221, 224
Meikyoku   143
Melodiya   126
Melody Maker   40, 48, 222
Melton, B.   10
memory places   22
Mengel, N.   66
Merurido   146
Mestre, J.   222–8
Miami Herald, The   48
Mia X
   "My FEMA People"   18
microhistory of capitalism   186–94
Midnight Oil   62
Mighty Diamonds   33
Miller, K.   75
Miller, W.   41–2
Minaj, N.   213
Minions: The Rise of Gru   2–3, 5–6 n.1
Minogue, K.   211
Mise Eire (I am Ireland)   160
Mitchell, J.   42
Mitchell Manufacturing Company   248
Mixtape   2
Molina, J.   233
Mondial Twist   227
Mondosinfonola   103
Monticone, K.   65

Moore, B.
   Top of the Pops   168, 169, 178
morality for aesthetics, substituting   78–81
Morgan, D.   31
Morin, E.   209
Morris Music   17
Mosshart, A.   235
Motoi, K.   145
Mott the Hoople   42
Mo'Wax   226
Mudhoney   10
Muldoon, R.   180
Münchinger
   Four Seasons   143
Murder Strikes Pink   181
Mushroom   24 n.23
music celebrity, in record store   237
Musikexpress   188
Muzak   105
Muzica Store, Bucharest   122, *123*, 130, 131

National Iranian Oil Company (NIOC)   167
Nelson, D.   62
New Musical Express   40, 61, 175, 178, 181
New Rose   224
New Towns Act of 1946   261 n.24
New York Times, The   137
Nichibei Chikuonki Kabushikigaisha (The Japan-American Phonograph. Mfg Co., Ltd.), *see* Nippon Chikuonki Shokai (Nipponophone Co., Ltd.)
Nichols, R.
   Roger Nichols & The Small Circle of Friends   146
Nicks, S.   47
Nigeria
   Jazzhole   134–9
   Musical Society of Nigeria (MUSON)   138
Nigerian Association of Gramophone Record Dealers (ANGARD)   136
Nigerian Recording Association   136
NIOC, *see* National Iranian Oil Company (NIOC)

Nippon Chikuonki Shokai
    (Nipponophone Co., Ltd.)   142
Nippon Columbia   143, 144
Nippon Gakki   143
Nipponophone   142
Nippon Polydor   142, 179
Nirvana   10
NME   48
No Fun   189
Nora, P.   22
Northern Lights   10
Northern Soul
    emergence of   200–1
    scene   201–3
Northern Soul   200
Nouvelle Vague   223
Nuthin But Fire Records   20–2, 24 n.34

Obasanjo, O.   137
Obey, E.   138
O'Brien, O.   158
O'Connell, M.   64
Okusanya, B.   136
Oldenburg, R.   20
One Nation Party   64
online marketplace   75–6
online music distribution   66
Onyekwelu, C. T.   136
OPEC, see Organization of the Petroleum
    Exporting Countries (OPEC)
Open Books + Records, Ft.
    Lauderdale   48–9
Organique   225
Organization of the Petroleum Exporting
    Countries (OPEC)   136
Ormonde   26
Osmonds   165
Otautahi Christchurch   176–7, 180
Our Price   75
Our Price Records   73
outsider communities   2, 160
Owens, C.   183

Padovan, M.   68
Pandora   135
Paphides, P.   252
Parker, D.   183

Partridge Family, The   43
Partridge Family Songbook, The   43
Pastels, the   147
Pat Egan's Sound Cellar, Dublin   153–4
Patton, T.   65
Paz, E.
    Dust & Grooves: Adventures in Record
        Collecting   2
Peaches, Ft. Lauderdale, Florida   48–9
Peaches Records and Tapes   20, 21
Peake, T.   178
Peck, O.   234
Peel, J.   178
"pen and display" method   243
Penn, D.   12
Pentagram Music   105
Pere Ubu   10
personal choice   245
Peterson, J.   10
Peterson, R. A.   146
Philips Record Club   177–8
Pickett, C.   49
Pied Piper House   145–7
Pig City (1970s–1980s), imports and
    alternatives in   61–3
Pin Group   185 n.10
Pink, S.   252
Pink Floyd   129, 178
Place de la Bastille   211
Plain Dealer, The   44
plaisir-jouissance distinction   79
pleasure   76–8
Plourde, C.   210, 215
Plugd   155
Plus One Records   65, 66, 68
political liberalism   64
Pop Rivets   189
Porto
    record stores, analysis of   99–104,
        100, 103
Porto Calling   115
Portugal
    biopolitics of resistance   110
    independent record stores (1998–
        2020)   109–18
    material and affective curation, in
        record stores   110–13

ritual practices, in independent record stores   113–16
(trans)local music scenes, in record stores   116–17
post-Katrina New Orleans
  aural public sphere   19–22
  bounce   18–19
  independent record store in   17–23
postwar crisis   122–5
Powderfinger   63
Powell, J. E.   37 n.3
  "Rivers of Blood" speech   37 n.3
practiced place   22
Preiserhöhung   188
Premier Music (formerly Polygram Records)   137, 175, 179
Presley, E.   53
Pretty Vacant   188, 189
Price, M.
  Homegrown   235
Pride, L.
  I'm Com'un Home in the Morn'un   207 n.16
Pringle, W.   178
professionalization   129–30
project music   18
Prout, B.   50–7
Pryce, G.   31
Pryor, R.   44
Puetz, K.   110
Punkhouse   191

Quare Groove   158, 159
Quarter Moon in a Ten Cent Town   44
Queen Street Mall   66

Race Relations Act   37 n.3
racism   3
Radio Caroline   220
Radio Eireann   162 n.2
Radio Luxembourg   154
Radio Nova   224, 225
RAF, *see* Red Army Faction (RAF)
rage   63
Rainbow record store, Harlem, New York   31
Raine, S.
  Authenticity and Belonging in the Northern Soul Scene   198
  Northern Soul Scene, The   198
Randall Park Mall   46
Randy's Record Shop, Gallatin, Tennessee   5
Rankin, C.   61–5
Rave Magazine   65
RCA   179, 247
Rebotini, A.   226
Record Bar   51–2
  cult of   50–7
  family values   52–4
  next generation, charge of   55
Recording Industry Association of America (RIAA)   247
Record Joynt   180
Record Revolution, Coventry Rd., Cleveland Hts., Ohio   40–2
Record Store Day   3, 5, 49, 69
record stores
  location of   105
  as places for shared knowledge   105–6
  ritual practices in   113–16
  as spaces of material and affective curation   110–13
  and tourism   104–5
  (trans)local music scenes in   116–17
  see also individual entries
Record Theater, Mayfield Rd.   44–7
Record Theater, S.O.M. Center Rd., Mayfield Hts., Ohio   44–7
"Record Village," Shibuya, Tokyo   141, 147, 148
Recovery   63
Red Army Faction (RAF)   186, 187
Reed, L.   178
Reggae Record Shops, Black London   26–37
Regurgitator   63
Reid, A.   30
Reish, G.   5
religious fanaticism   3
retromania   121
Revolver Records   105
Rhymes & Reasons   44

RIAA, *see* Recording Industry Association of America (RIAA)
Richard, C.   184
Ilie Tic   202
Riedel, F.   256
Rip Off Records   186–94
　hanging out and shopping, experience of   190–1
　new endeavors   191–3
　opening of   188–90
　rebranding   191–3
Riverside Traders   180
Robert, M.   211, 215–17
Roberts, K.   204
Rock&folk   220
Rocking Horse Records   60–9
Rodrigo, O.   82
Rollercoaster Records   155
Rolling Stones   41, 61, 127, 167
Rollins, H.   1
Romain BNO   225
Romania
　alternative record stores and 1980s   126–9, *128*
　Arts Committee   122
　Cold War openness   122–5
　history of record stores in   120–31
　new (old) record store   129–30
　postwar crisis   122–5
　professionalization   129–30
　Union of Composers   122, 126, 130
Ronstadt, L.   46
Roppongi WAVE   147, 148
Rotkotz   189
Rough Trade Paris   219–28
Royal, J.   126
Rudolph, C.   139
Rug, E.   223
Rundgren, T.   44
run-out groove   92–3

Safari, S.   167
Saison culture   147–8
Salisbury, J. E.   168
Sanden, V.   7, 10
Sattar   166
Saunders, S.   244

Schwartz, A.   9
Scott, B.   67
Screamfeeder   63
Searling, R.   203, 205
self-service record retail   241–9
　brief history of   242–4
　consumer choice, staging and extending   248–9
　customer's "inner desires," acknowledging   244–6
　display   246–8
　need to "buy with tour eyes"   246–8
Self-Service-The Greatest!   246
Seroka, H.   126
Setting the Record Straight   204
Sex Pistols
　"Anarchy in the UK"   62
　"God Save the Queen"   9
Sfinx   129
Shagari, S.   137
Shahr-e-Ketab   170
Shaker Square, Van Aken Rd., Cleveland, Ohio   42–4
shared knowledge, record stores as places for   105–6
Shazam   78–81
SheCan   170
Shepherd, R.   184, 185 n.9
Shôkai, A.   142
shubeen   28–9, 38 n.7
Sinatra, F.   42, 134
"Singles of the Week"   181
　Hollow Skai   188, 189
Skinny's Music   60–8
Skippy White's Records, Boston   4–5
Skyhooks   64
Slime   191
Smagghe, I.   223, 226
Small Wonder Records   180–2
Smith, N. W.
　Northern Soul Scene, The   198
Smith, W. H.   73
social distancing   235
social mobility   1, 135
sociocultural history   134
Sonic Sherpa   67–9
Sonic Youth

Master=Dik EP   63
Sony Music Entertainment   148
Soul Bowl Records   197–206
   disc jockeys   203–5
   literary review   198–9
   modern   203–5
Sound It Out   2
"Sound of Philadelphia"   202
Sounds   175, 178, 181, 187, 190, 192
Sound System culture, import of   30–3
South East Records   155
space-age ergonomics   259
specialist record shops   73–4
Spec's, S. Dixie Hwy, Coral Gables, Florida   48–9
Spec's, Town Center Mall, Glades Rd., Boca Raton, Florida   47–8
Spector, M.   47
Spence, S.
   Oar   7
Spillers Record Shop, Cardiff, Wales   3, 4
Spotify   78–81, 112, 135, 170, 171, 226
Spotlight   210, 214–16
Squirrels   48
Stafford, A.   61, 62
Stark, P.   10
Steen, C.   248
Stenagaard, W. L.   247
Stern   191
Stimmung (mood)   257–9
Stone, K.   36
Stone's Throw   44
Stooges   10
store planning   247
Straw, W.   81, 242
Street, J.   201
Student Liberation Front   24 n.23
Styles, H.   82
subcultural history   112, 113, 156
Suicide Commandos
   Time Bomb   12
Sumartojo, S.   252
Sumiya   146
Super Club N.V. Super Club   51
Supraphon   126
Swan, C.   157
Sweeney, M.   237

Swift, T.   82

Tabatô Records   117
Takahashi Sota   145
Takarajima   145
Takeichirô, M.   142
Tamla Motown Appreciation Society   200
Tangerine Dream   129
Tape Revolution (1971)   73
Tasjan, A. L.   49
taste   76–8
Taylor, E.   134, 139
Taylor, J.   167
   Walking Man   48
Taylor, S.   156
Teddy Boys   31
Tejuosho, K.   134
third spaces   20
Thirty-Three Record Shop   155
This England: The Wigan Casino   203
Thoroughbred   44
Three EPs, The   79
Thunders, T.   178
Time Off   65
T/K Records   9
Todd, A.   3
Tonight Show, The   236
Toowong Music Centre   62
Toshiba EMI   145
Touché Amoré   234
tourism, record stores and   104–5
touristification   98
Tower Records   146–8, 155
Townes Van Zandt: Live at the Old Quarter   47
Treehouse Records (formerly Oar Folkjokeopus, Oar Folk), Minneapolis   7–13
Trehus, M.   7
Trevaskes, R.   67
T-Rex   178
"Triggerman"   18
Triple J   63
Trouser Press   48
Tubb, E.   4, 5
Tubb III, E. D.   4–5

Twisted Wheel 202, 203

UB40 129
Udoji Commission 136
Udoji Public Service Review
    Commission 136
ugly 90-2
"Ultra Self-Service" 246
Uneeda Records 10
University of Canterbury Bookshop 178
Unknown Pleasures 183
Unterm Durchschnitt 186
urban geography 2, 252
Ursulescu, F.-S. 127
US Navy
    Operation Deep Freeze Antarctic
        research and exploration
        program 177

Van Der Graaf Generator 129
Vaseghi, K. (aka Kasra V) 169
Velvet Underground 10
Vere, W. 61, 66-8
VHF 185
Victor 142
Victor USA 142
Vidal, G. 211, 212, 214-17
Viguen 167
Village Voice 219
Vinyl Vintage 103
Virgin Megastores 75, 147
vocation 106-7
Vynehall, L. 236

Walden, P. 55
Walker, M. 204
Wall, T.
    Northern Soul Scene, The 198+
Warner Music 215
Warner-Pioneer 145
Warp 226
wax cylinder graphophone 141
Wayne, L.
    "Georgia . . . Bush" 18
    "Tie My Hands" 18
WEA 175, 179
Welskopp, T. 187

West Indian(s)
    arrival 27-8
    Britishness for 28
West Indian music
    blues dance 28-9
    cultural confluence of 26-37
    resilience 34-7
    resistance 34-7
    shubeen 28-9, 38 n.7
    Sound System culture, import
        of 30-3
What's In My Bag? (WIMB?) 230-7
    Home Edition 234-7
    introduction 231-2
    music celebrity 237
    Season 12 232-4
White, C. 254
"Why Can't I Be Like Other Girls" 45
Wiganization 203
Wigan's Chosen Few 203
Williams, H., Sr. 4
Willsteed, J. 61-2
Wilson, A.
    *Northern Soul: Music Drugs and
        Subcultural Identity* 198
Wilson, C. 78
Winstanley, R. 204
Wired 169
Wizard Import Records 62
Woman in Red, The 77
Wonder, S. 44
    "I Just Called to Say I Love You" 77,
        80
Wong, T.
    Tom the Great Sebastian 30, 31
Woodbine, Lord 38 n.6
Woodward, I. 116
    *Vinyl: The Analogue Record in the
        Digital Age* 75
World Health Organization 234
World Record International 177
Worley, M. 7
Wraparound Joy 44

Yamaha 144
Yamano Gakki 143
Yô-ban 141

Yoshirô, N. 146
Young, N.
  On the Beach 235
YouTube 135, 149
Y&T Records, Coconut Grove,
   Florida 48–9
Yunyû-ban 146, 148, 149
  culture, formation of 144–5
  yearning for 142–3

Zend Avesta 226
Zhivago 155
ZickZack Records 191–3
Ziegler, L. 77–8
Zonophone 136
Frank Z(iegert) 191

www.ingramcontent.com/pod-product-compliance
Lightning Source LLC
Chambersburg PA
CBHW051631230426
43669CB00013B/2252